DISCOVERY & EXPLORATION

154. "Founding of the Settlement", 1799.
Thomas Gosse's engraving of an heroic settlement of
New South Wales is the first Australian "history painting".

HORDERN HOUSE

RARE BOOKS • MANUSCRIPTS • PAINTINGS • PRINTS

DISCOVERY & EXPLORATION

77 VICTORIA STREET • POTTS POINT • SYDNEY NSW 2011 • AUSTRALIA
TELEPHONE (02) 9356 4411 • FAX (02) 9357 3635 • EMAIL books@hordern.com
Directors: Anne McCormick • Derek McDonnell
www.hordern.com

Cover illustration: from Garnier's account of the Mekong River Expedition (no. 383).

Hordern House
77 Victoria Street
Potts Point Sydney NSW 2011 Australia

Hordern House Rare Books Pty. Ltd. ACN 050 963 669
Directors: Anne McCormick, Derek McDonnell

First published in 1998 by Hordern House Rare Books Pty. Ltd.

National Library of Australia
Cataloguing-in-publication data.

Hordern House (Firm).
Discovery & Exploration.

Bibliography.
Includes index.
ISBN 1 875567 21 6.

1. Hordern House (Firm) - Catalogs. 2. Rare books - Australia - Catalogs. 3. Catalogs, Booksellers' - Australia. 4. Voyages and travels - Bibliography - Catalogs. 5. Australia - Discovery and exploration - Bibliography - Catalogs. 6. Oceania - Discovery and exploration - Bibliography - Catalogs.
I. McCormick, Anne. II McDonnell, Derek. III. Title.

016.91

Photography by Lucio Nigro.
Produced by Hordern House Publications, 77 Victoria Street, Potts Point 2011.

CONTENTS

PART ONE: *Early voyages, collections & utopias*

1. Early voyages....1-55
2. Collections, costumes, ethnography....56-85
3. Imaginary voyages & Utopias....86-102

PART TWO: *Cook to Bligh*

4. Captain Cook....103-146
5. Australian settlement....147-174
6. Australian exploration....175-201
7. La Pérouse & the search....202-212
8. Bligh & the Bounty....213-225

PART THREE: *World voyages*

9. General voyages....226-244
10. English voyages....245-271
11. French voyages....272-311

PART FOUR: *The East*

12. Far East & India....312-337
13. South-east Asia & East Indies....338-376
14. The Mekong River Expedition....377-388

PART FIVE: *The Pacific*

15. The Pacific....389-432
16. Hawaii....433-471
17. Ships & ship-building....472-481

Principal references cited

Author & title index

1: EARLY VOYAGES

ANSON'S VOYAGE

1. ANSON, George. WALTER, Richard, compiler. A Voyage Round the World... By George Anson, Esq; Commander in Chief of a Squadron of His Majesty's Ships, sent upon an Expedition to the South-Seas...

Thick quarto, with 42 folding engraved plates and maps; strong impressions of the plates, slight cracking to joints but a fine, clean and large copy in an attractive nineteenth-century binding of sprinkled calf, spine gilt in compartments; armorial bookplate, crest in spine-label. London, Printed for the Author, by John & Paul Knapton, 1748.

First and best edition of one of the most famous of all voyage books.

This is the official account of Anson's voyage. Based on Anson's own journal, and put together for publication by Richard Walters – starting a tradition of "ghost-writing" for voyage accounts that continues to this day – this narrative had an enormous popular success. For the mid-eighteenth-century reader, it was the epitome of adventure, and it was translated into several European languages and stayed in print through numerous editions for many years.

The voyage, a success in terms of ships plundered and treasure captured, was otherwise a disaster. Altogether six ships and nearly two thousand men set out in 1740, but only one ship and nine hundred men returned.

Borba de Moraes, p. 38; Hill, pp. 317-8; Kroepelien, 1086; Sabin, 1625. **$3500**

SCOTTISH ABRIDGEMENT

2. ANSON, George. WALTER, Richard, compiler. A Voyage Round the World...

Two volumes, duodecimo, with a folding engraved map; a very nice set in contemporary calf, red labels, spines numbered in gilt. Kilmarnock, J. Wilson, 1785.

A particularly scarce Scottish-printed abridgement of one of the most popular and enduring of all voyage accounts. **$1100**

CHAPBOOK ANSON

3. ANSON, George. The Voyage of Commodore Anson round the World.

Duodecimo, with six full-page illustrations; original half roan, spine worn at top. Dublin, William Folds, 1818.

Scarce Irish chapbook version of the famous voyage account, with a charming series of wood-engraved plates. **$285**

FIRST FRENCH EDITION

4. ANSON, George. Voyage autour du Monde... dans la Mer du Sud. Tiré des Journaux... par Richard Walter...

Quarto; title-page printed in black and red with engraved vignettes, 34 engraved maps and plates, mostly folding; a few edges frayed but a good copy in modern calf. Geneva, H.-A. Gosse, 1750.

Early French edition of the famous account of Anson's voyage (there was an Amsterdam edition, in a different French translation, the previous year).

Not in Hill or Kroepelien; Sabin, 1638. **$850**

LIFE OF ANSON

5. [ANSON] BARROW, Sir John. The Life of George Lord Anson...

Octavo, portrait frontispiece; contemporary half dark blue calf gilt. London, John Murray, 1839.

A finely-bound copy of the best early life of the famous voyager and naval supremo. With the bookplate of Admiral Sir Thomas Cochrane. **$975**

SPANISH IN THE EAST INDIES

6. ARGENSOLA, Bartolomé Leonardo de. Conquista de las Islas Molucas.

Folio, fine engraved title-page (allegorically representing the Spanish conquest of the Moluccas); light waterstain in a few corners and a small wormhole; an excellent copy in contemporary calf, spine and joints well restored; from the Peiresc library, with cipher in gilt on covers and in black at start of text. Madrid, Alonso Martin, 1609.

First edition: one of the most important books for the history of the Philippines, and essential for the history of Spanish exploration and trade in the East Indies, deliberately conceived to demonstrate Spanish colonial strength. Argensola – poet, historian, and friend of Cervantes – was commissioned by the Council for the Indies to write a history of the Spanish in the East Indies.

The result – "a work written with unusual elegance and judgement, becoming a man of letters..." (Penrose) – was this "history of the Spaniards in the East Indies, including their voyages of exploration, trade and their wars with the Portuguese, Dutch, English and natives of the region. The work relates primarily to the Philippines and the Moluccas (Spice Islands); however, sections deal with China, Java, Sumatra, Celebes, New Guinea, Ceylon and other areas. An account of the voyage of Sir Francis Drake is included, as is Pedro Sarmiento de Gamboa's voyage of exploration to the Strait of Magellan... This book is an essential work for the history of Spanish and Portuguese exploration in the East Indies" (Hill).

"Strictly speaking, Argensola's account is a history in a world setting of Spain's gradual progress in winning control over the Moluccas. But it is also a tract for the times and a graceful piece of political and dynastic propaganda... [which] proclaims that Spain still possessed the determination and strength to undertake conquests in distant places and to bring them off successfully. In his digressions on people and places, Argensola also adds significantly to the stockpile of information on Asia, especially the Moluccas, Java, Sumatra and Ceylon. His book also ties together neatly the affairs of Europe with struggles in overseas areas, for he sees the spice trade in its world-wide ramifications and makes his reader acutely aware of its immediate and potential interest for Japan and China. Through his accounts of the embassies sent from the East Indies to Europe he shows that diplomatic relations, not just trade, conquest, and missionizing could be carried on with insular Southeast Asia. And, here for the first time, the world had at hand an entire book centred directly on the Moluccas that could be read with profit and pleasure" (Lach).

This beautiful copy is from the library of the celebrated Provençal antiquarian, naturalist and collector Nicolas Claude Fabri de Peiresc (1580-1637), with his complex monogram on the binding. Peiresc's famous library has been the subject of several studies; A.R.A. Hobson points out (*French and Italian collectors and their bindings*, Roxburghe Club, 1953) that his bindings are almost certainly provincial rather than Parisian, and that at one time his household included two bookbinders.

Alden, 609/65; Hill, p. 475; Kraus, Drake, 33; Lach, III, pp. 311-2; Medina, Filipinas, 48 (& BHA 551); Palau, 16089; Retana, 67; Sabin, 1946; Steele, pp. 120-21. **$18,500**

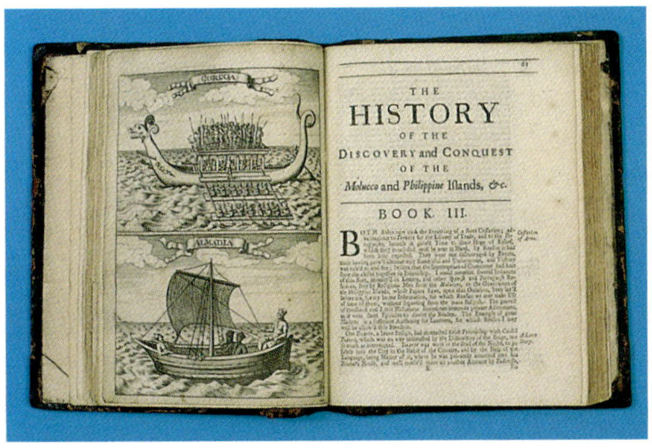

7. *Argensola*

FIRST ENGLISH EDITION

7. ARGENSOLA, Bartholomew de. The discovery and conquest of the Molucco and Philippine Islands containing their history, ancient and modern... With an Account of many other adjacent islands, and several remarkable passages through the Streights of Magellan, and in other Parts...

Quarto, with an engraved frontispiece, a map and two plates; slightly browned; contemporary panelled calf, rubbed, rebacked. London, 1708.

First English edition. Stevens, the translator for this edition, believed that the rarity of the original

6. *Argensola*

(Madrid, 1609: see previous entry) even in the seventeenth century was one of the reasons for its not being translated for a hundred years. Stevens also remarks in the preface that the book contains a great deal of information in addition to what is implied by the title: "This is frequent in Spanish books, wherein we generally find much more than the titles promise, contrary to what is practis'd with us...".

Hill, p. 475; Sabin, 1948; Steele, pp. 120-21. **$4750**

VOYAGE OF THE SPEEDWELL

8. BETAGH, William. A Voyage round the World. Being an Account of a Remarkable Enterprize, begun in the Year 1719, chiefly to cruise on the Spaniards in the great South Ocean...

Octavo, folding world map as frontispiece; a very good copy in modern panelled calf. London, T. Combes, 1728.

"One of the most important of the buccaneering

expeditions" (Hill). This is the first edition of this alternative account of the voyage of the *Speedwell*, presenting a substantially different account to that given by Shelvocke. Betagh, Captain of Marines on the voyage, casts doubt on the Shelvocke narrative wherever possible, claiming that it was untrue and that it cast an unfair light on the captain, John Clipperton. Betagh's narrative includes good material on Chile and Peru, with detailed observations on the techniques of gold and silver mining, as well as on Baja California and Macao.

Hill, p. 25; not in the Kroepelien collection. **$6850**

BYRON IN PATAGONIA

9. BYRON, John. The Narrative of the Honourable John Byron... containing An Account of the Great Distresses suffered by himself and his companions on the Coast of Patagonia, from the year 1740, till their arrival in England, 1746. With a description of St. Jago de Chile... also a relation of the loss of the Wager Man of War, one of Admiral Anson's Squadron.

Octavo, with a frontispiece; contemporary sprinkled calf, rubbed; old rebacking, spine worn at top; various owners' signatures including a presentation to "Arthur Cadogan from Genl. Fox". London, S. Barker, 1768.

First edition. "Admiral Byron's narrative of the loss of the *Wager* is one of the most thrilling accounts in the language, and supplied his illustrious descendant with many particulars for the shipwreck in Don Juan" (Sabin). Byron, later a noted Pacific explorer and eventually governor of Newfoundland, was known as "foul weather Jack"; he was a midshipman aboard the *Wager* and his narrative provides a notable supplement to the main account of Anson's voyage. He gives "an account of pillaging, treachery, and murder by the crew, most of whom claimed that since their pay ceased the moment the ship was lost they were no longer subject to military discipline and it was every man for himself. It told of Byron's own living off the sea, without shelter or clothing, on the 'most unprofitable spot on the globe of the earth'; of his going without a full meal for thirteen months; of his living three years in the lands of the Spaniards... and of his eventual return to England... almost five years after the shipwreck" (R.E. Gallagher, *Byron's Journal*, Hakluyt Society, 1964).

Hill, pp. 41-2; Sabin, 9730. **$2400**

BYRON & CLERKE TOGETHER

10. BYRON, John. The Narrative of the Honourable John Byron... Containing an account of the great distresses suffered by himself and his companions on the Coast of Patagonia... [bound with] CLERKE, Charles. A Voyage round the World, in His Majesty's Ship the Dolphin, commanded by the Honourable Commodore Byron... By an Officer on board the said ship. The second edition.

The two works bound together, octavo, the first with frontispiece, the second with three plates and advertisement leaf; very faint stain in a few foremargins, but both works good and tall copies, bound together in contemporary mottled calf, joints slightly cracked at top. London, 1768 and 1767.

A happy but unusual marriage of two works: the first edition of Byron's narrative of the Anson expedition (see previous item) with the second edition of the anonymous unauthorised account of Byron's subsequent command of the *Dolphin* on her circumnavigation of the globe, published anonymously but usually attributed to midshipman Charles Clerke who later sailed on all three of Cook's voyages.

Hill, pp. 41-2 & 310; O'Reilly-Reitman, 241; Sabin, 9730, 9732. **$5500**

BYRON CHAPBOOK

11. BYRON, John. The Narrative of the Honourable John Byron...

Duodecimo, engraved frontispiece (partly defective, remargined); frontispiece and title stained; some stains, modern half calf. London, A. Millar [et al.], 1788.

Scarce chapbook of the famous voyage account, not noticed by the standard bibliographies. **$400**

SPANISH MAP OF THE STRAITS OF MAGELLAN

12. [?CLERKE, Charles] Viage del Comandante Byron al rededor del mundo... con un nuevo mapa del Estrecho...

Small quarto, with a frontispiece and a coloured folding map; a fine copy in contemporary vellum. Madrid, Don Francisco Mariano Nipho, 1769.

The rare first Spanish edition, and a very desirable version of the book, not only for the considerable additional material provided by the translator Casimir de Ortega, but especially for the fine and detailed map of the Straits of Magellan by Juan de la Cruz Cano y Olmedilla, specifically commissioned for this edition. This large and handsome map is attractively coloured in outline, and shows the formidable extent of contemporary Spanish knowledge of the straits. The frontispiece to the book has the familiar image of an English sailor dwarfed by two Patagonian giants.

O'Reilly-Reitman, 245; Palau, 38229; Sabin, 9733. **$5500**

OLD-WORLD EXPLORERS

13. CALVERT, Ibert F. The Discovery of Australia.

Quarto, plates and maps; a good copy in the original cloth. London, Dean & Son, 1902.

Second, enlarged and best edition. Calvert worked tirelessly on the various theories relating to which nation of old-world explorers first landed in Terra Australis. Medieval maps are reproduced, and the early discoveries in Australian waters are well chronicled. **$325**

PRESENTATION COPY

14. COLLINGRIDGE, George. The Discovery of Australia. A Critical, Documentary and Historic Investigation Concerning the Priority of Discovery in Australasia by Europeans before the arrival of

Lieut. James Cook, in the "Endeavour", in the year 1770.

Quarto, with numerous maps (some folding), plates and text illustrations; a fine copy in half morocco gilt by Morrell. Sydney, Hayes Brothers, 1895.

A fine presentation copy, inscribed by Collingridge on the half-title to "my good friend D. Thomas", in a de luxe binding.

This was an important early contribution to the history of Australia and the pre-Cook discovery of the continent, and the first serious attempt at a geographical history of the continent. Although outmoded, it is a standard and basic work for the collector; Collingridge was able to claim "I believe I can truly say that I have read every book, and examined every map of real importance to the question, which has been produced in English, French, Spanish, Portuguese, Italian and Dutch".

Ferguson, 8465. $1285

PRE-COOK DISCOVERY

15. COLLINGRIDGE, George. The First Discovery of Australia and New Guinea. Being the Narrative of Portuguese and Spanish Discoveries in the Australasian Regions, between the Years 1492-1606...

Octavo, illustrations and maps (some coloured); a fine copy in the original cloth. Sydney, William Brooks, 1906.

An important book in the development of any history of pre-Cook discovery in Australia. $325

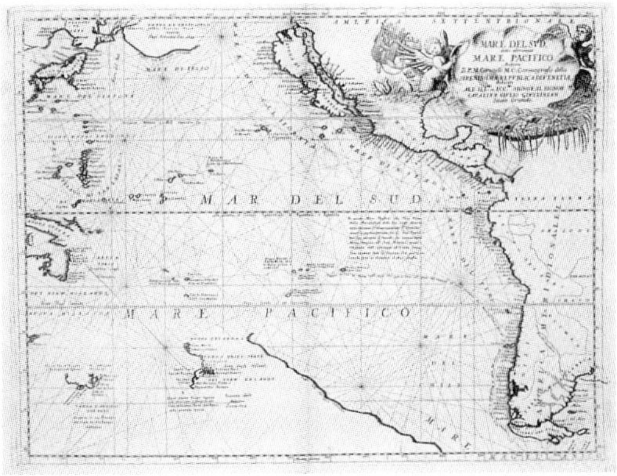

16. Coronelli

FINE PACIFIC MAP

16. CORONELLI, V.M. Mare del Sud detto altrimenti Mare Pacifico.

Engraved map, 445 x 600 mm.; in fine condition, mounted. Venice, (1696).

A fine Pacific map by the great Venetian cartographer. Coronelli's splendid map depicts the coast of America from California (shown as an island) to Terra del Fuego. New Zealand is shown as a land of enormous extent, with the eastern coastline marked "Piaggie non per anco conosciute" and a note of Dutch discoveries in 1654. Southern Tasmania has the Tasman discoveries noted, as does northern Australia which is joined to New Guinea.

Tooley, 350. $3750

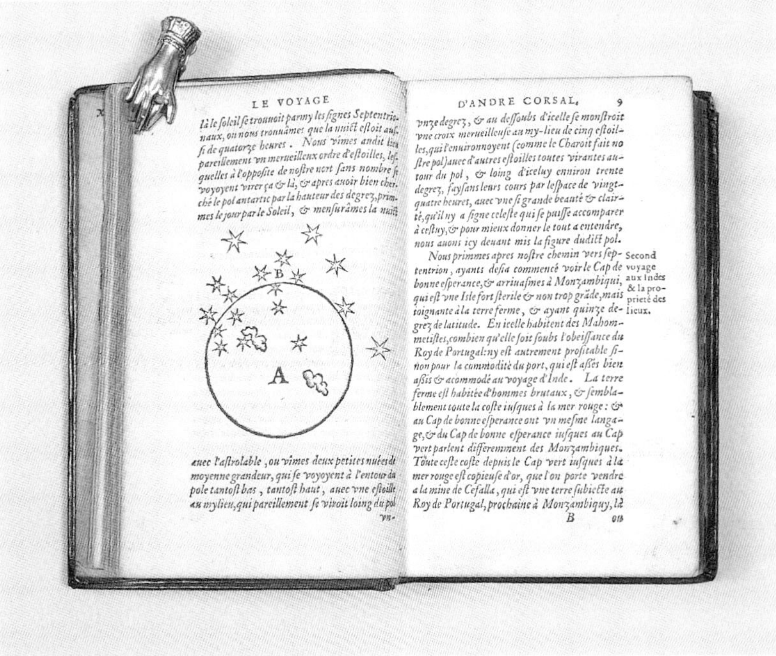

17. Alvares

THE SOUTHERN CROSS

17. [CORSALI] ALVARES, Francisco. Historiale description de l'Ethiopie... [with Corsali's letters about his discovery of the Southern Cross].

Small octavo, printed in italic and roman letter, with a woodcut of the stars of the Southern Cross and six woodcut plans of Ethiopian churches; a fine copy in sixteenth-century French calf, decorated with interlaced Greek lambdas. Antwerp, J. Bellere, 1558.

Including a narrative of great significance to Australia and the southern hemisphere: Corsali's description of his identification of the Southern Cross. The main purpose of the book is the eyewitness account, by its chaplain, of the first Portuguese embassy to reach Ethiopia in 1520-26, here in its first separate edition in French. First published in Portuguese in 1540, this was the earliest first-hand description of Ethiopia by a known European and of great importance for the history of the country, not least for its unequivocal identification of King David of Ethiopia as the Prester John of the Indies.

Alvares's narrative is preceded by the two letters of Andrea Corsali, included here because this Florentine traveller ended his days in Ethiopia. The first letter is of outstanding interest for Corsali's account of the Southern Cross, sighted by him on his voyage to India in 1515. His description and illustration of the constellation was the first to outline its shape in detail as a cross: after the publication of his *Lettera* the term "cross" or "crosiers" recurs frequently and in 1606, for example, Quiros, on his quest for the Southern Continent, instructed his captains to ascertain their position at night by the "crucero". The narrative also contains a tantalising reference to a continental land in the vicinity of New Guinea, which alone would make the *Lettera* an important element in the canon of pre-Cook discovery of Australia and the Pacific.

The first editions of Corsali's two letters appeared in Italian in 1516 and 1517 respectively and are both of utmost rarity; the important 1516 letter is known in only three copies.

Alden, 558/3; Atkinson, 117n; Voet, 53B. **$15,500**

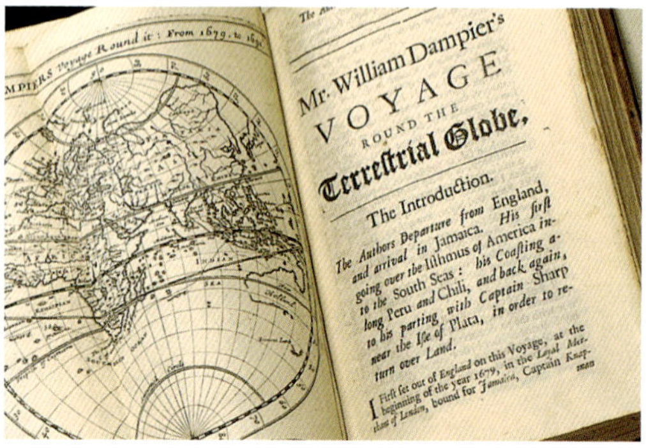

22. Dampier

DAMPIER'S FIRST VOYAGE

18. DAMPIER, William. A New Voyage Round the World. Describing particularly, The Isthmus of America, several Coasts and Islands in the West Indies... the Isle of Guam one of the Ladrones, Mindanao, and other Philippine and East-India Islands near Cambodia, China, Formosa, Luconia, Celebes, &c. New Holland...

Octavo, with five maps (four folding) and some woodcut images in the text; margins of several maps frayed (some loss to the edge of East Indies map restored), a few stains; contemporary panelled calf, joints repaired, new endpapers. London, James Knapton, 1697.

Dampier's first voyage account, describing adventures in the West and the East Indies, and including the first account of the first English landing in Australia. This is the second edition, corrected from the first edition of the same year. In 1688, "being now clear of all the Islands, we stood off South, intending to touch at New Holland, a part of Terra Australis Incognita, to see what the country would afford us... New Holland is a very large tract of Land. It is not yet determined whether it is an Island or a main Continent; but I am certain that it joyns neither to Asia, Africa, nor America...".

Dampier describes the appearance of the country and its natural produce, and includes a long description of the inhabitants. The folding "Map of the East Indies" shows a large proportion of the north-west coast of Australia and marks the point (present-day Dampier Land) where the expedition stopped before turning back to the East Indies.

See also catalogue number 345, Giolo, for Dampier's "Painted Prince", Jeoly.

Hill, p. 75; Wing, D162. **$1650**

THIRD EDITION CORRECTED

19. DAMPIER, William. A New Voyage Round the World...

Octavo, with five folding maps and some woodcut images in the text; a good copy in modern half calf. London, James Knapton, 1698.

Dampier's first voyage account: third edition corrected. There were a number of early editions of this important book, the seventh appearing in 1729 as volume 1 of his collected works (see below).

Hill, pp. 75-6; Wing, D163. **$1700**

THE CONTINUATION

20. DAMPIER, William. Voyages and Descriptions. Vol. II... A Supplement of the Voyage round the World... Two Voyages to Campeachy... A Discourse of Trade-Winds...

Octavo, four folding maps; contemporary calf, a few stains and a bit rubbed. London, James Knapton, 1699.

First edition of the continuation of Dampier's *New Voyage round the World*, described as Volume II on the title-page (and numbered 2 on the spine) but quite separately published. The first part is the most interesting, dealing with Dampier's experiences in the Far East and the East Indies.

Hill, p. 76; Wing, D165. **$900**

VOYAGE TO NEW HOLLAND

21. DAMPIER, William. A Voyage to New Holland, etc., in the Year 1699...

Octavo, with five maps and ten plates, the frontispiece map frayed at one edge; contemporary panelled calf, a few stains and marks but a good copy, joints rubbed but sound, gilt-panelled spine. London, James Knapton, 1703.

The important and rare first edition of the first deliberate English voyage to Australia, one of very few classic pre-Cook voyages to the continent. Dampier was the first Englishman to visit Australia when he stopped on the west coast in 1688, but he remained only a short time (publishing an account in 1697 – see above). In 1699, he set sail again on this voyage to New Holland. His account represented a third volume of adventures, and the title-page here (and the binder's label on the spine) refers to this as Vol. III though it was in fact published quite separately. The engraved plates – birds, fish and plants – are among the very earliest depictions of Australian natural history.

Widely regarded as the greatest English explorer and navigator before Cook, Dampier was also a popular and exciting writer. His books went through many editions – indeed he is still in print in one form or another – and copies of the first edition of this voyage to New Holland are now uncommon.

Hill, p. 76. **$5500**

THE COLLECTED DAMPIER

22. DAMPIER, William. A Collection of Voyages. In Four Volumes...

Four volumes, octavo, with many engraved plates and maps; an attractive set in period-style panelled calf. London, James and John Knapton, 1729.

An excellent set of the collected edition of Dampier. By 1729 Dampier's published work had reached a total of three volumes, and a fourth volume added by the publishers contains Funnell's account of the Dampier voyage as well as the narratives of Cowley, Sharp, Wood and Roberts. Dampier's publishers Knapton used the latest editions of each of the four volumes with a new general title-page to produce the collected edition as it appears here; thus the first volume is in its "seventh edition, corrected" form, while the others are represented by third or fourth edition texts.

Dampier's complete works represent a major body of Pacific description, and are of considerable importance for any study of the discovery and colonisation of the Pacific. The "four-volume Dampier" is generally considered the best edition of the various works, and is quite scarce today; this is a good set in an attractive binding.

Hill, pp. 77-78. **$14,000**

FIRST MAP OF AUSTRALIA

23. DE JODE, Gerard. Novæ Guineæ Forma & Situs...

Copper-engraved map, approx. 340 x 215 mm.; very good original condition, framed. Published as part of 'Speculum Orbis Terræ', Antwerp, 1593.

A rare and important sixteenth-century map, of great interest to early Australian and Pacific exploration. Considered by both the experts Schilder and Tooley to be the first printed map of Australia, De Jode's map predates by over a decade the first Dutch explorations in this area – Willem Jansz's voyage in the *Duyfken* in 1605 "to discover the great land Nova Guinea and unknown east and south lands". This map is thus either imaginary, or composed from scattered information, possibly of Portuguese origin, that would have been available to Europeans in the late sixteenth century.

"The speculation of a Southern Land had always preoccupied geographers. The existence of Terra Australis was further "proved" by reports by sailors who had been driven off course and sighted land far to the southward. This combined with a belief that Tierra del Fuego, sighted by Ferdinand Magellan's ships in 1520, was a headland of Terra Australis (as shown, and for the first time so called, by Oronce Finé on his world map of 1531). Yet Magellan's men thought the land they saw consisted of islands, and Francis Drake in 1578 "plainly discovered" that it was "no continent" but "broken islands" instead..." (*Landmarks of Mapmaking*).

Hondius showed a strait between New Guinea and a coastline of Terra Australis on his world map of 1590; this was evidently a lucky guess based on Sir Francis Drake's explorations, but was not finally proven until the voyages of James Cook nearly two hundred years later.

New Guinea had been named by the navigator Ortiz de Retes in 1545. The Latin inscription on the map states that it is uncertain whether New Guinea is an island or part of Terra Australis. The map also shows the Solomon Islands, and more than half of the engraved surface depicts the north of Australia. There are engravings of a native with bow and arrow attacking a gryphon, and a snake and a lion.

23. De Jode

A separate southern continent had appeared in world maps since the first part of the sixteenth century, but there is no earlier map that shows "an Australia" in such specific terms. De Jode maintained, in accompanying text, that "...after this region the huge Australian land follows which – as soon as it is once known – will represent a fifth continent, so vast and immense is it deemed...".

'Landmarks in Mapmaking', p. 247; Perry, p. 16; Schilder, map 13; Tooley, 385. **$15,500**

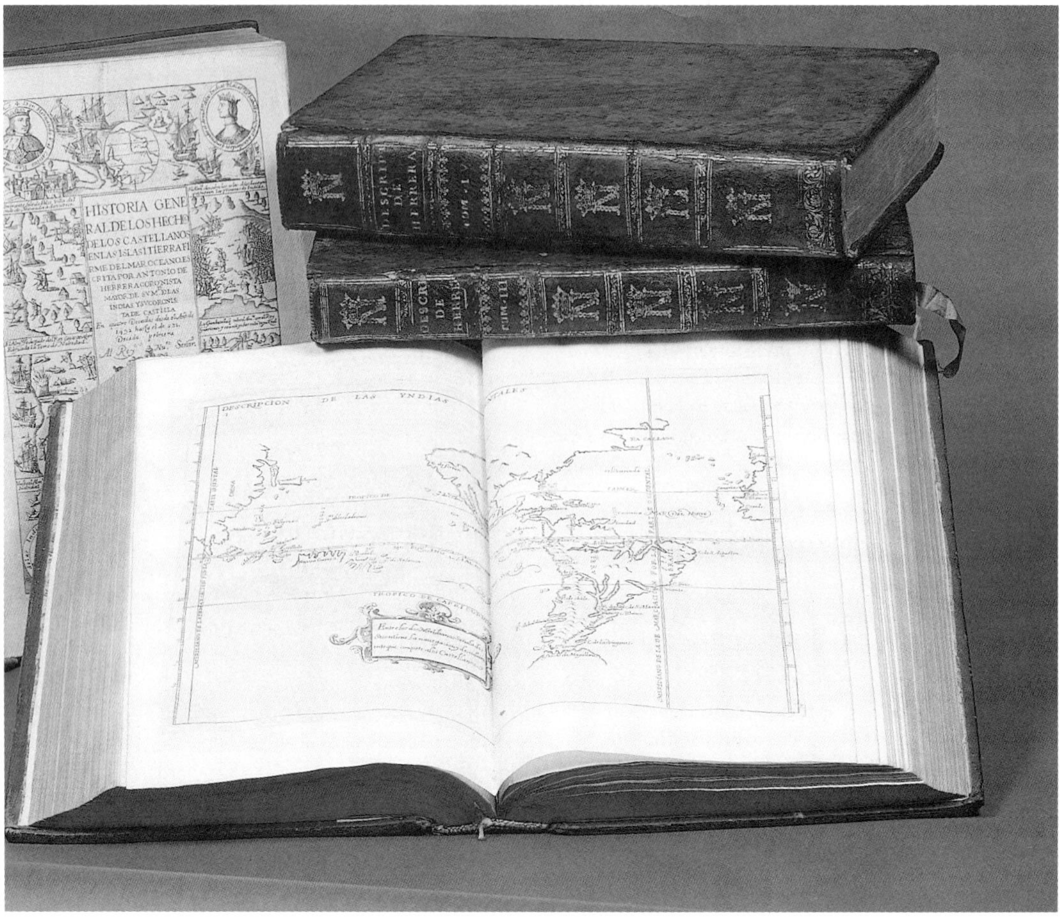
26. Herrera

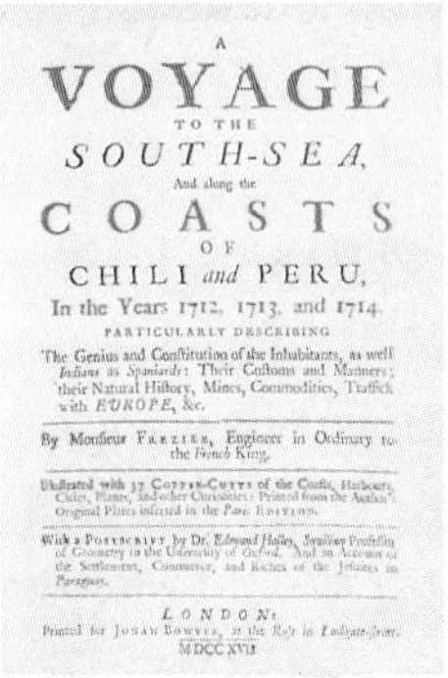
A

VOYAGE

TO THE

SOUTH-SEA,

And along the

COASTS

OF

CHILI *and* PERU,

In the Years 1712, 1713, and 1714.

PARTICULARLY DESCRIBING

The Genius and Constitution of the Inhabitants, as well *Indians* as *Spaniards*: Their Customs and Manners; their Natural History, Mines, Commodities, Traffick with *EUROPE*, &c.

By Monsieur FREZIER, Engineer in Ordinary to the *French* King.

Illustrated with 37 COPPER-CUTTS of the Coasts, Harbours, Cities, Plants, and other Curiosities: Printed from the Author's Original Plates inserted in the *Paris* EDITION.

With a POSTSCRIPT by Dr. *Edmund Halley*, Savilian Professor of Geometry in the University of Oxford. And an Account of the Settlement, Commerce, and Riches of the Jesuites in *Paraguay*.

LONDON:

Printed for JONAH BOWYER, at the Rose in *Ludgate-Street*.

MDCCXVII

24. Frezier

PACIFIC COAST OF SOUTH AMERICA

24. FREZIER, Monsieur. A Voyage to the South-Sea, and along the Coasts of Chili and Peru...

Quarto, with 22 maps and plans and 15 engraved plates and views; an attractive copy in contemporary calf, gilt, spine gilt in compartments with double labels; spine worn at top. London, Jonah Bowyer, 1717.

Rare: the first English edition, with a new postscript by Edmund Halley dealing with South American navigation. Frezier sailed under the Horn to reach the west coast of South America, and his description was much used by later navigators. He gives good descriptions of Chile and Peru. One of his odder claims to fame is the introduction of the strawberry to Europe from Chile. It is illustrated here.

Borba de Moraes, p. 329; Hill, p. 115; Sabin, 25926. **$9250**

DUTCH DISCOVERIES

25. HEERES, J.E. The Part Borne by the Dutch in the Discovery of Australia 1606-1765...

Folio, with two large folding coloured charts (some small tears) and 17 maps and figures in the text; original green cloth, a little rubbed but a very good copy. London and Leiden, 1899.

First edition, with text in Dutch and English on opposite pages. Published by the Royal Dutch Geographical Society to commemorate its twenty-fifth anniversary, this represented a major step forward in the study of the early discovery of Australia. Heeres was also the editor of the first publication of Tasman's journal.

Ferguson, 10355. **$725**

HERRERA'S DECADES

26. HERRERA, Antonio de. Descripcion de las Indias Occidentales... [*with*] Historia General de los hechos de los Castellanos en las Islas i Tierra Firma del Mar Oceano [Decades 1-8]...

Nine parts bound together in four volumes, folio, with engraved title-pages (incorporating portraits of missionaries, generals, governors, and soldiers, views of America, etc); with fourteen engraved maps; eighteenth-century mottled calf, a bit rubbed and joints cracking but firm, some neat restorations; crowned monogram "N" in compartments of spines; stored in two fitted quarter calf boxes. Madrid, Imprenta Real, 1730.

A most attractive set of the second and best edition of Herrera's *Decades*. This massive work, one of the primary accounts of the early Spanish conquest of the New World, was originally published in Madrid between 1601 and 1615 in eight "decades" spanning the years 1492 to 1554. Only the first three were ever translated, although the work as a whole was "of the greatest importance for the history of the conquests, colonization and progress of the Spaniards in America..." (Hill).

The *Descripcion* that accompanies the *Decades* is also of great significance: it first appeared in 1622, in several languages, and contained the first true publication of the journal of the Le Maire-Schouten expedition around the world (see Schilder, pp. 34-6). Sailing through Lemaire Strait, and rounding Patagonia via Cape Horn, this voyage had an enormous effect on future navigation, helping to put an end to the idea of the imaginary Southland. The series of fourteen maps that accompany the *Descripcion* includes one of the Pacific with the Straits of Magellan clearly mapped and another of the East Indies with good details of the Solomon Islands and of the north coast of New Guinea in both.

This excellent set has on its spines repeated crowned "N" monograms – which we have not been able to identify. This copy was once confidently described as having come from Napoleon's library, but the monograms certainly date from the time of binding, and the binding is simply too early. Ollivier does not record any other suitable candidates among the French book-collecting aristocracy. A number of leaves in this set are bound out of order (not surprisingly given the complexity of the book's make-up) but all are present. A full collation is available on request.

Alden, 730/117-188; Borba de Moraes, p. 401; Hill, p. 145; Sabin, 31541 & 31546 ("deservedly regarded as the best edition"); Palau, 114287-8; Wagner 'Spanish South-west', 12K & 12L. **$17,500**

SOUTHERN HEMISPHERE MAP

27. HONDIUS, Henricus and Johannes JANSSON. Polus antarcticus.

Handcoloured engraved map, 435 x 495 mm.; mounted. Amsterdam, 1650.

This classic map of the Southern Hemisphere centred on the South Pole was published in many editions during the mid-seventeenth century. Tasman's discoveries were not shown until 1657, so this version shows the extent of knowledge of the continent just before Tasman. It shows the discoveries of Pieter Nuyts on the south coast of Australia, as well as the earlier history-making Dutch landfalls on the west Australian coastline.

The decorative borders here are filled with historical detail from penguin-hunting to bizarre native costumes, and this is a fine example of the decorative presentation shown in several important Dutch maps of the period.

Schilder, map 44 (a 1639 version). **$3250**

FIRST DUTCH VOYAGE
TO THE EAST INDIES

28. [HOUTMAN] LODEWYCKSZOON, Willem. Prima pars descriptionis itineris navalis in Indiam Orientalem...

Quarto, folding engraved plate, engraved map on the title-page, 48 engraved maps and illustrations; several woodcut coastal profiles and charts in the text; neat library-stamps on title; an excellent copy in old boards, respined in calf. Amsterdam, Cornelis Nicolaus, 1598.

The first Dutch voyage to the East Indies: this is one of two original descriptions of the voyage that survive (the other is attributed to Pieter Stockmans and others – see below). This narrative is based on the original journal of Lodewyckszoon, one of the officers of the fleet, who sailed aboard the *Mauritius* – and so differs substantially from the other account.

Houtman's voyage was the very first trading fleet sent by the Dutch to the East Indies: four ships were sent out in 1595 under the command of Cornelis de Houtman and Pieter Dirckz, three of which returned to the Netherlands in 1597. Their venture prompted the departure of several further fleets, culminating in the establishment of the VOC in 1602.

The most significant geographical achievement of Houtman's voyage was that by sailing round Java it proved that it could not be part of the supposed southland (see Schilder, *Australia Unveiled,* ch. VI). The map on the title-page of this account shows the oceans and coasts navigated between Europe and the East Indies, including the promontory of a southern continent, with the names "Beach", "Lucach" and "Maletur", in close proximity to an island, "Java Minor".

Lach points out that the book "contains the first continuous account of Bali in any European language, Its prosperity is described in... great detail...". The long and detailed description of Java is probably derived from Portuguese accounts, and "contains information about Portuguese trade and forts in the Moluccas where De Houtman's ships did not visit. Perhaps it came from Pedro de Tayde, the Malaccan-born Portuguese pilot who was so helpful to De Houtman's men in Bantam. In any case, it has not been found in any of the printed Portuguese sources".

This is the first edition in Latin of this narrative: editions in Dutch and French were also published in Amsterdam in the same year. The engraved maps and illustrations are common to all three editions and are listed by Tiele (pp. 125-7), who notes that the folding plate of the bazaar at Bantam (present in this copy) is nearly always lacking.

Alden, 598/60; Lach III, pp. 438-9 & pp. 1302-59 passim (illustrated in section following p. 1380); Tiele, 510; Tiele-Muller, 112; not in Landwehr. **$22,850**

28. Lodewyckszoon

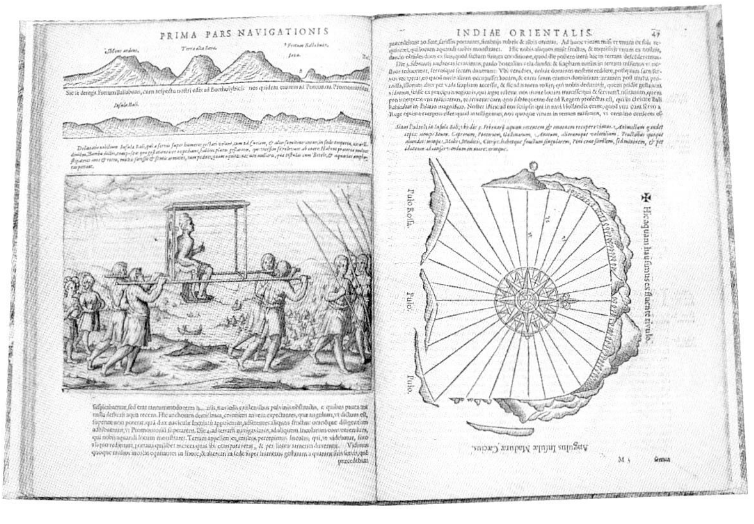

29. Stockmans

ALTERNATIVE ACCOUNT OF THE HOUTMAN VOYAGE

29. [HOUTMAN] STOCKMANS, Pieter and others (attributed). Diarium Nauticum Itineris Batavorum in Indiam Orientalem...

Oblong folio, woodcuts, 15 plates and maps; large map of Bali, woodcut of birds and 24 coastal profiles; some early underlinings in the text; contemporary vellum, slightly stained, an excellent copy. Arnhem, Johan Jansonius, 1598.

The first Dutch voyage to the East Indies: this is one of two original descriptions of the voyage that survive (the other is by Lodewyckszoon – see above). Lach, who points out that "Firsthand reports of insular Southeast Asia arrived in the Netherlands with Cornelis de Houtman's fleet in August, 1597", also notes that Rouffaer and Ijzerman have conjectured that "the author or more likely authors [of this narrative] had apparently sailed on the "Hollandia"; one of them was possibly a midshipman named Pieter Stockmans...".

This very rare and important book gives a day-by-day narrative of this highly significant adventure, the first of what was to be an almost endless series of voyages made by the Dutch in the most ambitious and successful of all colonial ventures. The history of the VOC and their trading from bases in the East Indies effectively begins here, with a speculative voyage commissioned by a group of nine major Amsterdam merchants in 1595-1597. "In 1598, the year after its return, no fewer than twenty-five ships were sent out by merchants of the provinces of Holland and Zeeland alone..." (Lach).

The narrative, richly illustrated with maps, views and plates, includes the often lacking Appendix, a pictorial pilot with woodcuts illustrating the coasts of the Cape of Good Hope, the waters of S. Bras, the Capes of St. Just and Augusta, and parts of the coasts of Madagascar, Sumatra and Java, as well as a fine full-page engraved map of the island of Bali.

This is the first edition in Latin and thus the first international edition; a Dutch-language edition appeared in the same year, while a different, shorter version in Dutch had appeared the year before (entitled *Verhæl vande reyse* rather than *Journæl vande reyse*, it evidently derived from the same original journal; it was known to Tiele at the end of the nineteenth century in a single copy, and Alden in turn could cite only the Nederlandsch Historisch Scheepvaart Museum copy in 1980). Three issues of this Latin version are known: this version for distribution by an Arnhem publisher, another with the Paris address of Adrianus Périer, and – most often seen – an issue with the Middleburg address of Bernardus Langenese. This excellent copy of the Arnhem issue in original binding has the ownership inscription of Caspar a Mekerad, dated 1600, on the title-page.

Alden, 598/47; Lach, pp. 437-8; Rouffaer and Ijzerman, 'De eerste schipvaart der Nederlanders naar Oost-Indië onder Cornelis de Houtman', The Hague, 1915-35, II, pp. xxii-xxx; Tiele, 505n; Tiele-Muller, 108n; not in Landwehr. **$28,500**

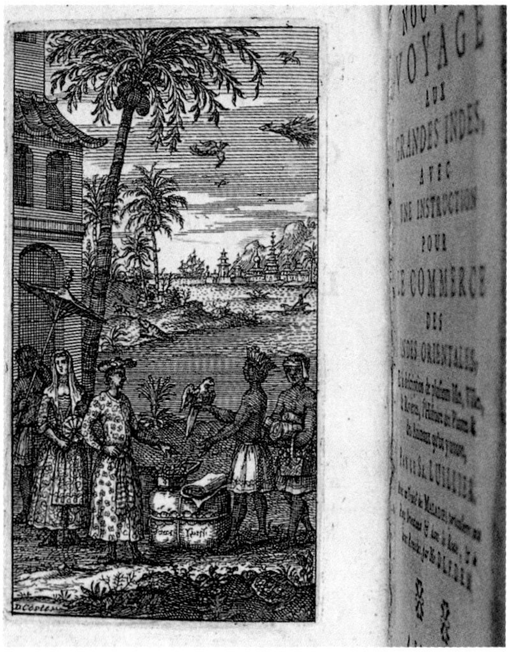

30. Luillier-Lagaudiers

EARLY TRAVELS TO INDIA

30. LUILLIER-LAGAUDIERS, —. Nouveau voyage aux Grandes Indes avec une instruction pour le commerce des Indes Orientales, Et la description de plusieurs Isles, Villes & Rivières, l'Histoire des Plantes & des Animaux qu'on y trouve...

Small octavo, engraved frontispiece, title in black and red; small stain in some outer margins well clear of text, otherwise a fine copy in contemporary French calf gilt. Rotterdam, Jean Hofhout, 1726.

Rare account of early travels to India, concentrating on Bengal and Pondicherry, and with a section on trade and commerce in the East Indies. This edition has an additional chapter on diseases encountered en route to the East Indies and their treatment ("par Mr. D.L.F.D.E.M."), which the James Ford Bell catalogue notes only in the Rotterdam edition of 1742.

See James Ford Bell, L526-8. **$4850**

MAGELLAN FROM MANUSCRIPTS

31. [MAGELLAN] PIGAFETTA, Antonio. Relation du premier voyage autour du monde par Magellan. 1519-1522...

Large octavo, numerous illustrations; unopened in the original wrappers, a few spots around spine but an excellent copy. Anvers, Paris, 1923.

A handsome narrative of Magellan's circumnavigation, edited by Denucé from the two manuscripts in the Bibliothèque Nationale, Paris and the Phillipps manuscript (now in the Beinecke Library). **$375**

MAGELLAN & THE EAST INDIES

32. [MAGELLAN] PENROSE, Boies. A Link with Magellan. Being a Chart of the East Indies...

Large quarto, double-page colour facsimile; a fine copy in the original boards, uncut. Privately Printed, 1929.

A handsome publication, very scarce on the market: only 150 copies were printed for private distribution by Penrose, one of the great collectors of voyage material and a considerable scholar in his own right. This examination of a Spanish manuscript chart in his possession points to Magellan and to general Portuguese knowledge as sources of information for the impressive knowledge of the East Indies and New Guinea shown in the manuscript. **$325**

VOYAGES TO TERRA AUSTRALIS

33. MAJOR, R.H. Early Voyages to Terra Australis, Now called Australia...

Octavo, with six maps; a good copy in contemporary calf, rebacked; gilt stamp of the New Zealand General Assembly on front cover. London, 1859-61.

A well-bound copy of this work on pre-Cook exploration in the Pacific which assembles important and often little-known documents, mainly translated from the Dutch and Spanish. It covers the period from the beginning of the sixteenth century to the Cook discoveries and includes narratives by Torres, Dampier and several Dutch explorers.

Ferguson, 12195. **$1425**

KING MANUEL'S LETTER

34. MANUEL, King of Portugal. Epistola potentissimi ac invictissimi Emanuelis Regis Portugaliæ & Algarbiorum, &c. De victoriis habitis in India & Malacha, ad Leonem X Pont. Maximum.

Small quarto, 4 pp. with drop-title, roman letter; in fine condition in recent quarter morocco. "Rome, Jacob Mazochius, 9 August 1513" [?in fact Venice, c. 1513].

King Manuel's letter to Pope Leo X announcing the spread of Portuguese dominion in the East Indies after the victories at Goa and Diu, and the capture of Malacca, effectively sealing Portuguese control of the South-east Asian trade: as the emporium for spices from the Moluccas and goods from China its strategic value was immense. Malacca became the base for Portuguese penetration of the archipelago and northward into the China Sea.

This copy of the letter is in a previously unknown edition, probably the second or the third. The first edition appeared in Rome in 1513 with the imprint of Jacob Mazochius. Our edition has the same colophon and date but is in a smaller type, on two instead of six leaves. It has been pointed out that no Rome printer of this period is recorded as employing a type this small; there is a strong possibility that this may be a piracy, in which case one should look to Venice as the likeliest place of printing.

Within a year the letter had been reprinted not only in Rome (Streit, 380), but also in Strasbourg and Vienna (*ibid.*, 380n), and it also circulated in German translation (*ibid.*, 381 & 382). These various editions attest to immense contemporary curiosity in the geopolitics of the new world.

"It was not until 1511 that the Portuguese captured the great entrepot of Malacca and began to penetrate farther eastward. News of Albuquerque's victory became known in Europe in the spring of 1513, and Manuel quickly dispatched letters to Rome informing the papacy of Portugal's progress. Manuel's speedy proclamation of the capture of Malacca had certain specific objectives apart from his understandable desire to let the world know of his distant conquests. The question had already been raised in Spain whether the Moluccas were not within the Spanish demarcation, and a project was being considered in 1512 for advancing a claim to and taking possession of the Spice Islands. Ever watchful as to the security of his demarcation claims, Manuel immediately set to work to obtain papal support and recognition of Albuquerque's advances into south-eastern Asia. And, it was Manuel's good fortune that a pope had been elected in March 1513 who was likely to receive the news cordially.

"Pope Leo X (Giovanni de' Medici) was the second son of Lorenzo the Magnificent and the real ruler of his family and the Republic of Florence. The traditional good relations between Portugal and the Florentine merchants and navigators certainly did no harm to Manuel's cause in the eyes of the urbane Medici pope. Moreover, a strong case could be made in Rome for confirming Manuel's claims inasmuch as the Portuguese were regularly defeating Muslims and constantly enlarging the field of possible Christian missionary enterprise. In fact, public celebrations of thanksgiving were held in Rome shortly after the news of Albuquerque's Conquest of Malacca arrived" (Lach).

JFB, M129 (Rome edition); Lach, I, pp. 166-7; Streit, 'Bibliotheca Missionum', IV, 380-82. **$78,500**

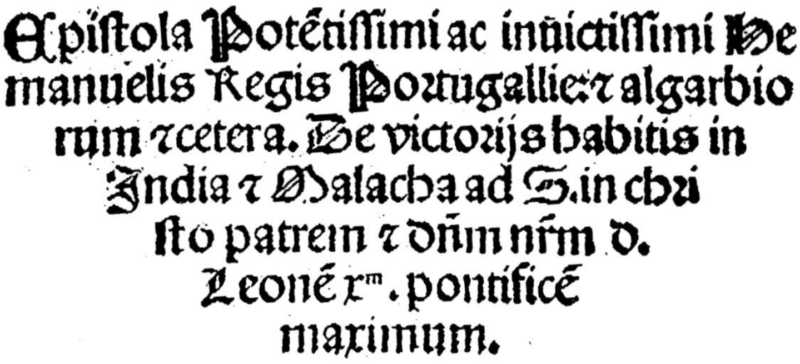
Epiſtola Potētiſſimi ac inūictiſſimi De manuelis Regis Portugallie: ⁊ algarbio rum ⁊ cetera. De victorijs habitis in India ⁊ Malacha ad S. in chri ſto patrem ⁊ dn̄m nr̄m D. Leonē x^m. pontificē maximum.

34. King Manuel

35. *Pomponius Mela*

ANTIPODEAN BALANCE

35. MELA, Pomponius. Pomponii Melæ de situ orbis libri III... curante Abrahamo Gronovio.

One volume bound in two, octavo, with a folding engraved plan and a number of engraved vignettes; a fine copy in eighteenth-century red morocco, flat spines gilt, double olive morocco labels, gilt edges, marbled endpapers. Leiden, Samuel Luchtmans, 1722.

A most attractive copy of this standard text of the ancient geographer Pomponius Mela, edited by Gronovius; this copy probably from the Lamoignon library, in typically attractive binding; later in the library of the National Geographic Society.

Pomponius Mela's view of the shape of the world incorporated a huge land mass surrounding the south pole – the "early notion of a fifth continent and that notion's important historical development as an intermingling of fact and fiction. The phantom of *Terra Australis*, an unknown south-land, haunted the minds and maps of cosmographers for more than two millennia... The landmasses of the northern hemisphere needed a counter-balance in the south. These speculations... were still influential in the eighteenth century until the time of James Cook. It was felt that an undiscovered southern continent had to exist because the known land masses of the southern hemisphere were not sufficient to balance those of the northern half of the globe..." (Schilder, *Australia Unveiled*, pp. 7-8). **$4000**

MORTIER'S EASTERN HEMISPHERE

36. MORTIER, Pierre. Partie Orientale du Monde qui contiennent l'Europe, l'Asie et Afrique...

Engraved map, in contemporary colour; 593x 502 mm.; mounted. Amsterdam, Covens and Mortier, c. 1730.

A rare map of the eastern Hemisphere, extending eastwards from South America to New Guinea: an earlier version first appeared in the *Neptune François* produced in Paris in 1693. The world map on Mercator's projection was issued in two sheets (this is the eastern sheet) under the general title *Carte Generale de Toutes les Costes Du Monde*; each sheet also had its own title in an oval cartouche.

Mortier re-engraved the *Neptune* after returning to Amsterdam and published several counterfeit versions. A slightly larger version of the map was engraved by Baltasar Ruyter in 1703, with additional changes to the Asian sheet, and this state was re-issued in the 1730s with the imprint of Covens and Mortier as here. The northeast coast of Asia has far more detail and unidentified promontories extending from the coast. Australia is charted according to the discoveries of Tasman's voyages of 1642-43 & 1644. Many of the early Dutch discoveries in Australia are shown including those of Carstensz (1623), de Wit (1625), Hartog (1616) and Houtman.

Shirley 559, pl. 389 (the earlier edition). **$3250**

37. *Nodal*

RELACION
DEL VIAGE,
QUE POR ORDEN DE SU MAGESTAD,
Y ACUERDO DE EL REAL CONSEJO
DE INDIAS,
HICIERON LOS CAPITANES
BARTHOLOME GARCIA DE NODAL,
y Gonzalo de Nodal, Hermanos, naturales de Pontevedra,
AL DESCUBRIMIENTO
DEL ESTRECHO NUEVO DE SAN
Vicente, que hoy es nombrado de Maire, y
Reconocimiento del de Magallanes.
REIMPRESSO DE ORDEN
DEL Sr. D. JOACHIN MANUEL DE VILLENA
y Guadalfaxara, Marquès del Real Theforo, Cavallero del Orden de S. Juan, del Consejo de S.M. Gefe de Esquadra de la Real Armada, y Presidente de la Real Audiencia, y Casa de la Contratacion á las Indias:
EN UTILIDAD DEL HOSPICIO DE LA SANTA Charidad de la Ciudad de Cadiz.
LLEVA AÑADIDO
LAS DERROTAS DE LA AMERICA OCCIDENTAL de unos Puertos á otros, que diò á luz el Theniente de Navio de la Real Armada Don Manuel de Echavelar.
CON LICENCIA EN MADRID:
Y Reimpresso en Cadiz por *Don Manuel Espinosa de los Monteros*, Impressor de la Real Marina, Calle de S. Francisco.
Donde se hallarà.

STRAITS OF MAGELLAN

37. NODAL, Bartolome García de. Relación del Viage... al descubrimiento del estrecho nuevo de San Vicente, que hoy es nombrado de Maire, y reconocimiento del de Magallanes...

Small quarto, with a folding map; a fine copy in contemporary quarter calf (labelled as one volume in a series), later marbled sides. Cádiz, Manuel Espinosa de los Monteros, [1766 or 1769].

Second edition, and very scarce: the original 1621 edition is noted by Sabin as "one of the rarest books of its class". No date is given on the title-page of this edition; the licence is dated 1766, while the map is dated 1769. The second part is new to this edition: separately paginated, it has its own title-page ("Instruccion exacta, y util de las derrotas, y navegaciones, que se executan en todos tiempos en la America Septentrional ... Sacala a luz D. Manuel de Echevelar ... Año de 1753. Con Licencia: en Cadiz en la Real Imprenta de Marina, Calle de San Francisco").

This voyage marked the Spanish reflex reaction to the Dutch discovery of the Le Maire Strait (called San Vicente by the Spanish); it had been discovered by Le Maire and Schouten during their 1616 voyage in search of the Southern Continent. The Dutch success alarmed the Spanish – but they were also interested in a safer means of passage into the South Sea; they therefore sent out two caravels under the command of the brothers Bartolome García de Nodal and Gonzalo de Nodal, to report on the Le Maire and Schouten discoveries.

"This was the first circumnavigation of Tierra del Fuego. The whole voyage, out and back, had taken only nine months and twelve days: 'a period which, in the present state of navigation, would be reckoned very short for the performance of such a voyage, and was then unprecedented' (Burney) – many a Dutch and English ship took longer just to reach Cabo Virgines. The ships had never once separated, and except for one man hanged at Rio for plotting mutiny, there was not a single death. Rapid as the voyage was, it was not scamped; the Nodals brought back careful and clear sailing instructions, and paid unusual attention to tidal observations. Their conduct of the voyage was indeed a model of decision and efficiency; but they fell victims to the general slackness which was creeping over Spanish naval affairs: the Spain of Philip III was not that of Philip II" (Spate II, p. 26).

Borba de Moraes, p. 616; Medina, 'Hispano-Chilena', 468; Palau, 99486; Sabin, 55395-6. **$12,500**

NODAL MANUSCRIPT

38. [NODAL, Bartolomé Garcia de] Two manuscript documents in Spanish concerning the appointment of the explorer Nodal to command an expedition to the Straits of Magellan.

Folio, manuscripts in ink on paper; pp. [4]; [6], stitched together in black paper wrappers. Spain, c. 1618.

A remarkable pair of voyage manuscripts, once in the Phillipps collection: the King of Spain's appointment of Nodal to explore the Straits of Magellan, with detailed instructions for the conduct of the expedition.

These two documents are contemporary copies (one may possibly be a draft) of the instructions given in the name of Philip III of Spain to Nodal in August 1618, and offer a remarkable glimpse of the careful planning and organisation which preceded the departure of his expedition. The maintenance of discipline is stressed, as is the need to work with his brother and deputy, Gonzalo, and chief navigator, Diego Ramirez de Arellano; the need to proceed by the most direct route, avoiding distractions; the careful supervision of the pilots and recording of navigational observations, including the making of fair copies of charts; the embarking of stores for ten months; and the importance of the two ships staying together. Directions are given for navigating the approaches to the Strait of Magellan and the supposed location of the Strait of Le Maire is indicated; possible tidal and meteorological conditions are described. In essence the documents reveal that little was left to chance and that Nodal was despatched well briefed and with clearly defined procedures and objectives, which the actual course and conduct of the voyage reveal were closely and successfully followed. These instructions were not printed in Nodal's account of the expedition (Madrid, 1621; or Cadiz, c. 1769 – see above).

These manuscripts come from the collection of Sir Thomas Phillipps (MS 13306, Sotheby's, 27 June 1938, lot 35 (£11); a pencil note states "Puttick, 24 May 1854, lot 7371", at which sale the pair was presumably acquired by Phillipps). **$24,000**

FACSIMILE-ATLAS AND PERIPLUS

39. NORDENSKIOLD, Adolf. Facsimile-Atlas... [*and*] Periplus an Essay on the Early History of Charts and Sailing Directions.

Two volumes, folio, with 111 double-page maps and 184 maps and figures in the text; original half roan, the spines slightly worn. Stockholm, 1889-1897.

A pair of seminal cartographic works. The *Facsimile-Atlas* is a detailed and indispensible study, which includes numerous beautifully-produced facsimiles, chapters on Ptolemy's atlas and the various editions of it, ancient non-Ptolemaic maps, the first New World maps, and traces the history of cartography down to the 'modern' period of Apianus, Gastaldi, Mercator and Ortelius.

The *Periplus*, intended as a companion work, contains chapters on pre-Ptolemaic cartography, maps and sailing directions of the middle ages, portolans, and the discovery and charting of Africa, Asia and the Pacific to the time of Tasman.

Besterman, 1153. **$5750**

THE SERIES OF FACSIMILE MAPS

40. NORDENSKIOLD, Adolf. Facsimile-Atlas...

Folio, with many maps; original half roan, worn. Stockholm, 1889.

The atlas by itself (see above) – a detailed and indispensible study of early mapping, with a beautiful series of facsimile maps. **$3250**

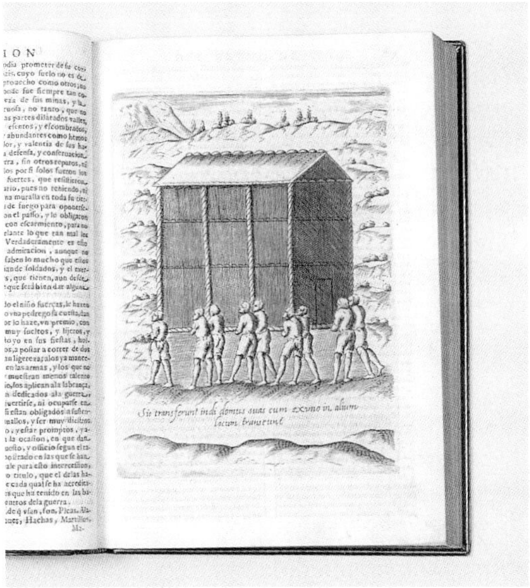

41. Ovalle

BEST EDITION OF OVALLE

41. OVALLE, Alonso de. Historica Relacion del Reyno de Chile...

Small folio, with 14 plates, 21 portraits, a folding map, and 18 full-page woodcuts; the map backed; a fine copy in a handsome Zaehnsdorf binding of full dark olive morocco, gilt. Rome, Francisco Cavallo, 1646.

Very rare: the best edition of the book, the first edition in Spanish; an Italian translation of 1646 did not include the portraits, and the English edition a century later was limited to the first five books of the original. The series of portraits of Chilean governors and Spanish-American dignitaries appears only here.

This was the first general history of Chile. The map shows a good knowledge of the Straits of Magellan, but combines solid geography with fancies about creatures to be found there. It is reproduced by Duviols (pp. 36-7) with a note about the long-running myth of men with tails that may have started with Mandeville, here continued by Ovalle – who also shared with Marco Polo some interesting ideas about apemen.

Of particular interest is the fine engraving of the stars of the Southern Cross or *cruzero*, which Ovalle notes (in later translation) are "indeed very beautiful, and shine with great liveliness. I suppose that those who have not seen them in their own Place and Situation, would be glad to see a Draught of them...". Corsali's original depiction of the early sixteenth century had reappeared halfway through the century in Eden's *Decades*, but not in many other printed sources.

The plate illustrates Ovalle's chapter XX, devoted to the stars of the south. Ovalle quotes De Bry – "the Learned of Our Nation, who have sailed on the South Sea... do affirm without dispute that those Stars which are near the Antartick Pole are more in number, and brighter and bigger...[and especially that] the Stars of the Constellation of the *Cruzero*... their Splendours and Beauty is extraordinary...".

Alden, 646/112; Medina, BHC 118; Palau, 207397; Sabin, 57972. **$22,500**

QUIROS ON MENDANA

42. [QUIROS] DE MORGA, Antonio. The Philippine Islands, Moluccas, Siam, Cambodia, Japan, and China, at the Close of the Sixteenth Century.

Octavo, engraved frontispiece, folding plate; half morocco binding, faded spine. Hakluyt Society, 1868.

Antonio De Morga's original work was printed in Mexico in 1609, and is impossibly rare. He was the first auditor of the Audiencia of Manila, and compiled this work from original documents to which he had access. Of particular importance is the inclusion – here for the first time in English translation – of Fernandez de Quiros's account of Alvaro de Mendaña's second voyage.

Ferguson, 12806. **$585**

EARLY APPEARANCE OF QUIROS

43. [QUIROS] MEGISER, Hieronymus. Septentrio Novantiquus, oder die newe Nortwelt...

Small octavo, title in black and red; with 11 (of 12) engraved plates, two of them slightly defective; somewhat browned throughout; old marbled boards. Leipzig, Henning Grosse, 1613.

Rare first edition of this collection of voyages, "never before translated into German": this is an example of changing tastes as an increasingly wide readership required the expensive, large-format travel books (like De Bry) to be adapted to an economical pocket size.

This copy lacks plate 11 (Samoyed dog-sled) and has two imperfect (Iceland map and plate 6, Pygmies), but the important final section of the book (pp. 445-473) is complete: this contains an abridged translation of Quiros's recently published *Memorial* of his voyage to the southern continent. This is one of the earliest appearances of the narrative in a general travel book, predating its first French and English printings, among others.

The other sections of the book include material on Iceland and Greenland, whales and other sea-animals decribed by Venetian navigators, Hariot's account of Virginia, Hudson's search for the northwest passage, and so forth.

Sabin, 47383; not in the James Ford Bell catalogue. **$5850**

TORRES AND PRADO

44. [QUIROS] STEVENS, H.N. and G.F. BARWICK. New Light on the Discovery of Australia as revealed by the Journal of Captain Don Diego de Prado y Tovar.

Octavo, maps and plates, two folding maps; original blue linen cloth. London, Henry Stevens, 1930.

A fine copy of this important publication, following the modern rediscovery of the original journal of the voyage of Torres and Prado, with a detailed account of the discovery of Torres Strait and parts of northern Australia. They had separated from Quiros at Espiritu Santo, and sailed to Manila. This is the independent issue of the book (it also appeared as Hakluyt Society Series II, Vol. lxiv). **$385**

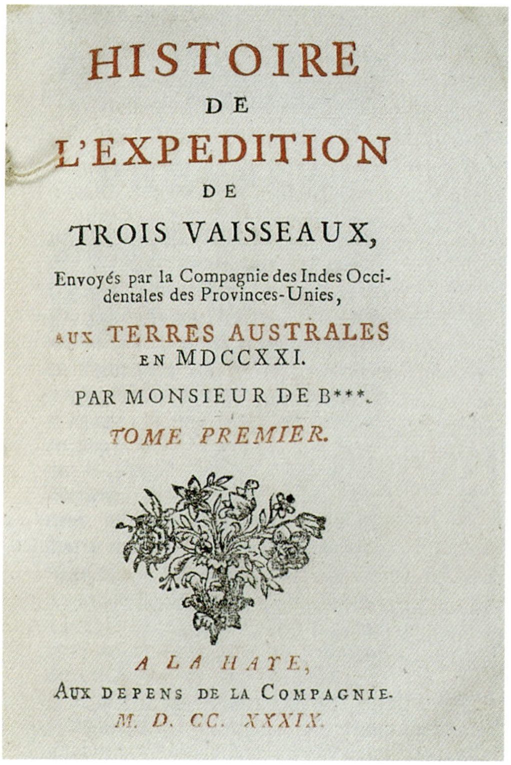

45. Behrens

THE DUTCH CHASING QUIROS

45. [ROGGEVEEN] [BEHRENS, Carl Friedrich] Histoire de l'expedition de Trois Vaisseaux, envoyés par la Compagnie des Indes occidentales des provinces-Unies, aux Terres Australes en MDCCXXI. Par Monsieur de B *...**

Two volumes in one, small octavo; title-pages in black and red; a fine copy, completely uncut, in modern boards. The Hague, la Compagnie, 1739.

A very scarce book: this is the account of the Roggeveen voyage, one of the most important of the Dutch voyages, ranking with those of Le Maire, Schouten and Tasman. It set out to rediscover the part of Terra Australis which Quiros had led the exploring world to believe existed, but in fact added substantially to the Dutch record of Polynesian discoveries, with Easter Island and Samoa being described here for the first time. There were altogether three contemporary narrative versions of the voyage, in Dutch, French and German: all are rare. This first French – and generally preferred – edition of Behrens's account was based on the 1737 German version, and very likely translated by Behrens himself, who travelled under Roggeveen as commander of marines, though some authorities have attributed the translation to De Brosses.

Borba de Moraes, 95; Kroepelien, 70; Mackaness, pp. 41-2; O'Reilly-Reitman, 230; Sabin, 4379; Taylor, 77; Tiele, 933n. **$7850**

ONE OF 200 COPIES

46. ROTZ, Jean. The Maps and Text of the Boke of Idrography presented by Jean Rotz to Henry VIII now in the British Library. Edited by Helen Wallis

with a Foreword by Viscount Eccles.

Large folio, with a coloured illustration on the title, 16 double-page coloured plates, and 11 illustrations in the text; original brown half-morocco, top edges gilt; a fine copy. Oxford, Printed for Presentation to the Members of the Roxburghe Club, 1981.

First edition: one of only 200 copies for sale of this magnificent production, reproducing in full colour Rotz's celebrated charts of the coastlines of the world. Rotz's *Boke* is the earliest surviving work of Dieppe hydrography to show "the Londe of Java", called by later Dieppe hydrographers "Java-la-Grande", and provides some of the strongest evidence in favour of an early Portuguese discovery of Australia, discussed by Helen Wallis in the Introduction ("Java-la-Grande: the first sight of Australia"). Rotz's life, times and works are also examined in detail and the plates are preceded by a transcription of the *Boke*. **$2850**

PORTUGUESE EXPANSION IN THE EAST INDIES

47. SAN ROMAN DE RIBADENEYRA, Antonio. Historia general de la Yndia Oriental. Los descubrimientos y conquistas, que han hecho las armas de Portugal, en el Brasil, y en otros partes de Africa, y de la Asia; y de la dilatacion del Santo Evangelio por aquellas grandes provincias desde sus principios hasta el año de 1557.

Quarto, with an engraved title and two engravings in the text; light waterstains to some upper margins (touching headlines on a few leaves towards the end of the book), small marginal repair and some minor soiling, but an excellent copy in contemporary limp vellum, spine lettered by hand. Valladolid, Luis Sanchez, 1603.

First edition of this significant history of Portuguese expansion in Asia and the East Indies. As Lach points out, it is "a genuine history based on the best printed sources available in Europe... it merits attention because it put the general history of overseas activities into the Spanish language and so brought the best of the Portuguese chroniclers to a European audience...". Lach goes on to compare the book with Herrera (see catalogue number 26) as "another popular book which brought certain of the Spanish conquests in the East, especially in the Moluccas, to a broad public...".

San Román lists at the start of the book the principal authorities that he consulted (including Marco Polo, Galvao, and Alvares on Ethiopia); this bibliography forms an invaluable guide to the sources of knowledge of the eastern world available in Europe at the beginning of the seventeenth century. It is cited as "List A" (the earliest) in F.M. Rogers's *Europe informed: an exhibition of early books which acquainted Europe with the East* (Harvard, 1966).

Borba de Moraes notes, quoting Salva, that "copies in good condition and containing the engraved frontispiece are rare"; the fine engraved title by Juan Baptista Morales, often defective, is present here in good condition.

Alden, 603/100; Lach, III, p. 322; Palau, 293611 ("Obra importante"); Retana, 55; Sabin, 76188. **$15,000**

47. San Roman de Ribadeneyra

STRAITS OF MAGELLAN

48. SARMIENTO DE GAMBOA, Pedro. Viage al Estrecho de Magallanes... En los años de 1579 y 1580...

Quarto, with three folding plates; a fine large copy, completely uncut, with ample margins, and one of the large copies, printed on thicker paper, as noted by Sabin; small library stamp of the Foreign Office on title-page, with unobtrusive withdrawn stamp on verso; modern half calf. Madrid, Imprenta Real, 1768.

The first printing – from the manuscript in the Royal Library, Madrid – of this highly important sixteenth-century description of the Straits of Magellan.

Sarmiento (who had earlier sailed with Mendaña) had been in Peru when Francis Drake attacked: he was sent to the Straits of Magellan to intercept the *Golden Hind*, but Drake had by then decided to continue on a circumnavigation. Sarmiento instead explored and mapped the Strait, and sailed for Spain to convince the government to form settlements there. A sixteen-ship fleet, with Sarmiento as governor of the colonists, sailed from Cadiz in 1581 – but the settlement failed, and even Sarmiento fell into English hands and was a prisoner for five years.

This was, as J.H. Parry signals (*The Discovery of South America*, London, 1979, pp. 285-9) "the most serious and thorough exploration of Magellan's Strait in the sixteenth century – indeed the most thorough exploration ever made before the voyage of the *Beagle* in the nineteenth century". In fact the officers of the *Beagle* were lost in admiration for Sarmiento's explorations, but also for his vainglorious attempts at settlement. Phillip Parker King concluded: "This was the first, and perhaps will be the last, attempt made to occupy a country, offering no encouragement for a human being; a region, where the soil is swampy, cold, and unfit for cultivation, and whose climate is thoroughly cheerless...".

Hill, p. 267; Sabin, 77094. **$6000**

THE DUTCH IN THE EAST INDIES

49. SCHOUTEN, Wouter. Reys-Togten naar en door Oost-Indien; in welke, de voornaamste Landen...

Small quarto, two parts in one volume; engraved title and portrait and 43 engraved plates by C. Decker, repair to inner edge of one plate with slight loss; contemporary sprinkled calf, front flyleaf cut away, binding a little rubbed but a good copy; early bookplate. Amsterdam, Andries Van Damme, 1708.

An extensive account of travels in the East Indies from 1658 to 1665, with a fine series of engraved views of the various regions, including Batavia, Bantam, Ceylon, Malacca and Amboyna, with a number of depictions of military engagements by land and sea. This is the second edition: the first had appeared in 1676. There were several later versions.

Alden, 708/116; Tiele, 990; see James Ford Bell, S186. **$4500**

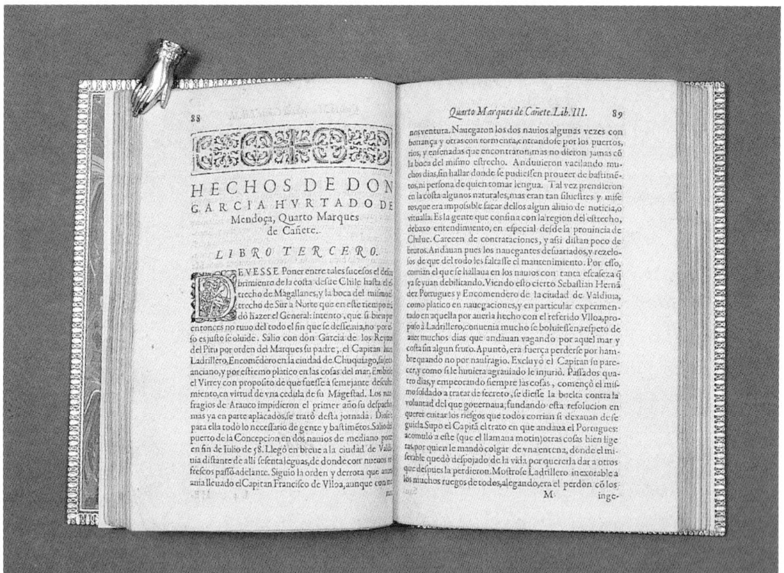

50. Suarez de Figueroa

THE VERY RARE FIRST ISSUE

50. SUAREZ DE FIGUEROA, Cristobal. Hechos de Don García Hurtado de Mendoça, Marqués de Cañete...

Small quarto, with an engraved title, occasional spotting; modern morocco, a fine copy. Madrid, Imprenta Real, 1613.

First edition, the first of two issues (a second issue appeared three years later). This has always been a rare book, and at least two famous collectors had trouble finding it: Alexander Dalrymple states in his *Historical Collection* that he had "not been able to procure a copy of Figueroa" until, when his first volume was in the press in 1769, he acquired a copy "from Spain" and was able to add the final section to his book; and Rex Nan Kivell wrote to Bjarne Kroepelien (quoted in the Kroepelien catalogue): "I was delighted to get your list of books and am very envious of them, particularly... your Suárez de Figueroa... [which] I have been hunting for twenty years".

The text, an account of the deeds of Don García Hurtado de Mendoça (captain-general of Chile and from 1590-96 viceroy of Peru), was commissioned by his family; it describes the wars against the Araucano Indians in Chile and Don García's rule as viceroy; the incursions of Drake, Cavendish and Hawkins are touched upon.

Most importantly however, it is a prime source for the history of Mendaña and Quiros. Mendaña's discovery of the Solomons in 1568 is narrated (pp. 228-37), and there is a long account of his second voyage (pp. 238-90), which was despatched from Peru in 1595 to colonise the Solomons and to continue the search for the Southern Continent. Although Mendaña failed to rediscover the Solomons and died along with many of his crew – the chief pilot, Quiros, finally guiding the survivors to safety in the Philippines – his second voyage is especially significant marking the first discovery of an important Polynesian island group, the Marquesas (named "Las Marquesas de Mendoça" after the viceroy) and led directly to Quiros's own expedition of 1605 and his celebrated but fruitless quest for Terra Australis Incognita.

This is one of two principal contemporary printed sources for Mendaña's momentous voyage, the other being Quiros's report to Don Antonio de Morga published in the exceedingly rare *Sucesos de las Islas Filipinas*, Mexico, 1609.

Suárez de Figueroa's account is based on the manuscript *Narrative* of his friend Quiros, who had proposed him to the King in 1610 as chronicler for the proposed expedition to the Southern Continent; Celsus Kelly has established that Suárez de Figueroa "incorporates as his own many phrases, paragraphs and even at times entire pages, except for minor alterations, from the 'Narrative' of Quiros on the Mendaña 1567 and 1595 voyages. For this he must surely have received Quiros's permission since they were both in Madrid at that time...".

Alden, 616/111; Kelly, 'Calendar', 301; Kelly, "The Narrative of Pedro Fernandez de Quiros", in 'Historical Studies Australia and New Zealand' IX, 34, May 1960; Kroepelien, 1260n.; Medina, 'Hispano-Chilena', 41; Palau, 323906; Sabin, 93312. **$55,000**

A NEW WORLD NOT YET KNOWN TO THE ENGLISH

51. [TASMAN] NARBOROUGH, John, and others. An Account of Several Late Voyages and Discoveries: I. Sir John Narborough's Voyage to the South Sea... II. Captain J. Tasman's Discoveries on the Coast of the South Terra Incognita. III. Captain J. Wood's Attempt to Discover a North-East

Passage to China. IV. F. Marten's Observations made in Greenland...

Octavo, folding "Chart of the western and southern oceans", folding chart of Terra del Fuego, polar map and 19 engraved views; a good copy in contemporary calf, rubbed, old rebacking, one joint splitting but firm. London, D. Brown, 1711.

"Of the greatest importance to an Australian collection, as it contains one of the earliest accounts of Abel Janszoon Tasman's famous voyage of 1642 from Batavia, in which he discovered Tasmania and New Zealand and visited Tonga and Fiji" (Hill). The account of Narborough's voyage in the *Batchelour* through the Strait of Magellan and into the Pacific was widely read by later navigators including the survivors of the *Wager* (part of Anson's fleet), who used this account for their own navigation through the passage.

The editor, Tancred Robinson, exclaims of Tasman's voyage: "tis the Discovery of a new World, not yet known to the English, 'Tis probable by Abel Jansen Tasman's Navigation, that New Guinea, New Carpentaria, and New Holland, are a vast prodigious Island, which he seems to have encompass'd in his Voyage...".

This is the second and best edition, with an additional chart of the western and southern oceans which did not appear in the first version. The editor gives a resumé of earlier voyages, including those of Quiros, Drake and Magellan, and makes a plea for scientific exploration of the globe; lamenting "that the English nation have not sent with their Navigators, some skilful Painters, Naturalists, and Mechanists, under publick Stipends and Encouragement as the Dutch and French have done...".

Hill, pp. 1-2 (first edition only). **$4850**

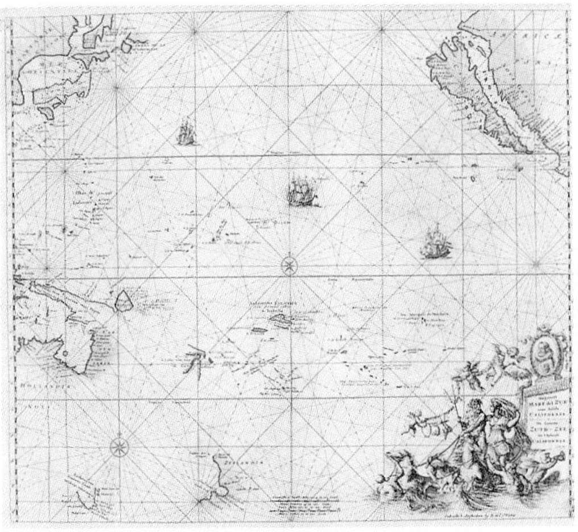

52. De Wit

TASMAN'S DISCOVERIES

52. [TASMAN] WIT, Frederick de. Magnum Mare Del Zur cum insula California...

Engraved map, 485 x 560 mm.; contemporary handcolouring, framed. Amsterdam, [1680].

A splendid map of the Pacific. Frederick de Wit, an engraver and map-seller who settled in Amsterdam in 1648, was to become one of the leading Dutch map-makers of the seventeenth century. His firm's "maps were distinguished by their excellent craftsmanship, exactness and beauty" (Schilder).

De Wit worked from the discoveries made by Abel Tasman in 1642, so that the discoveries on the north coast of Australia are recorded in detail. On the west coast of America California is still shown as an island. The map is typically well-decorated in the style of the best seventeenth-century Dutch publications: a fine cartouche incorporates a figure of Neptune and shows early ships under full sail, as well as a portrait of Magellan, the first person to cross the Pacific – the name that he himself bestowed on the great ocean.

Clancy & Richardson, pp. 90-91; Tooley, plate 101. **$4750**

THE BONAPARTE MAP

53. [TASMAN] WIEDER, Dr. F.C. Tasman's Kaart Van Zijn Australische Ontdekkingen 1644 "De Bonaparte-kaart".

Quarto text, with numerous illustrations, accompanied by the large separate colour map; fine in quarter cloth, the map unbound as issued. The Hague, Nijhoff, 1942.

Controversial study of the Bonaparte map in the Mitchell Library, with a fine facsimile of the map. Wieder was taken to task for misunderstanding the map's legend to mean that it was actually by Tasman; it is now generally agreed that the cartographer derived his information from an imperfect source itself based on authentic originals. "Although the Bonaparte map was neither drawn in 1644, nor by Tasman, it is, nevertheless, a very important document in the history of navigation and can be reckoned among the most famous and the most beautiful maps produced by Dutch cartographers..." (Schilder).

Schilder, p. 148. **$485**

QUIROS'S OWN ACCOUNT

54. TORQUEMADA, Juan de. Primera [-tercera] parte de los viente I un libros rituales I Monarchia Indiana, con el origen y guerras, de los Indios Ocidentales de sus poblaciones, descubrimiento, conquista, conversion y otras cosas maravillosas de la mesma tierra...

Three volumes, folio, with engraved titles and a folding map; contemporary mottled calf, wear to heads and tails of spines, with some loss to extremities, some slight worming to inner joints, but overall an attractive copy in original condition. Madrid, Nicolas Rodriquez Franco, 1723.

Second and best edition. The map (which did not appear in the 1615 Seville edition) centres on the Pacific, and shows the Chinese coast, the Philippines, Solomons, New Guinea and "Tierra Austral" to the west and Mexico and South America to the east. The preface mentions the extreme rarity of the first edition – only three copies were known in Madrid in 1723 – and explains that most copies had apparently been lost in a shipwreck, presumably en route to Mexico.

The book is especially important in the literature of Pacific exploration because of Torquemada's account

54. Torquemada

of Quiros's voyage of 1605, in which he sailed from Callao to locate the fabled Southern Continent. The earliest extensive description of the expedition to be printed, this remained the only reliable source available to navigators, geographers and historians until the nineteenth century. Celsus Kelly ("The narrative of Pedro Fernandez de Quirós" in *Historical Studies, Australia and New Zealand* IX/34, May 1960) has shown that Torquemada had access to Quiros's own account of the voyage as well as to the journal of Munilla, the commissary of the Franciscans who was also on the voyage. In fact Torquemada probably met Quiros in Mexico City in 1607, and again in Madrid in 1613. Kelly shows too that Torquemada probably also interviewed two other members of the expedition, the ensign Pedro López de Sojo and the sergeant Pedro García de Lumbreras.

Torquemada's book is important for many other reasons, not least for its account of Legazpi's expedition to the Philippines (see Retana, 226). Hill describes the book as the "key work on the early history of Spanish North America, particularly Mexico, the Southwest and California… the primary source of Vizcaíno's 1596 and 1602 expeditions and voyage of discovery to California. It contains Father Antonio de la Ascensión's letter which relates to Vizcaíno's voyage during which most of the bays, headlands and rivers, including San Diego, were renamed. Torquemada also deals at length with Juan de Oñate's conquest of New Mexico, and briefly with the expeditions of Coronado, Alarcón, Espizo and Niza… The Mexican historian and statesman Lucas Alamán called Torquemada the Livy of New Spain…" (Hill).

Alden, 725/195; Hill, p. 291; Medina, BHA, 2491; Palau, 335033; Sabin, 96212; Wagner, 'Spanish Southwest', 18a.

$19,750

DUTCH DISCOVERIES ON
THE AUSTRALIAN NORTH COAST

55. VALENTIJN, François. Kaart van de Zuyd-Ooster Eylanden van Banda…

Engraved map, 450 x 560 mm.; mounted. Amsterdam, c. 1726.

An important map, showing the earliest Dutch discoveries on Australia's northern coast (named *Nova Hollandia of Het Zuyd Land*) and including Arnhem Land. The islands in the Arafura Sea, and the south coast of New Guinea and Timor Island are shown.

Based on Tasman's own drafts, this map shows not only his discoveries but adds those made by Van Colster in 1623 (the north-east corner of Arnhem Land), with those of Pietersz in 1636, including Melville Island and Dundas Strait. The map was prepared for François Valentijn's monumental five-volume work (see catalogue number 83).

Tooley, 1269. **$2200**

2: COLLECTIONS & ETHNOGRAPHY

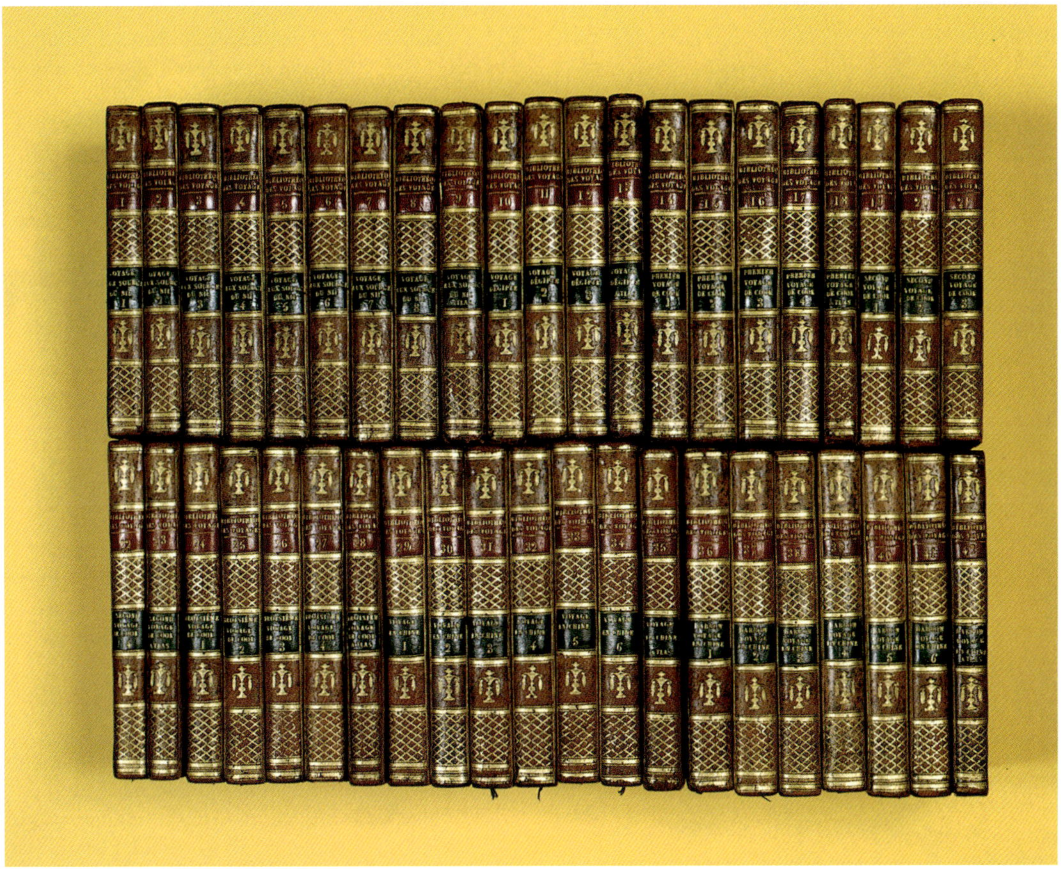

57. Breton.

COSTUMES OF THE WORLD

56. BEDELET, Leonie C. Le Monde en Estampes: Types et Costumes des principaux Peuples de l'univers…

Oblong quarto, with 24 coloured lithographs with explanatory text; original morocco-backed gilt-decorated cloth, all edges gilt, silk doublures; a fine copy. Paris, Amedée Bedelet, c. 1858.

Very scarce; this handsome work on world costumes, derived from voyage accounts and intended mainly for children, contains a splendid series of coloured lithographs by J. Boquin, after drawings by Leloir and Fossey. They depict the native costumes of various peoples of the world, including Europe, Asia, Africa, North and South America (with a rendering of American Indians negotiating with a fur trader, while a buffalo hunt proceeds in the background), and the Pacific Islands, including Australia (with a depiction of Aborigines "pillant des débris de naufrage").

Not in Ferguson. **$4500**

MINIATURE VOYAGES: 42 VOLUMES

57. BRETON, J.B.J. Bibliothèque portative des voyages.

42 volumes, 16mo., including seven atlas volumes with engraved plates and folding charts; some isolated spotting but a fine set in contemporary marbled calf gilt, gilt spines, red and green labels, gilt dentelles, all edges gilt. Paris, Lepetit, An VII (1799)-1807.

An exquisite set: this delightful French miniature collection of voyages comprises Bruce's voyage to the source of the Nile (nine volumes), Norden's voyage to Egypt (four volumes), Cook's three voyages (fifteen volumes), Macartney's travels in China and Tartary (seven volumes) and Barrow's account of the Macartney expedition to China (seven volumes).

Of particular interest is the complete set of Breton's edition of Cook (1804) in fifteen volumes, three of them atlases, each of which contains a folding map and finely engraved plates, miniature replicas of those published in the official account.

See Beddie, 66; Muir, 952. **$9850**

FOUNDLINGS, BEGGARS & CRIMINALS

58. BROSSES, Charles de. Histoire des Navigations aux Terres Australes… Contenant ce que l'on scait des moeurs & des productions des Contrées découvertes jusqu'à ce jour; & où il est traité de l'utilité d'y faire de plus amples découvertes, & des moyens d'y former un établissement.

Two volumes, quarto, with seven folding maps, with the 'asterisk' leaves; contemporary French calf, a fine copy. Paris, chez Durand, 1756.

One of the most important general works dealing with early voyages to the Pacific, which aimed to stimulate French discovery and colonisation of the South Seas. It contains an account of all voyages to the south, beginning with the second Vespucci expedition of 1502 and going up to 1747, including the voyages of Magellan, Drake, Schouten, Tasman and others. "An extremely important and thorough collection of voyages, and one of the outstanding works relating to the early history of Australasia" (Hill). De Brosses's text was later silently used as the basis for Callander's highly influential *Terra Australis Cognita* of 1766-8, and in this form it had much to do with British plans for colonial expansion by transportation.

"Extremely important. Here de Brosses suggests that France should colonise Australia, virtually predicting the basis of its final settlement by maintaining that the colonisation should be by France's 'foundlings, beggars and criminals'. This is somewhat scarce, but with patience copies can still be obtained", wrote Rodney Davidson in 1970; the book is nowadays actually quite rare on the market. This copy is complete with the "asterisk leaves", so-called because they are numbered 437*-450*: they contain the details of the discoveries of the Dutch in Australasia, and are not always present. For the Australian collector they are particularly important. The maps include those of Australia, New Guinea, and north-eastern Australia which make the book particularly significant.

Davidson, pp. 34-5; Hill, pp. 34-5; Kroepelien, 132; O'Reilly-Reitman, 93. **$7000**

FIRST GERMAN EDITION

59. BROSSES, Charles de. Vollständige Geschichte der Schiffarthen nach den noch gröstentheils unbekanten Südländern…

Two volumes, quarto, with six folding engraved maps; a very good copy in the original plain grey boards, printed blue paper labels on spines. Halle, Johann Justinus Gebauer, 1767.

First German edition, based on the original French version of 1756, and one of the most important general works dealing with early voyages to the Pacific; the translation, preface and notes are the work of J.C. Adelung.

Kroepelien, 133; O'Reilly-Reitman, 95; Sabin, 8389; not in the Hill collection. **$3600**

EARLY SOUTH SEA DISCOVERIES

60. BURNEY, James. A Chronological History of the Discoveries in the South Sea or Pacific Ocean.

Five volumes, quarto, with 28 engraved maps (16 folding), 13 plates and six text woodcuts; some staining to endpapers and a few leaves of text at the end of the first volume; spotting to endpapers but a good set in contemporary half morocco, spines renewed. London, Luke Hansard, 1803-1807.

First edition. "The most important general history of early South Sea discoveries containing practically everything of importance on the subject" (Hill); Burney's great compilation "must always form the basis of historical research for early voyages and discoveries throughout the Pacific" (Hocken).

The collection covers more than 250 years of Pacific exploration prior to that of Cook, including Spanish, Dutch, French and English voyages, with a general "History of the Buccaneers of America", and concludes with Bougainville's voyage to Tahiti.

Burney, who had sailed with Cook as lieutenant during the last two voyages, received encouragement from Sir Joseph Banks (to whom the work is dedicated) and enjoyed free access both to Banks's magnificent library of books and manuscripts, and to Dalrymple's collection of scarcer Spanish books. Whenever possible, he relied on manuscript accounts, generally comparing them with printed narratives for purposes of style. The collection contains much that is nowhere else accessible, and will always remain one of the chief authorities for the history of the geographical exploration and discovery of the Pacific.

Burney himself explains (in volume 5) that "the termination of this present work is adapted to the commencement of voyages in another collection [i.e. that of Hawkesworth], which with the addition of M. de Bougainville's voyage round the world, follow as an immediate sequel, without any chasm being left, to the Discoveries here related…". This is therefore the essential precursor to the series of voyage publications, starting with Wallis, Carteret, and Cook, that began the new age of discovery.

Ferguson, 372; Hill, p. 40; Hocken, pp. 30-34; O'Reilly-Reitman, 104; Sabin, 9387. **$16,500**

VOYAGES TO THE TERRA AUSTRALIS

61. CALLANDER, John. Terra Australis Cognita; or, Voyages to the Terra Australis, or Southern Hemisphere, during the Sixteenth, Seventeenth, and Eighteenth Centuries…

Three volumes, octavo, three folding maps; neat restoration to several upper fore-corners, map in vol. 2 repaired at fold with some loss, bound without the advertisement leaf in vol. 1; contemporary calf gilt, slightly worn, double labels (one chipped); contemporary ownership inscriptions. Edinburgh, A. Donaldson, 1766 [-1768].

First edition: an important collection of voyage material, gathered partly as background for an early, serious proposal for European settlement of the 'Southern Continent'. This is an example of the scarce issue with the variant dedication to Sir Laurence

Dundas in volume 1 (replacing the dedication to Charles Townshend who had died in 1767).

Callander's collection – based in part on the French compilation by de Brosses (see catalogue number 58) – contains forty-one voyage narratives, including Magellan, Drake, Sarmiento, Cavendish, Quiros, Schouten, Narborough, Dampier, Funnell, Woodes Rogers, Shelvocke and Roggeveen among others. He argues for the establishment of a penal settlement in the Pacific, to support the further exploration and settlement of New Holland, and claims the inevitability of English occupation of New Holland by virtue of superior sea power.

Hill, p. 367. **$8250**

COOKE'S EAST INDIA SCENERY

62. COOKE, George. Scenery of the East India Islands, Etc. A Series of Views of the Most Remarkable Localities Celebrated for Picturesque Beauty of Historical and Other Associations. Exquisitely Engraved by Cooke, from Original Drawings.

Oblong quarto, 26 copper-engraved plates; small wormholes in some lower margins affecting a few captions; original stiff paper buff wrappers, lettered on the front cover within a decorative border, cloth spine; covers slightly browned and rubbed at edges, some plates with light spots and marginal tears, but generally a very good copy. London, Longman, Hurst, c. 1811-13.

A very rare collection of voyage plates, almost all of them South Seas views. The plates are engraved by George Cooke after Hodges, Webber, Lesueur, and other voyage artists, and were ultimately used to illustrate Pinkerton's huge *General Collection of the Best & Most Interesting Voyages & Travels* (see catalogue number 79).

Cooke, a prominent line-engraver in the early nineteenth century, is best known for his engraved topographical views of England. He engraved many of Turner's drawings for the *Picturesque Views on the Southern Coast of England.* In 1826 he began his favourite work, *Views in London and its Vicinity*, which ended with the twelfth number before his death; and in 1833 he produced *Views of the Old and New London Bridges* after drawings by his son Edward William Cooke. He was one of the original members of the Society of Associated Engravers.

The series of engravings includes views of Tahiti, Hawaii, New Zealand, New Guinea and the Marquesas, and has a fine view of Sydney and the mouth of the Parramatta River after the French artist Lesueur. This separate issue of the engravings, predating the completion of Pinkerton's publication, is surprisingly rare: we have found no record of it in any of the bibliographies we have consulted including Ferguson, the National Maritime Museum Catalogue, Hill, Kroepelien, Lada-Mocarski, O'Reilly-Reitman, the National Union Catalog, and the British Museum Catalogue. **$8750**

62. Cooke

COOKE'S CHINA AND TURKEY

63. COOKE, George. Scenery of China, Turkey, Etc. A Series of Views of the Most Remarkable Localities...

Oblong quarto, 25 copper-engraved plates; original buff wrappers, lettered on the front cover within a decorative border, cloth spine; covers slightly browned and rubbed, some plates with light spots and marginal tears, generally a good copy. London, Longman, Hurst, c. 1811-13.

First edition: a selection of engravings including views of Nankin, Pekin, the Great Wall of China, the Plain of Troy, Siphanto, the mountains of Caucasus and various Turkish scenes ultimately used to illustrate Pinkerton's *General Collection of the Best & Most Interesting Voyages & Travels* (1808-1814). Like Cooke's *Scenery of the East India Islands* (see above), this separate issue of the engravings, predating the completion of Pinkerton's publication, is surprisingly rare: it is not recorded in the Blackmer catalogue despite its obvious interest, while NUC locates a single copy. **$3600**

THE DELUXE ISSUE LIMITED
TO FIFTY COPIES

64. DALRYMPLE, Alexander. An Account of the Discoveries made in the South Pacifick Ocean.

Quarto, illustrated with a frontispiece portrait, seven folding maps and plates; hand-bound in full blue calf, custom-made solander case. Sydney, Hordern House for the Australian National Maritime Museum, 1996.

The special issue. Only 50 signed copies were prepared of this deluxe issue of this publication, the only reprint to have been made of this famous book, first published in 1767. The original, one of the rarest of all eighteenth-century publications about the Pacific, and in particular Australia, is known in only 22 copies world-wide.

In 1996 the Australian National Maritime Museum invited Dr. Andrew Cook of the British Library, the

64. Dalrymple

leading expert on Alexander Dalrymple, to assess the importance of this work and to contribute the introductory essay for this publication. In his scholarly essay Dr. Cook examines in detail the influence that this work had on the *Endeavour* voyage of Captain Cook, as it was one of the only listed references mentioned by Banks or Cook on *Endeavour*. Dalrymple, working from manuscripts and original data in archives across the world, had prepared a summary of maritime discoveries made in the Pacific region up to 1764, and his work was of considerable significance to the success of Cook's voyage. The essay is accompanied by a Foreword by Dr. Kevin Fewster, Director of the Australian National Maritime Museum, who has also signed and numbered each of the 50 deluxe copies.

The original issue of the book is faithfully facsimilised, including the charts and views, and the complete work is a most handsome limited edition, printed on fine quality paper, in eighteenth-century style. **$550**

DALRYMPLE'S COLLECTION

65. DALRYMPLE, Alexander. An Historical Collection of the several Voyages and Discoveries in the South Pacific Ocean...

Two volumes in one, quarto, with 17 engraved maps and plates, five of which are folding; contemporary calf, rebacked. London, Printed for the author, 1770-71.

First edition of a work of far-reaching importance by the leading English hydrographer. Passionately involved in the argument over the possible existence of a southern continent, Dalrymple partially translates here some twelve original accounts which support his belief in its existence. His collection begins with Magellan's voyage of 1519, and the Spanish accounts that he translates include Mendaña's voyage to the Solomon Islands in 1595, and that of de Quiros in 1606. The Dutch accounts include those of Le Maire, Schouten, Tasman, and Roggewein. Dalrymple's long introduction on trade and his "investigation of what may be farther expected in the South Sea" carefully expound his belief in the existence of a "Great Southern Continent".

Dalrymple had produced an analysis of voyages to the Pacific in 1764, but its issue was evidently very restricted (see previous item). Before that, he had been offered the command of the *Endeavour* voyage, but partly because of his insistence on being given an Admiralty commission, the command went instead to Cook. His disappointment is hinted at in the remarkable "undedications" of this work: to Byron "who discovered scarcely anything but Patagonians" and to Banks who "infatuated with female blandishments forgot for what he went abroad and hastened back to amuse the European world with stories of enchantments...".

Davidson, pp. 36-7; Hill, p. 71; Hocken, pp. 7-9; Holmes (first edition), 32; Kroepelien, 245; O'Reilly-Reitman, 97. **$11,000**

FIRST FRENCH EDITION

66. DALRYMPLE, Alexander. Voyages dans la mer du Sud, par les Espagnols et les Hollandois. Ouvrage traduit de l'Anglois de M. Dalrymple, par M. de Fréville.

Octavo, with three folding maps; a fine copy in contemporary French calf gilt. Paris, Saillant & Nyon, Pissot, 1774.

First French edition of the famous Dalrymple collection, translated by the influential de Fréville, who also translated the official accounts of Cook's voyages. His own book *Histoire des Nouvelles Découvertes faites dans la Mer du Sud*, also published in 1774, summarised contemporary knowledge of the Pacific, and his work in general had a significant influence on late-eighteenth-century French voyagers.

Hill, p. 73; Sabin, 18338. **$4200**

EARLY COLOUR LITHOGRAPHY

67. ENGELMANN, Godefroy and G. BERGER. Porte-feuille géographique et ethnographique contenant des planches pour l'explication... du globe terrestre... par le procédé nouveau de la Lithographie.

Quarto, frontispiece and 88 plates and maps, lithographed and all but two in original handcolouring; an excellent copy in contemporary French calf, ornately decorated in gilt. Mulhaus and Paris, G. Engelmann, 1820.

A beautiful series of highly-coloured plates of peoples of the world: the first major educational use of lithography. The images are largely derived from voyage accounts, and of particular interest are those of the Pacific. They include a splendid tableau of the Aborigines of New Holland partly based on images from the Baudin voyage, a reworked Death of Cook

67. Engelmann

in Hawaii, a fine plate of New Zealand Maoris based on Parkinson, as well as the familiar images of the dancing girls of Ulieta and the famous tattooed man of the Marquesas. There is also a folding map of Oceania. A series of plates of ethnic types includes one with five figures all of which derive from Cook or Baudin, including an Australian Aborigine and an Hawaiian warrior.

This wonderful and rare colourplate educational extravaganza, originally accompanied by a volume of text not present with this copy, is an early publication by one of the great names of lithography. Godefroy Engelmann was effectively the pioneer of the printing process which would revolutionise printing: he "remania complètement toute la partie mécanique de cet art dont Senefelder est l'inventeur" (Hoefer). The book was published over a four-year period, coming off the press in twenty-five 'livraisons'. Engelmann's biographer Lang notes that "Cet ouvrage a été annoncé par les Affiches de Mulhausen no. 52 du 30/12/1815 et citée par le Journal de la Librairie no. 631 du 24/12/1816": the earliest livraisons are thus among the very first lithographic printings of any note.

This important book is unrecorded by both Ferguson and Beddie despite its relevance both to Australia and to Captain Cook. In fact it seems not to have been recorded by any of the major travel bibliographers.

Gumuchian, 2413; Lang, 'Godefroy Engelmann', 33. **$6500**

THE BUCCANEERS

68. ESQUEMELING, Joseph. The History of the Bucaniers of America...

Duodecimo; original bookseller's calf-backed boards, worn. Glasgow, Robert Duncan, 1785.

Chapbook edition of the classic pirate account of Esquemeling (Exquemelin). First published in Amsterdam in 1678, this account reappeared in innumerable editions over several hundred years in Holland, Germany, England, France and Spain, and it has been said that "perhaps no book of the seventeenth century in any language was ever the parent of so many imitations and the source of so many fictions...".

The exact details of the life of Esquemeling have never been firmly established. He was born around 1645, probably in Harfleur, France. "In 1666 he was engaged by the French West India Company and went to Tortuga, where he stayed for three years. There he enlisted with the buccaneers, probably as a barber-surgeon, and remained with them until 1674. Shortly afterwards he returned to Europe; but he was later once again in the Caribbean" (National Maritime Museum, *Piracy and Privateering*). After his pirating days were over he settled in Holland, possibly because he was a Huguenot.

See National Maritime Museum, 'Piracy and Privateering'.

$550

FRENCH DISCOVERY

69. FLEURIEU, Charles Pierre Claret de. Découvertes des François, en 1768 & 1769, dans le sud-est de la Nouvelle Guinée. Et Reconnoissances postérieures des mêmes Terres par des Navigateurs Anglois qui leur ont imposé de nouveaux noms; Précédées d l'Abrégé historique des Navigations & des Découvertes des Espagnols dans les mêmes Parages...

Quarto, with twelve folding charts; an excellent copy in contemporary quarter blue calf. Paris, Imprimerie Royale, 1790.

First edition. Fleurieu, the leading light in the early history of French exploration in the South Seas, gives accounts of the voyages of Mendaña, Quiros,

Carteret, Bougainville, Cook and others. The maps, based on actual discoveries, are used to illustrate Fleurieu's theories, most of which ultimately proved to be correct.

The great work on the French discovery of the Solomon Islands, this was also a direct result of the voyage of the First Fleet to New South Wales: Phillip's *Voyage*, first published in 1789, had included the journal of Lieutenant Shortland on his return voyage in the *Alexander* transport from Botany Bay to England, during which he coasted along a group of large islands which he named "New Georgia". Fleurieu, writing in a period of intense international rivalry over Pacific discoveries, denies that discovery and promotes those of Louis de Bougainville and Jean de Surville. He draws on unpublished manuscripts as well as the printed narratives of Cook, Bougainville, Phillip, Mendaña, Quiros and others.

There are also interesting comments on the La Pérouse voyage. Fleurieu announces the receipt of journals from as far as Botany Bay, and in discussing the Great Ocean Chart (a fragment of which is published here), says publication is to be delayed until his later discoveries can be inserted.

Ferguson, 81; not in the Hill collection; Kroepelien, 436; Sabin, 24748. **$6750**

FIRST ENGLISH EDITION

70. FLEURIEU, Charles Pierre Claret de. Discoveries of the French in 1768 & 1769 South-East of New Guinea...

Quarto, with 12 folding charts; a very good copy in recent half maroon calf, marbled boards. London, John Stockdale, 1791.

First English edition. The translator notes that La Pérouse has still not turned up and "the apprehension for his loss increases daily".

Ferguson, 105; Hill, p. 105; Kroepelien, 437; Sabin, 24749. **$6750**

IN CONTEMPORARY CALF
WITH THE CADIZ LEAVES

71. HAKLUYT, Richard. The Principal Navigations, Voiages, Traffiques and Discoveries of the English Nation, made by sea or over-land, to the remote and farthest distant corners of the Earth, at any time within the compasse of these 1500 yeeres...

Three volumes in two (as usual), small folio, woodcut initials and head and tail pieces; slight soiling to titles; original plain dark calf bindings, spines numbered in gilt, red leather labels; slight rubbing, spines and joints wearing but the covers firmly held; contemporary bookplates of Charles Lord Maynard; unrestored and preserved in a quarter morocco case. London, George Bishop, Ralph Newberie and Robert Barker, 1599-1600.

A splendid unsophisticated copy of one of the classics of travel literature and the first English collection of voyages. Hakluyt's collection will always be the primary source for the history of early English exploration, as well as one of the gems of Elizabethan letters. "It is difficult to overrate the importance and value of this extraordinary collection of voyages" (Sabin); "an invaluable treasure of nautical information which has affixed to Hakluyt's name a brilliancy of reputation which time can never efface or obscure" (Church).

This is the second edition, greatly expanded from the single-volume original version of 1589 and effectively a new work – "the first edition contained about 700,000 words, while the second contained about 1,700,000... This was indeed Hakluyt's monumental masterpiece, and the great prose epic of the Elizabethan period. In design it was similar to the first edition: the first volume concerned voyages to the north and north-east; the second volume, to the south and south-east; the third volume, to America. All sections were expanded; the first two were approximately doubled and the American part was almost tripled.

"Much that was new and important was included: the travels of Newbery and Fitch, Lancaster's first voyage, the new achievements in the Spanish Main, and particularly Raleigh's tropical adventures. At first sight the expanded work appears a vast, confused repository, but closer examination reveals a definite unity and a continuous thread of policy. The book must always remain a great work of history, and a great sourcebook of geography, while the accounts themselves constitute a body of narrative literature which is of the highest value in understanding the spirit and the tendencies of the Tudor age" (Penrose, p. 318).

This copy has the second version of the volume 1 title-page, dated 1599, in its first state (with line 7 ending 'yeres'); but it retains the genuine original printing of the Cadiz leaves, which are often only present in later reprints or facsimiles. These leaves were suppressed after the fall from royal favour of the Earl of Essex, who returned to England from Ireland without leave in 1599. The original title-page (dated 1598) had made mention of Essex's "famous victorie atchieued at the citie of Cadiz", and so it was quickly replaced with the present version (dated 1599) which makes no mention of Cadiz.

Normally, the seven Cadiz leaves were simply removed from the end of the first volume. It would seem, therefore, to be unexpected that the second version of the title should be present with the suppressed Cadiz leaves in their original printing; however, Quinn noted that twenty-two per cent of the institutional copies examined took this form, and proposed two possible explanations: that the cancelled title-page must have been added to copies from which the Cadiz leaves were not excised, or that sets of leaves which had been excised were later reattached to copies which had the new title-page. This latter theory is supported by the condition of this present copy – in contemporary binding, in unsophisticated condition – in which the first page of the Cadiz leaves is lightly soiled, evidencing that the section has sat somewhere waiting to join the rest of the text. But it could not have sat separately for long, since our binding is of the period, and the Cadiz leaves have definitely not been inserted after

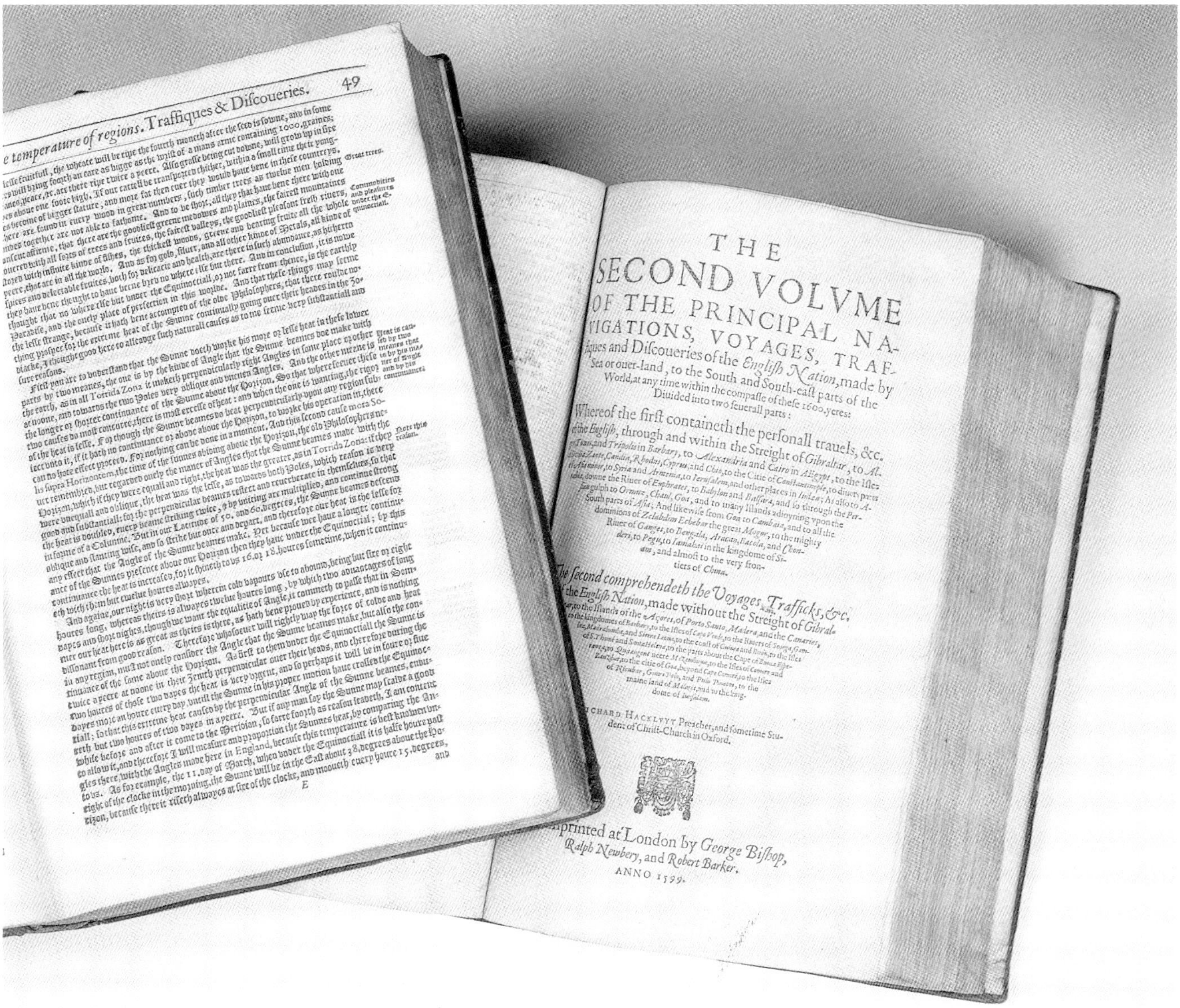

71. Hakluyt

the time of binding.

The world map that was intended to accompany the third volume was only actually issued with a handful of copies, and is almost always missing; Church was able to trace just thirteen copies of the map, a figure that Quinn's census for *The Hakluyt Handbook* could only increase to nineteen out of the total of two hundred and forty copies of the book identified, most of them held by institutions.

Quinn remarks that even allowing for the ravages of time this "survival rate is sufficiently low to raise the possibility that not all copies were equipped with the map, either because it was made available after many sets had been sold (which would mean that its date might be later than 1599), or because it was an optional extra supplied at additional cost" (p. 496).

Borba de Moraes I, pp. 328-9; Church, 322; James Ford Bell, H10; Quinn, p. 490; Printing and the Mind of Man, 105; STC, 12626a; see also A. Payne, 'Richard Hakluyt and his Books', Hakluyt Society Annual Talk for 1996. **$42,500**

JUVENILE COLLECTION

72. HALL, Captain Basil. Fragments of Voyages and Travels, including anecdotes of a naval life: chiefly for the use of young persons. First [- Third] Series.

Nine volumes, duodecimo; a nice set in contemporary half calf with double labels, a few stains and a bit rubbed. Edinburgh, Robert Cadell, 1832-34.

A full set of all three series of this attractive juvenile collection of voyage accounts: this is the second of several editions. "This set is usually listed as Hall's chief work, in which he describes his adventures around the world, including his experiences with the allied forces at the Battle of Coruna in 1809, and his cruise to Ireland and Madeira in the *Endymion*. Visits to the Cape of Good Hope, the Comoro Islands, Ceylon, Borneo, etc., are described in detail, and much information about the South Seas and the Monsoons of the tropics is presented..." (Hill).

Hill, p. 439. **$775**

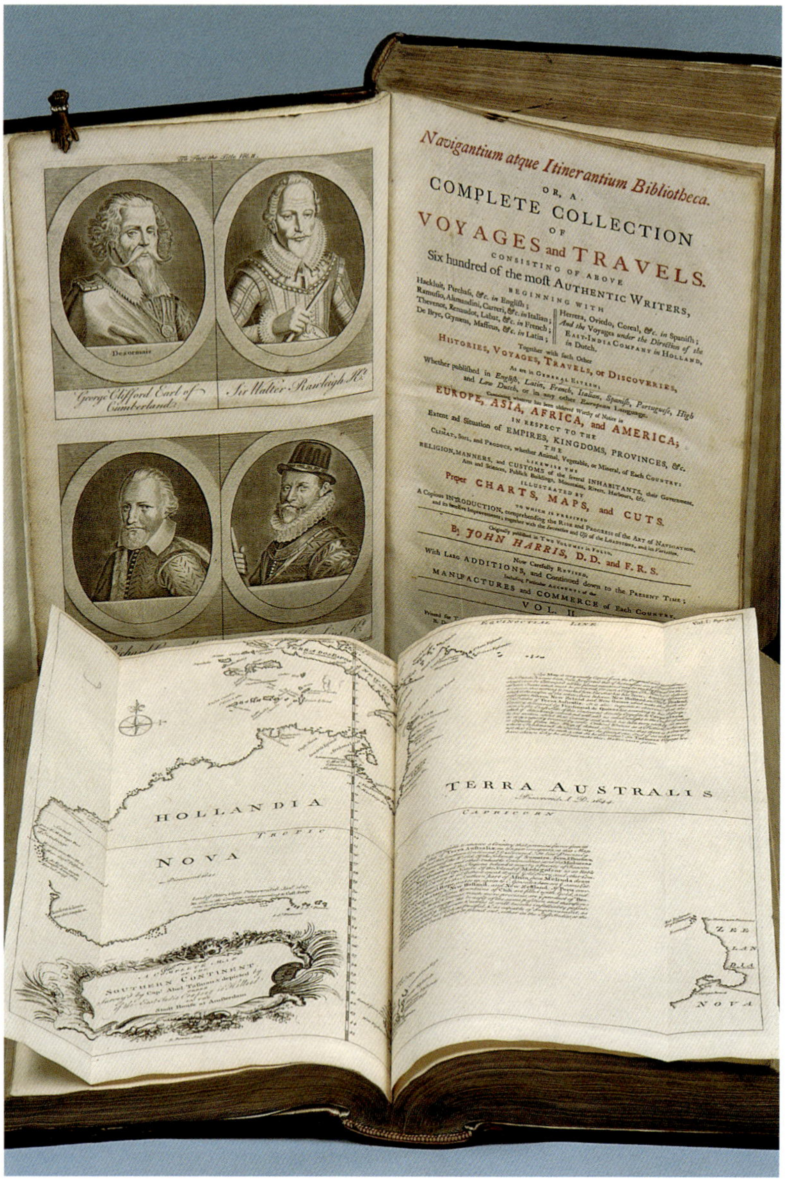

73. Harris, Navigantium

WITH THE TASMAN MAP

73. HARRIS, John. Navigantium atque Itinerantium Bibliotheca… A Complete Collection of Voyages and Travels.

Two volumes, folio, with 61 engraved maps and plates (several folding) by or after Emmanuel Bowen; a fine copy in contemporary mottled calf gilt, joints and labels expertly renewed. London, T. Osborne and others, 1764.

A handsome copy of the third edition, one of the preferred issues of the book with the full complement of maps, including the Bowen/Tasman map of Australia which did not appear in the first edition.

Harris's great collection of travels was first published in 1705 in slighter form, but extensively revised by John Campbell for the second and third editions: he added new voyages – many of Australasian interest – and had new maps prepared by Emmanuel Bowen, the most notable being "A Complete Map of the Southern Continent surveyed by Capt. Abel Tasman"; this is one of the earliest English maps of Australia.

Campbell actively encouraged further voyages to the imperfectly known Southern Continent in continuation of the work of those (including de Quiros, Pelsært, Tasman and Dampier) whose narratives he published. "He recommended an expedition to Van Diemen's Land, and a voyage to New Guinea by which means… 'all the back coast of New Holland, and New Guiney, might be thoroughly examined; and we might know as well, and as certainly, as the Dutch, how far a Colony settled there might answer our Expectations'…" (Williams and Frost, *Terra Australis to Australia*, Melbourne, 1988, p. 28).

Alden, 744/116; Davidson, pp. 37-8; Hill, pp. 137-8; Lada-Mocarski, 3; Perry, p. 60; Sabin, 30483; Schilder, 'Australia Unveiled', map 87; Tooley, 241. **$14,000**

COSTUMES OF THE WORLD

74. HARRIS, John, publisher. The Costume, Manners, and Peculiarities of Different Inhabitants of The Globe, calculated to instruct and amuse The Little Folks of All Countries.

Octavo, title and 16 numbered leaves, printed on one side only and each containing a half-page handcoloured woodcut accompanied by a poem related to the illustration; modern green cloth, some repairs throughout. London, John Harris, c. 1830.

This charming and naïve work includes, in addition to the usual nationalities, "Otaheitians", "North American Indians", and "Negroes". This is a very rare book, not appearing in any of the standard bibliographies except Gumuchian (which lists the 1821 first edition and a printing of 1831). NUC locates a single copy of what seems to be our printing as well as one copy of an 1822 edition. **$1185**

NORTH AMERICAN INDIANS. 1

ALL hail! Indian chieftain, thy object I see;
The calumet held in thy hand
Proclaims that a peace is concluded by thee,
A peace which is likely to stand.
The murderous tomahawk leave with thy wife,
Nor raise it again to destroy human life.
Thine arrows and bow o'er thy shoulders are slung
The axe in thy girdle may rest;
Of peace thy beloved has gratefully sung,
In peace may thy offspring be blest!
That offspring now cradled in moss, but design'd
To witness the blessings of peace on mankind.

74. Harris, Costume

THEATRICAL PANORAMA

75. HARVEY & DARTON. A Geographical Panorama Exhibiting characteristic representations of the Scenery and Inhabitants of various Regions.

Theatrical panorama, with ten large backdrop scenes and 23 cut-out foreground scenes (all handcoloured lithographs on card, hand-lettered in pen & ink on the verso of each), and two mahogany pillars with brass attachments, contained in the original mahogany box, 260 x 300 mm., with sliding illustrated lid, the underside of which is grooved for placement of scenes; lacking two pasteboard pieces which form proscenium arch and one back scene (VII: village on Norway/Sweden border), with two duplicate back scenes and two prop cards, lid illustration somewhat darkened, but generally in superb original condition; with the original instruction booklet (octavo, 23 pp.; original titling wrappers, a little dusted and foxed). London, Harvey and Darton, c. 1840.

Delightful theatrical panorama based on scenes from around the world. Each of the nine scenes – from Arabia to the Arctic to Tahiti – is brought to life with a landscape backdrop, in front of which parade the exotic inhabitants and animals native to the area. It is designed as a small theatre, and the audience is instructed to place the background and foreground scenes into the slits of the upturned lid and illuminate the whole from the sides with two candles; thus creating the nineteenth-century version of a three-dimensional movie.

Home entertainments of this kind, particularly those which transported the armchair traveller to newly discovered and infrequently visited lands, were very popular in the eighteenth and nineteenth centuries. This particular example is of an extremely high quality: the box and the props are of mahogany and each of the background landscapes and the foreground scenes is exquisitely coloured by hand. It would have always been an item of luxury.

After sampling the delights of Arabia, Hindostan, the Arctic circle with the Perry expeditions of the 1820s, Mont Blanc and China, one arrives at a scene of particular interest, "A View in the Society Islands". The scene is recreated with the assistance of several well known characters from John Webber's drawings of Cook's third voyage. In front of the backdrop is the "A Young Woman of Otaheite Dancing". The instruction booklet explains that "the dancer wears upon her head a cap composed of braids of human hair… the hairs are so nicely joined, that Sir Joseph Banks had pieces of it above a mile in length, without the appearance of a knot". The male figures playing the drums have been directly drawn from Webber's plate "A Human Sacrifice, in Morai, in Otaheite". The other figure is "The young woman of Otaheite, bringing a Present" of Webber; "the cloth, the manufacture of the country; it is made from the bark of trees… This mode of wrapping cloth round the bearer, is only practised when a considerable present is to be made".

Although a couple of props and one backdrop scene are lacking, this is a remarkably intact theatrical panorama, of exceptional interest for its inclusion of scenes from Cook's Third Voyage. **$24,500**

75. Harvey & Darton

HENRY'S HISTORICAL ACCOUNT

76. HENRY, David. An Historical Account of all the Voyages round the World, performed by English navigators… Faithfully Extracted from the Journals of the Voyagers…

Four volumes, octavo, with five folding maps and 44 engraved plates; old marbled boards, uniformly rubbed, neatly respined in calf; an attractive set. London, Printed for F. Newbery, 1773-4.

Scarce voyage anthology, largely relating to Australia and the South Seas. The last two volumes contain an account of Cook's first voyage, mostly based on his journal and that of Parkinson, the official artist who died during the course of the voyage. Other voyages described in the collection include those of Drake, Cavendish, Cowley, Dampier, Edward Cooke, Rogers, Clipperton and Shelvocke, Anson, Byron, Wallis, Carteret, and Bougainville, while the Appendix contains the *Journal of a Voyage to the North Pole* by Phipps.

The early 1770s were a prolific period for the English explorers, and the date of this publication coincides with one of the high points for London publishers of travel accounts. As a result, several of the accounts in this collection are of particular interest as being strictly contemporary narratives rather than later abridgements by hack-writers. The Cook material, for example, is published in the same year as the official account of the first voyage, as is also true for the accounts of Wallis and Carteret, while Bougainville's account had appeared for the first time in English only a year earlier; Phipps's *Voyage* did not in fact appear independently until a year later.

The two most significant English expeditions to Australia – Dampier and Cook, respectively the first Englishman to reach Australia and the discoverer of the east coast – are fully treated here, and the collection as a whole gives a clear picture of English activities in the South Seas with Australia as the central achievement.

Beddie, 655; Hill, p. 142; Sabin, 31389. **$3200**

MOLL'S SYSTEM OF GEOGRAPHY

77. MOLL, Herman. A System of Geography: or, a New & Accurate Description of the Earth in all its Empires, Kingdoms and States. Illustrated with history and topography, and maps of every country, fairly engraven on copper, according to the latest discoveries and corrections…

Two parts in one volume, folio, with an engraved additional title and 43 maps, letterpress title printed in red and black; occasional marginal dust-soiling, a few leaves browned; contemporary panelled calf, joints slightly cracked. London, Printed for Timothy Childe, 1701.

First edition of this universal geography. Although his *System of Geography* is essentially a factual compilation, Moll had close links with Daniel Defoe, Woodes Rogers and other proponents of schemes to exploit the riches of the South Sea. Towards the end of the work Moll gives as much information as was then available to English writers and readers about "Nova Guinea", "Terra del Fuogo [sic]" and the "Southern Unknown Region, or Terra Australis Incognita". Describing this latter, he "distinguish'd the Parts of it by the several Names given it by Pilots and Captains, who have sail'd by them…". Under this head he mentions New Zealand, Van Diemen's Land, New Holland, "Carpentaria", and "La Tierra Austral del Espiritu Santo" (of Quiros).

Alden, 701/184; Phillips, 536. **$6750**

77. Moll

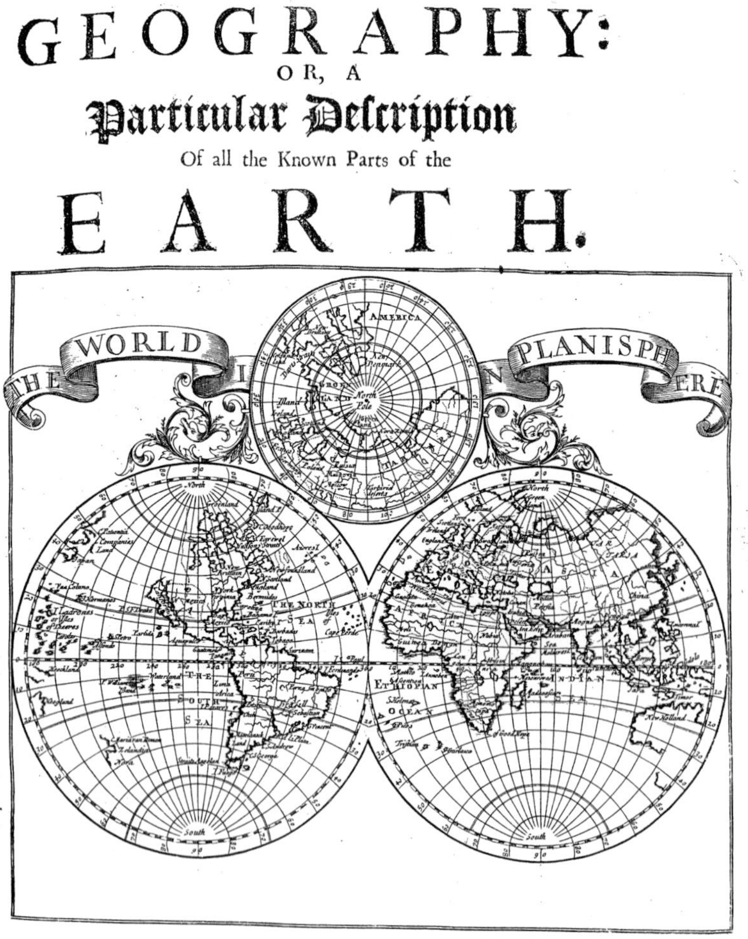

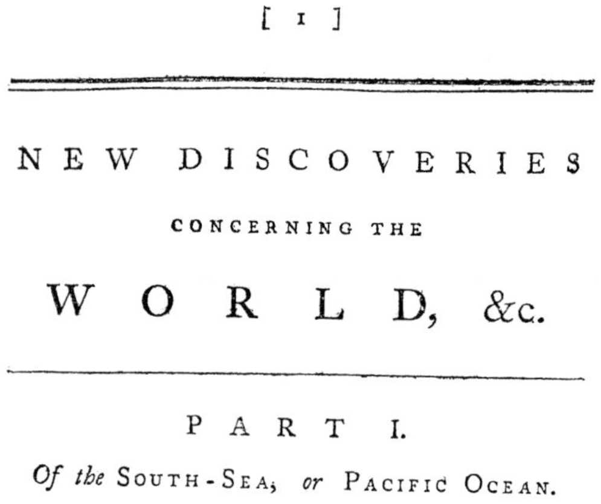
[1]

NEW DISCOVERIES
CONCERNING THE
WORLD, &c.

PART I.
Of the SOUTH-SEA, *or* PACIFIC OCEAN.

78. New Discoveries

BETWEEN A CANOE AND AN ICEBERG

78. NEW DISCOVERIES. New Discoveries Concerning the World and its Inhabitants.

Octavo, with two folding plates and two folding maps; old quarter pebblegrain cloth and marbled boards, spine and edges rubbed but a very good copy. London, J. Johnson, 1778.

First edition of this early compendium of information about Australia, New Zealand and the South Seas. The accounts of Cook's first two voyages are assembled from a variety of sources, including Hawkesworth, Parkinson, Forster and Cook himself. This gives a good overall picture of the voyages, while the material on the second voyage in particular, published during Cook's absence on the third voyage and before any news of his death, is of a good early date.

The various discoveries in the South Seas, and especially those of Cook, are put into a context of particular significance for Australia, by comparison with the voyages of Mendaña, Quiros, Tasman, Le Maire, Schouten, Dampier, Roggeveen and others. One of the engravings bears more than a passing resemblance to the celebrated Dufour panorama, the printed French wallpaper which illustrates Cook's and others' discoveries; in the plate here, a marvellous composite of people, places and animals discovered in the South Seas, we see a "Kangaroo" and a "Woman of New Holland", as well as "A Canoe of O-Taheitee" which, defying climate, is separated from *Endeavour* only by "A Floating Ice-Island, numbers of which were seen towards the Southern Frigid Zone".

Beddie, 9, 1220, 1297; Hill, p. 210; Kroepelien, 893. **$2850**

17 VOLUMES QUARTO

79. PINKERTON, John. A General Collection of the best and most interesting Voyages and Travels in all parts of the world; many of which are now first translated into English. Digested on a new plan.

17 volumes, quarto, with numerous engraved maps and plates; contemporary half diced russia, spines ornately panelled in gilt between raised bands; the marbled sides mostly rubbed, some worn, and the joints generally splitting (as often with russia leather of the period); nonetheless an acceptable set. London, Printed for Longman, Hurst, Rees and Orme, 1808-1814.

An elaborate collection of voyage accounts, containing details of virtually every major voyage to date, often for the first time in English. "Of great value for its texts" (Hill), Pinkerton's *General Collection* also contains one of the earliest bibliographies on Australasia and the Pacific. This 255-page catalogue found in volume 17 lists voyages by geographical location, and includes sections on world voyages, Australasia, New Holland, and Polynesia; followed by a detailed analytical index, certainly one of the most ambitious ever attempted, comprising in its 472 pages everything from Aa to Zwellendam.

The author's summary in the same volume discusses the achievements of Cook from a contemporary perspective: "The Voyages of Cook may therefore be regarded as forming an illustrious epoch... the old catalogues of pictures and statues, with trifling adventures by sea and land, which were called books of travels, have sunk into obscurity before the new and important works". Pinkerton concludes his remarks on Cook with the note that "...an English colony was sent to Australia, which will infallibly diffuse illuminations over that remote quarter of the world" (p. xxxi).

This set – whose bindings, though worn, are attractive – has the label of J.R. Hogg, Bookseller, Madras, who advertises "Books neatly bound": this may well be a Madras binding in the English style of the period.

Beddie, 77; Hill, p. 236; Sabin, 62957. **$5250**

PORTUGUESE & SPANISH VOYAGES

80. RICHARDSON, W., publisher. A General Collection of Voyages and Discoveries, made by the Portuguese and Spaniards, during the fifteenth and sixteenth centuries...

Quarto, two engraved views, four maps and two portraits (as issued); an excellent copy, completely uncut in the original boards, joints and edges rubbed; boxed. London, W. Richardson [etc.], 1789.

First edition of an important collection of early voyages, neatly anthologising most of the major Spanish and Portuguese early navigators, including all the major discoverers such as Columbus, da Gama, Vespucci, but with a number of lesser-known accounts as well.

There has always been some confusion regarding the maps issued in this book, as the printed list does not exactly correspond with the maps as found in all copies. The map of "India" called for is in fact a map of "The East Indies", and no map of the West Indies was issued. Those that should be present, and are in this copy, are therefore "Chart of the Western Coast of Africa", "Map of Africa", "Map of the East Indies", and "Map of the Canary Islands". Additionally there are portraits of Henry the Navigator and Albuquerque, and views of Madeira and Tenerife.

Hill, p. 429; Mendelssohn (1979) II, p. 322; Sabin, 26875. **$5500**

80. Richardson

THE INTERESTING AND ENTERTAINING VOYAGES OF THE CELEBRATED

GONZALEZ AND VAZ,
GONZALEZ ZARCO,
LANZEROTA,
DIOGO GILL,
CADA MOSTO,
PEDRO DI SINTRA,
DIOGO D'AZAMBUZA
BARTHOLOMEW DIAS,
VASCO DE GAMA,
VOYAGES TO THE CANARY ISLANDS
VOYAGES OF COLUMBUS,
NINO AND GUIERRA,
OJEDA AND VESPUSIUS
CORTEREAL,
ALVAREZ CABRAL,
FRANCIS ALMEED,
ALBUQUERQUE,
ANDREA CORSALI,
VOYAGE TO ST. THOMAS,
VOYAGE OF DE SOLIS, PINZON, &c.
VOYAGE OF JOHN PONCE,
GRIJALVA,
NICUESSA,
CORTES,
OJEDA AND OCAMPO,
MAGELLAN.

WITH OTHER VOYAGES,
TO THE EAST-INDIES, THE WEST-INDIES, ROUND THE WORLD, &c.

Adorned with COPPER-PLATES, MAPS, &c.

81. Stuart

GEOGRAPHICAL ANTHROPOLOGY

81. STUART, Martinus. De Mensch zoo als hij voorkomt op den bekenden Aardbol…

Six volumes, octavo, coloured plates; original half calf, rebacked. Amsterdam, J. Allart, 1802-1807.

A fine set of the first edition with a splendid series of forty-one plates of different ethnic types, printed in colours and finished by hand. This famous anthropological work was "the first Dutch treatise on geographical anthropology… Quite a few plates are after the travel accounts of Captain Cook, La Pérouse, Carver, Levaillant, etc…" (Landwehr). A German version appeared a few years later and another Dutch version came out in the 1830s.

Landwehr, 447; Lipperheide, Ma9; Teile, 1065. **$4250**

MARITIME GEOGRAPHY

82. TUCKEY, James H. Maritime Geography and Statistics, or, A Description of the Ocean and its Coasts, Maritime Commerce, Navigation &c. &c. &c.

Four volumes, octavo; contemporary ownership inscriptions, an excellent set in later half green morocco. London, Black Parry & Co., Booksellers to the Hon. East India Company, 1815.

A wide-ranging work on maritime geography in all its aspects, which describes the principal voyages round the world, including those of Cook, La Pérouse, Broughton, and Vancouver. It also contains sections on America, New Spain (including California), the Northwest coast, Polynesia, New South Wales and Van Diemen's Land.

The author is better known for his *Account of a Voyage to Establish a Colony at Port Phillip* published in 1805 in which he described the attempt at settlement in Victoria after the voyage in the *Calcutta.*

Ferguson, 628; not in the Hill collection. **$750**

ENCYCLOPEDIA OF THE EAST INDIES

83. VALENTIJN, François. Oud en Nieuw Oost-Indiën…

Five volumes, folio, bound as eight volumes; with an engraved frontispiece, engraved dedication leaf, engraved portrait of the author, list of subscribers, errata; with a magnificent series of about 400 engravings, comprising maps, plans, portraits, views and natural history subjects (many folding), and engraved text illustrations; the portrait of Mattheus de Haan which is often missing here loosely inserted (he was still governor until 1729); folding printed tables; a lovely set in nineteenth century Dutch red morocco, gilt ornaments, spines panelled in gilt. Dordrecht and Amsterdam, 1724-1726.

A beautiful set of the great "encyclopedia of the Dutch East-Indies", "a work of great importance and extreme detail", with a superb visual record of early voyages in the form of its extensive and splendid series of engraved illustrations. This scarce and important collection covers not only Western Australia and the East Indies but also eastern Asia in considerable detail, and there is much too on the Philippines, Ceylon and the Cape of Good Hope.

The publication of Valentijn's book reflects an intense interest in the "Southland" in the early eighteenth century. To satisfy this audience, Valentijn was able to publish extensive narratives of the expeditions of Abel Tasman (1642-1643) and Willem de Vlamingh (1696-97), and to produce a number of images from these expeditions – which are among the very few visual records of Australia and the South Pacific prior to the Cook voyages.

Valentijn includes with the Tasman narrative a series of small charts adapted from the navigator's journals, along with a number of coastal profiles, which are the earliest views or plans of the Tasmanian and north Australian coasts, while the illustrations to the Vlamingh section include the famous view of the Swan River in Western Australia and the earliest depiction of the black swan. A general map of Tasman's voyages begins the section, with a fine depiction of the outline of the Australian continent established by him. There are also several New Zealand charts and views.

Willem de Vlamingh's voyage to the west coast of Australia was the initiative of Nicholas Witsen, Burgomaster of Amsterdam and Director of the VOC. The objective was not only scientific but directly linked to Dutch commercial activities. William Eisler and Bernard Smith point out (in *Terra Australis: the Furthest Shore*, AGNSW, 1988) that Witsen was an enthusiastic collector of paintings and artifacts, and insisted that a painter accompany the expedition to record the rarities found: "the only surviving drawings are the watercolour sketches of the Western Australian coast done by Victor Victorszoon… Witsen received from New Holland… some brambles of wood and a small chest containing shells collected on the beaches, fruits, plants etc… No specimens of land animals were brought back to Holland. Several black swans were captured but died on the return voyage…". Valentijn also remarked on the "rare and beautiful shells from the beaches of the Southland to be found in the cabinets of Simon Schynvoet of Amsterdam…".

Valentijn (1666-1727) spent several years in the East Indies, and made much use of official archives and documents (many of which were destroyed at Batavia during the Second World War) in compiling his collection. The great naturalist Georg Eberhard Rumpf helped him with natural history materials.

Cordier, 'Indosinica', 927-930; Landwehr, 467; Mendelssohn IV, p. 59; Nissen, ZBI 4213; Tiele II, 1121. **$58,500**

83. Valentijn

204 FINE COSTUME PLATES

84. WAHLEN, Auguste. Moeurs, Usages et Costumes de tous les peuples du monde…

Four volumes, large octavo; with a total of 204 finely handcoloured costume plates; contemporary quarter crimson roan, flat spines ornately gilt. Brussels, Librairie Historique-Artistique, 1844.

Scarce and beautiful work on the costumes of the world, finely illustrated with a wonderful series of highly coloured full-page plates. It is not recorded by Ferguson, although the fourth volume covers Oceania and includes the Aborigines of Jervis Bay and Kangaroo Island, and the exotic peoples of Pitcairn Island, New Guinea, Timor, Java, New Zealand and Hawaii. There are also images familiar to us from published voyage accounts such as the Tahitian dancer after John Webber from the account of Cook's third voyage, and portrait profiles of Maoris based on Sydney Parkinson's images published in the official account of the first voyage.

As usual, the plates in the publication do not exactly correspond with the lists published within the work. The table of contents describes 185 plates (actually 188, but an erratum shows that only 185 were issued) and subscribers could also order a series of twelve special supplementary plates – as here – making a total of 197 plates. This copy however has 204 plates. A full collation is available on request. **$4750**

84. Wahlen

COMPARATIVE ANTHROPOLOGY

85. [WILKES EXPEDITION] PICKERING, Charles. The Races of Man: and their geographical distribution.

Quarto, with a coloured folding map and 12 handcoloured plates; a good copy in contemporary calf gilt, spine and edges worn, joints repaired. Boston, Charles C. Little and James Brown, London, John Murray, 1848.

Large paper copy: one of only 150 copies published of the "Unofficial" issue; the "Official" issue, limited to 70 copies, was also dated 1848 but was not in fact distributed until 1849. This quite pioneering work in comparative anthropology appeared as the ninth volume of the official account of the United States Exploring Expedition commanded by Wilkes.

Haskell, 33. **$2400**

3: IMAGINARY VOYAGES & UTOPIAS

VIAGEM
ao interior da
Nova Hollanda,
OBRA MORAL, CRITICA, E RECREATIVA,
DEDICADA
A
Sua Magestade Fidelissima
A RAINHA.
Por V. J. A.

LISBOA: 1841.
Typ. de Vicente Jorge de Castro & Irmão,
Calcada de S. João Nepomuceno N.° 16.

Vende-se nas Lojas de Livros de Orcel, defronte dos Martyres — Bordalo, na Rua dos Capellistas — e Viuva Henriques, na Rua Augusta.

86. Aguiar

ARMCHAIR TRAVELLING BY A PORTUGUESE CIVIL SERVANT

86. [AGUIAR, Vasco José de] Viagem ao interior da Nova Hollanda, obra moral, critica e recreativa... Por V.J.A.

Three volumes bound in one, small octavo; a few spots but an attractive copy in contemporary (probably original) quarter calf, slight wear. Lisbon, Vicente Jorge de Castro e Irmão, 1841.

The only edition of a little-known imaginary voyage to Australia, not dealt with by most of the commentators on the genre. This strange piece was written by a Portuguese civil servant who probably never even left Lisbon. It was his only book: he died in 1855.

The anonymous hero of the narrative arrives in New South Wales from Liverpool, accompanied by his manservant, in 1836. He spends several years in the colony, and travels extensively in the interior. Fulsome descriptions of the social life of Sydney, and the beauties of Australia are used to paint a picture of an utopian new world full of hope and enterprise, in contrast with the evils of life in Europe.

Ferguson, 3133; Innocêncio, VII, 405. $2850

UTOPIAN CIVILISATION IN THE INTERIOR

87. FOX, Lady Mary. Account of an Expedition to the Interior of New Holland. Edited by Lady Mary Fox.

Duodecimo; an excellent copy, with the half-title but without the advertisement leaves found in some copies, in original grey boards, green cloth spine with printed paper label (slightly chipped). London, Richard Bentley, 1837.

The scarce first edition of this fantastical tale, "more authentic, while at the same time quite as incredible as anything Swift wrote..." (Friederich). An 1835 expedition leaves Bathurst for the Australian interior, where the exploring party comes across a group of English-speaking settlers, part of an utopian civilisation of three or four million descendants of sixteenth-century refugees from the Reformation wars, intermixed with whole tribes of Aborigines.

Lady Fox was a woman of liberal views and throughout the novel she satirically condemns racism and the convict system in Australia: "a new

ACCOUNT
OF
AN EXPEDITION
TO THE INTERIOR OF
NEW HOLLAND.
EDITED BY
LADY MARY FOX.

LONDON:
RICHARD BENTLEY, NEW BURLINGTON STREET,
Publisher in Ordinary to His Majesty.
1837.

87. Lady Fox

settlement with convicted criminals, – to form a new nation of the scum and refuse of mankind, – appeared to them so preposterous, that for some time they could not help supposing they must have misunderstood their informants".

Ferguson, 2264; Friederich, pp. 69-70. $985

ORIGINAL WRAPPERS

88. FOX, Lady Mary. Account of an Expedition to the Interior of New Holland.

Octavo; a very good copy in the rare original wrappers, worn. London, Richard Bentley, 1849.

The surprisingly scarce second edition.

Ferguson, 5040. $685

THIRD EDITION REVISED

89. FOX, Lady Mary. The Southlanders: an account of an Expedition to the Interior of New Holland.

Octavo; a fine copy, with the half-title and 4 pp. advertisements, in the original brown cloth. London, John W. Parker, 1860.

Third edition, revised and with an additional "humorous poem", "The Land of Contrarieties". The preface notes that "This little Work appeared several years ago. It was compiled by more than one person; all of whom are not now living…". This remark is interesting as the book has often in the past been attributed to Archbishop Whateley as well as to Lady Fox.

Ferguson, 9717. $525

EARLIEST ANTIPODEAN UTOPIA

90. [HALL, Joseph] Mundus alter et idem sive Terra Australis ante hac semper incognita longis itineribus peregrini academici nuperrime lustrata. Auth: Mercurio Brittanico…

Small octavo, with five folding maps (one in well-produced facsimile, small defect to corner of another); contemporary calf, spine neatly restored. "Frankfurt, the heirs of Ascanius de Rinialme" [in fact London and Hanau, c. 1606].

"One of the earliest, if not the earliest of the fictitious voyages set in Australia… an extremely rare work and seldom offered for sale" (Davidson). This supposed description of Terra Australis, with a highly fanciful cartography of the continent in five eccentric maps, was written as a satire on the vices and follies of mankind.

"Hall's 1607 *Mundus* gave us the earliest utopia to be set in the Antipodes. It would be more correct to describe it as a dystopia because it is not a blueprint for a perfect society but a place where life is disagreeable or unappealing. The genre has a long history: we have seen it in our own century in *Brave New World* and *1984.* But *Mundus* is a lighter work, a satire located in Terra Australis… [which] is divided into four regions. Crapulia is a land of gluttons where even the birds eat so much that they are too heavy to fly, and where the fattest men are picked as rulers… Viraginia is a republic of women

90. Hall

where there are no laws and everything is discussed. But it is hardly a blueprint for a feminist collective: all the women talk at the same time and 'none doth give ear, but each one yells as if she were horn-mad'…" (Dunmore, *Utopias and Imaginary Voyages*, Canberra 1988).

The bibliography of this significant precursor of *Gulliver's Travels* and close relation of More's *Utopia* is complex: see the article by J.M. Wands in *Papers of the Bibliographical Society of America*, 74, 1980, pp. 1-12. This is the usual form in which the book is seen, an intermediate issue combining the sheets from the London printing of 1605 and the Hanau printing of 1607.

Gibson, 702; STC, 12685.3. **$1500**

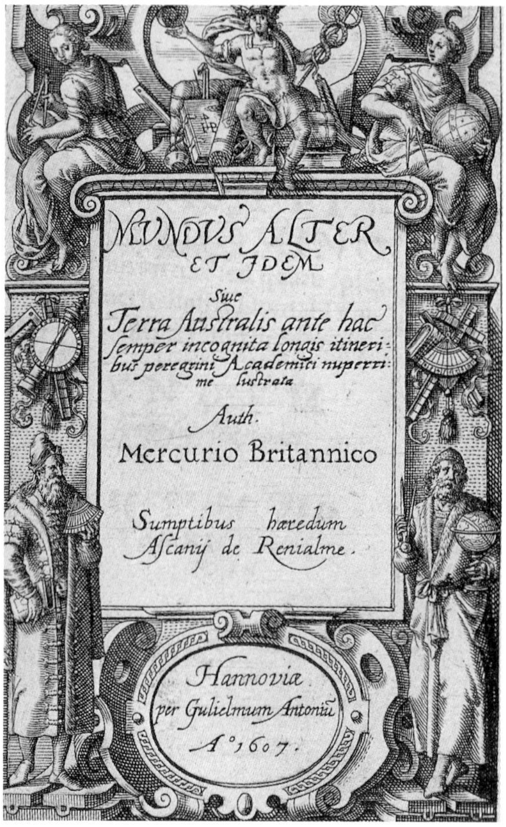

91. Hall

EARLY EDITION

91. HALL, Joseph. Mundus alter et idem sive Terra Australis…

Duodecimo, with five folding maps; an attractive copy in contemporary stamped vellum. Hanau, Gulielmus Antonius, 1607.

One of the early editions of the book which comprise a London printing of 1605, this Hanau printing of 1607, and an intermediate issue combining the sheets from the two. There were several later versions.

Gibson, 702; STC, 12685.3 (note). **$6250**

92. Hall

THIRD EDITION

92. HALL, Joseph. Mundus alter et idem sive Terra Australis…

Duodecimo, with five folding maps; library stamp on title, a couple of marginal tears and slightly defective last leaf but a very good copy in old vellum. Utrecht, Ioannes Wæsbergius, 1648.

Third edition of this rare imaginary voyage to a Terra Australis utopia. **$1200**

LIVING IT UP IN THE SOUTH SEAS

93. [LA PEROUSE] DECOUVERTES. Découvertes dans la Mer du Sud. Nouvelles de M. de la Peyrouse, jusqu'en 1794. Traces de son passage trouvées en diverses isles et terres de l'Océan pacifique; grande isle peuplée d'émigrés français.

Octavo, neat old library stamps at start; some foxing and a number of leaves stained, but overall an excellent uncut copy in original plain wrappers, spine repaired; boxed. Paris, chez Everat, c. 1795.

First edition of this rare La Pérouse fantasy, published in either 1795 (Ferguson, Sabin, Kroepelien) or 1798 (Gove, Library of Congress catalog); the only other edition was a Dutch version of 1799.

The disappearance of La Pérouse had led to

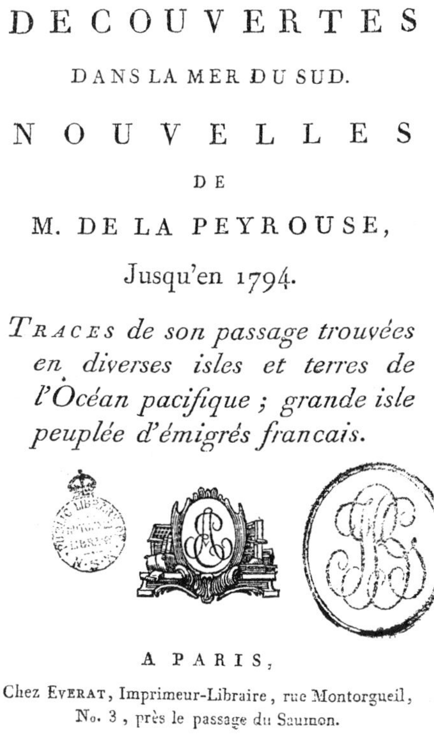
DECOUVERTES
DANS LA MER DU SUD.
NOUVELLES
DE
M. DE LA PEYROUSE,
Jusqu'en 1794.
TRACES de son passage trouvées en diverses isles et terres de l'Océan pacifique ; grande isle peuplée d'émigrés francais.
A PARIS,
Chez EVERAT, Imprimeur-Libraire, rue Montorgueil, No. 3, près le passage du Saumon.

93. Découvertes

intense speculation in France, and more than one writer used the mystery as a starting-point for a semi-utopian description like this of an island where La Pérouse and his crew were leading an idyllic life as refugees from revolutionary France. The South Seas paradise is used in metaphorical contrast to the horrors of the Revolution. "Very rare and… a desirable addition either to a collection of La Pérouse items, or to a library of Pacific voyages…" (Davidson).

Ferguson, 225; Gove, pp. 397-8; Kroepelien, 283. **$3250**

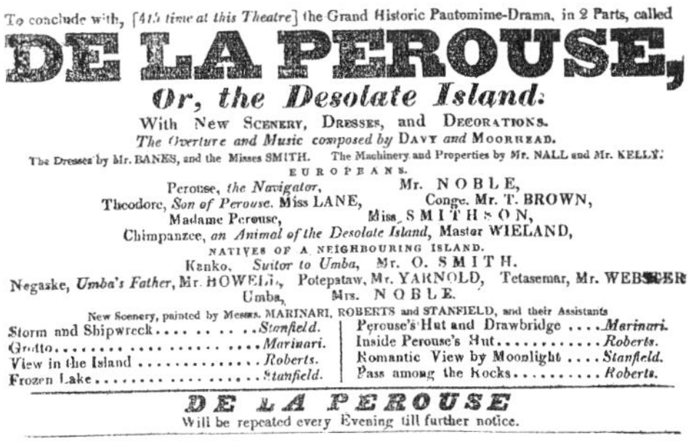
To conclude with, [4th time at this Theatre] the Grand Historic Pantomime-Drama, in 2 Parts, called
DE LA PEROUSE,
Or, the Desolate Island:
With New SCENERY, DRESSES, and DECORATIONS.
The Overture and Music composed by DAVY *and* MOORHEAD.
The Dresses by Mr. BANKS, and the Misses SMITH. The Machinery and Properties by Mr. NALL and Mr. KELLY.
EUROPEANS.
Perouse, *the Navigator,* Mr. NOBLE,
Theodore, *Son of Perouse.* Miss LANE, Conge. Mr. T. BROWN,
Madame Perouse, Miss SMITHSON,
Chimpanzee, *an Animal of the Desolate Island,* Master WIELAND,
NATIVES OF A NEIGHBOURING ISLAND.
Kanko, *Suitor to Umba,* Mr. O. SMITH.
Negaske, *Umba's Father,* Mr. HOWELL, Potepataw, Mr. YARNOLD, Tetasemar, Mr. WEBSTER
Umba, Mrs. NOBLE.
New Scenery, painted by Messrs. MARINARI, ROBERTS and STANFIELD, and their Assistants

Storm and Shipwreck	*Stanfield.*	Perouse's Hut and Drawbridge	*Marinari.*
Grotto	*Marinari.*	Inside Perouse's Hut	*Roberts.*
View in the Island	*Roberts.*	Romantic View by Moonlight	*Stanfield.*
Frozen Lake	*Stanfield.*	Pass among the Rocks	*Roberts.*

DE LA PEROUSE
Will be repeated every Evening till further notice.

94. Theatre Royal, Drury Lane

LA PEROUSE IN A SNOW STORM

94. [LA PEROUSE] PLAYBILL: THEATRE ROYAL, DRURY LANE. …De La Perouse, Or, the Desolate Island. With New Scenery, dresses, and decorations…

Folio broadside, 325 x 190 mm.; in fine condition. London, 1825.

Rare: European fascination with the whereabouts of the lost Pacific explorer often resulted in utopian explanations for his disappearance. This La Pérouse pantomime had actually been written and performed before Dillon discovered the whereabouts of La Pérouse's wreck in the New Hebrides in the late 1820s. This would explain the advertised "new scenery" by Mr Pitt, including "snow storm", "snow landscape" and "frozen lake"! **$485**

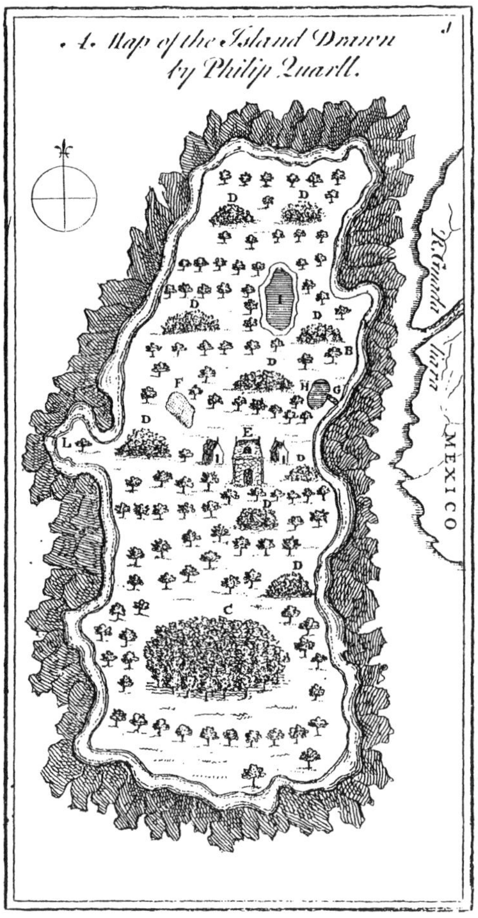

95. Longueville

WITH THE BIBLE AND BUNYAN

95. LONGUEVILLE, Peter. The Hermit: or, the Unparallel'd Sufferings and Surprising Adventures of Mr. Philip Quarll, an Englishman: who was lately discovered by Mr. Dorrington, a Bristol Merchant, upon an uninhabited Island in the South-Sea…

Duodecimo, with a frontispiece (Quarll and his monkey servant) and a map; quarter calf antique, a nice copy. London, J. Wren, 1768.

A fictitious account of life on a South Sea island, with a map of the island, Longueville's *Hermit* is the most famous of the imitations of *Robinson Crusoe*, and the most often printed; it first appeared in 1727. Gove lists over sixty editions and George Crabbe in *The Parish Register* accorded it a place alongside the Bible and Bunyan in most country homes. It was read by the young Coleridge with a deep impact on his poetic imagination (see Lowes, *The Road to Xanadu*, pp. 414-6).

As John Dunmore pointed out in his lecture *Utopias and Imaginary Voyages* (NLA, 1987), "Charles Lamb reports that schoolboys in his day were so taken in by the story that they planned to run away in search of the lonely island on which the hero had been ship-wrecked. It is not easy to appreciate in these days of atlases and instant communication how easy it was for readers to accept or to wonder about the probability of such wondrous tales of travel presented as factual". The "facts" were actually taken from genuine accounts of voyages such as Magellan and Dampier.

The publishers of this "fourth edition" also issued a "fourth" in 1763; the original edition appeared in 1727 (Alden, 727/149).

See Gove, p. 264; this edition not in Sabin (see 66949-52).
$375

THE HERMIT IN 1822

96. [LONGUEVILLE, Peter] The English Hermit; or, Life and Adventures of Philip Quarll, who was discovered by an English merchant on an uninhabited island in the South-Sea, where he had lived upwards of fifty years without any human assistance.

Duodecimo, frontispiece and engraved title; a good, neat copy in modern brown cloth. London, Dean and Munday, 1822.

One of very many versions. Gove lists over sixty editions, but does not note this printing.

See Gove, p. 266. **$245**

The English Hermit, or surprising Life and Adventures of Philip Quarll,

Philip Quarll and his Monkey discovered by Mr Dorrington.

London,
Printed for A.K. Newman, & Co. Leadenhall Street,
and
Dean & Munday, Threadneedle Street.

96. Longueville

THE EDGE-PARTINGTON COPY

97. [LONGUEVILLE, Peter] The English Hermit; or the Adventures of Philip Quarll... with several additions, by Mary Elliott.

Duodecimo, with a folding frontispiece and another plate; a sweet copy in the original printed wrappers, boxed. London, William Darton, 1822.

The Edge-Partington copy of this scarce edition of the popular imaginary voyage, first printed in 1727.

Gove, p. 266. **$625**

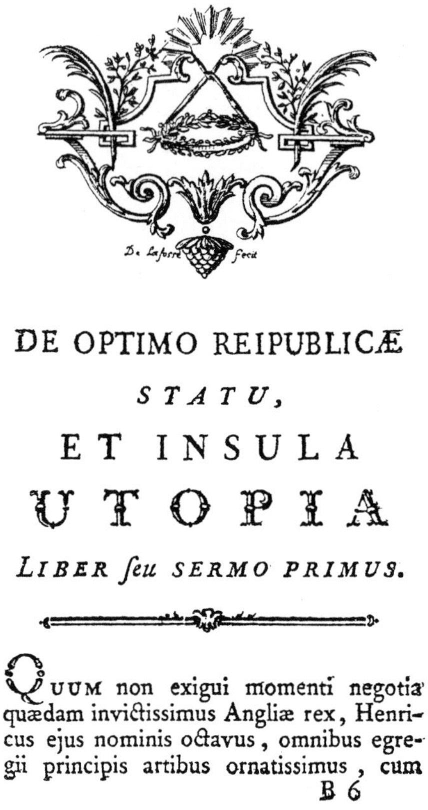
DE OPTIMO REIPUBLICÆ
STATU,
ET INSULA
UTOPIA
LIBER seu SERMO PRIMUS.

QUUM non exigui momenti negotia quædam invictissimus Angliæ rex, Henricus ejus nominis octavus, omnibus egregii principis artibus ornatissimus, cum
B 6

98. More

UTOPIA

98. MORE, Sir Thomas. De optimo reipublicæ statu, deque nova insula Utopia.

Duodecimo, with a frontispiece; a very nice copy in contemporary straight-grained blue morocco, gilt. London and Paris, Barbou, 1777.

An attractive edition of the most famous of all utopias, *Utopia*. More's philosophical conceit was responsible for an immense imitative or derivative literature that demonstrates what chords the book struck. As Oscar Wilde remarked, "A map of the world that does not include Utopia is not worth even glancing at, for it leaves out the one country at which Humanity is always landing. And when Humanity lands there, it looks out, and seeing a better country, sets sail. Progress is the realisation of Utopias". **$550**

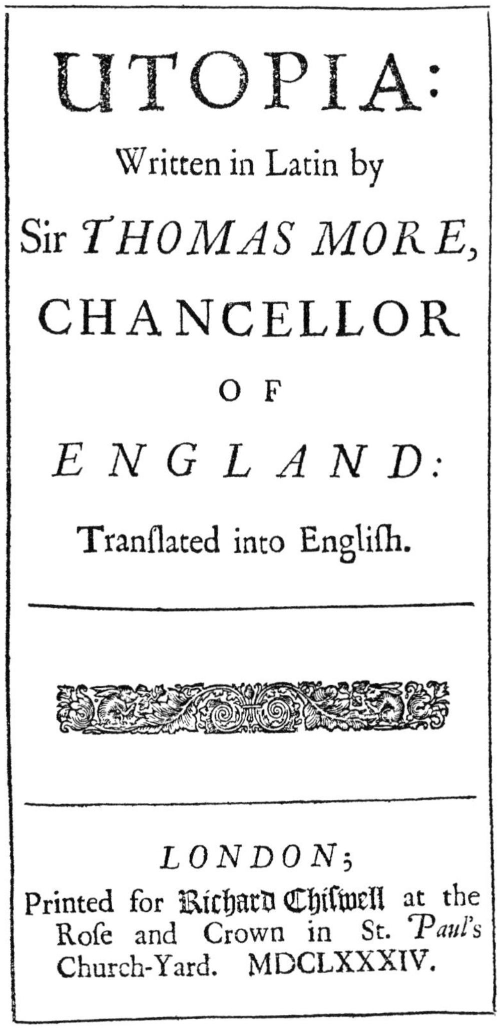

UTOPIA:

Written in Latin by

Sir *THOMAS MORE*,

CHANCELLOR

OF

ENGLAND:

Tranflated into Englifh.

LONDON;

Printed for Richard Chifwell at the Rofe and Crown in St. *Paul's* Church-Yard. MDCLXXXIV.

99. More

BURNET'S TRANSLATION

99. MORE, Sir Thomas. Utopia: written in Latin... Translated into English.

Octavo, with the initial and final blanks; contemporary ownership inscription obliterated on title; old calf, some surface wear but a very good copy. London, Richard Chiswell, 1684.

First edition of the translation by Gilbert Burnet, Bishop of Salisbury.

Wing, M2691. $550

FLYING INDIANS

100. [PALTOCK] PLAYBILL: Theatre, Beverley. Second night of Peter Wilkins... an entirely New Spectacle, founded on Fanciful Adventures... Peter Wilkins; or, the Flying Indians...

Folio broadside, 185 x 440 mm.; old folds, in excellent original condition. Theatre, Beverley, 21 May, 1828.

An attractive playbill based on *Peter Wilkins,* the famous imaginary voyage by Paltock, in which men take to the air to travel the globe. First published anonymously in 1750, it was an exceptionally popular book. Paltock's fanciful tale had a great literary impact: Southey called the winged people "the most beautiful creatures of the imagination that ever were devised", and both Coleridge and Shelley drew poetic inspiration from the story.

Here, brought to the stage in the north of England, it is used as a vehicle for some evidently tremendous stage-effects: "Indian Lake and Landscape, with the Descent of Two Gawries, or Flying Women... Romantic Rocks, Intermixed with Cascades and Waterfalls. Terrific Hurricane and Land Storm... approach of the Flying Islanders... Extraordinary Ascension of Peter Wilkins... to the Land of Flight". $375

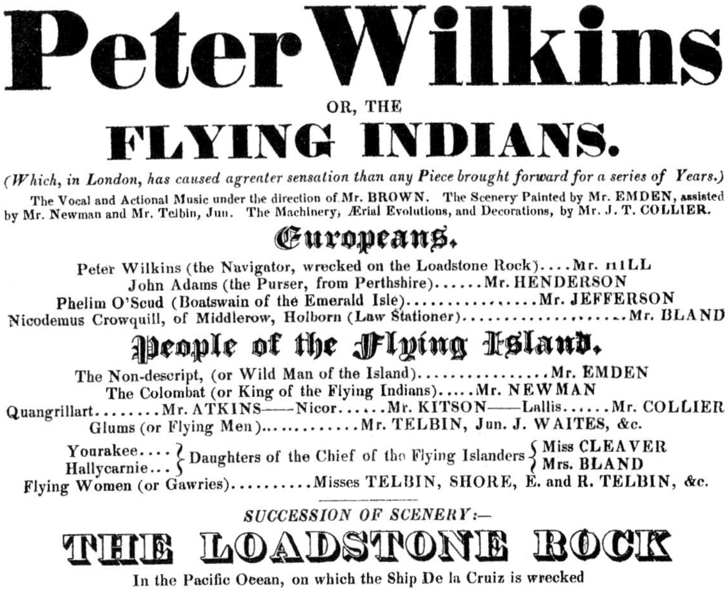

Peter Wilkins

OR, THE

FLYING INDIANS.

(Which, in London, has caused agreater sensation than any Piece brought forward for a series of Years.)

The Vocal and Actional Music under the direction of Mr. BROWN. The Scenery Painted by Mr. EMDEN, assisted by Mr. Newman and Mr. Telbin, Jun. The Machinery, Ærial Evolutions, and Decorations, by Mr. J. T. COLLIER.

Europeans.

Peter Wilkins (the Navigator, wrecked on the Loadstone Rock)....Mr. HILL
John Adams (the Purser, from Perthshire)......Mr. HENDERSON
Phelim O'Scud (Boatswain of the Emerald Isle)................Mr. JEFFERSON
Nicodemus Crowquill, of Middlerow, Holborn (Law Stationer)....................Mr. BLAND

People of the Flying Island.

The Non-descript, (or Wild Man of the Island)................Mr. EMDEN
The Colombat (or King of the Flying Indians).....Mr. NEWMAN
Quangrillart........Mr. ATKINS—Nicor......Mr. KITSON—Lallis......Mr. COLLIER
Glums (or Flying Men).............Mr. TELBIN, Jun. J. WAITES, &c.
Yourakee.... } Hallycarnie... } Daughters of the Chief of the Flying Islanders { Miss CLEAVER { Mrs. BLAND
Flying Women (or Gawries)..........Misses TELBIN, SHORE, E. and R. TELBIN, &c.

SUCCESSION OF SCENERY:—

THE LOADSTONE ROCK

In the Pacific Ocean, on which the Ship De la Cruiz is wrecked

100. Theatre, Beverley

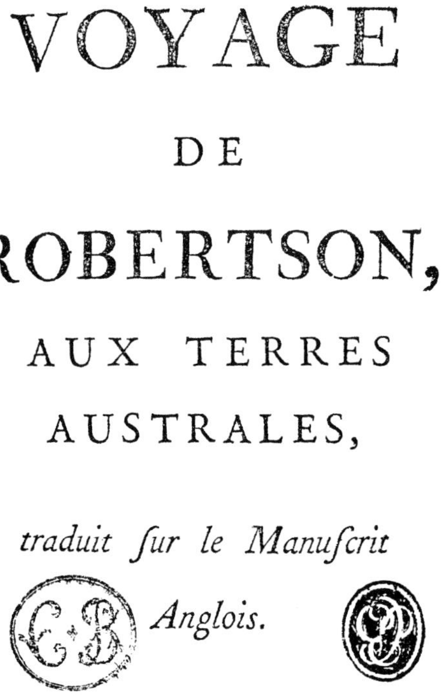

VOYAGE

DE

ROBERTSON,

AUX TERRES

AUSTRALES,

traduit fur le Manufcrit

Anglois.

101. Robertson

WRECKED IN AUSTRALIA

101. ROBERTSON, pseudonym. Voyage de Robertson, aux Terres Australes, traduit sur le manuscrit Anglois.

Duodecimo, two old library-stamps on title; a nice copy

102. Morin's Gulliver

in contemporary French calf, spine gilt, light wear to spine. "Amsterdam" [Paris], 1767.

A scarce and amusing imaginary voyage, which describes a large, fertile land – Terra Australis.

This pretends to have been translated from an unpublished English manuscript. Robertson is supposed to have sailed with Drake on the *Elizabeth* in 1585, to have been given command of a ship with a crew of ten, and to have been wrecked on the Australian coast, which he then explored.

As with other imaginary voyages, the new land supposedly discovered is used as a metaphorical contrast to the country in which the account is to be read – in this case, as often, France.

Gove, pp. 355-6. **$1200**

FINE FRENCH COLOURED PLATES FOR GULLIVER

102. [SWIFT] MORIN, Edmond (illustrator). Voyages de Gulliver en dessins par E. Morin.

Oblong octavo, handcoloured lithograph title and 30 handcoloured lithographs; fine in original green patterned cloth, some slight edge wear. Paris, chez A. de Vresse, n.d., c. 1850.

An exquisite and brilliantly illustrated children's version of Swift's classic. Written in 1726, *Gulliver's Travels* was an immediate success and has become one of the most frequently reproduced works in the English language; together with *Robinson Crusoe* it is the most reproduced tale for children. This version from the mid-nineteenth century is one of the most beautiful and imaginative pictorial renderings of Swift's story. Edmond Morin, a well-known illustrator of the period, perfectly captures the miniscule Lilliputians, the giants of Brobdingang, the absent-minded inhabitants of the floating island of Laptua and noble horses of the utopian land of the Houyhnhnms. His drawings have a vigour and unerring perspective which makes this one of the most desirable of all juvenile editions of *Gulliver.*

Gumuchian, 5484 (illustrated pl. 258). **$2850**

II: *Cook to Bligh*

4: CAPTAIN COOK

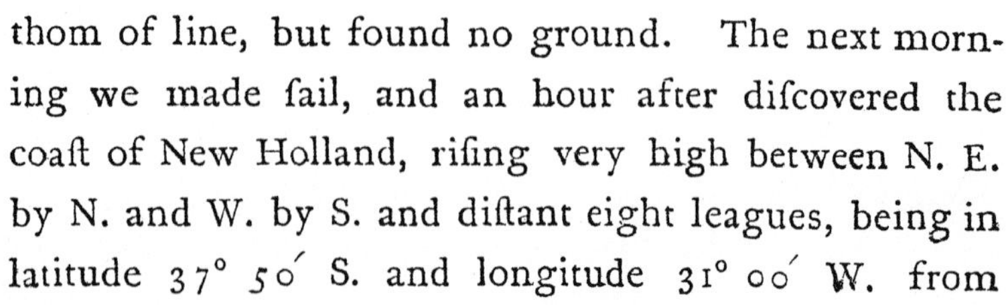
thom of line, but found no ground. The next morning we made ſail, and an hour after diſcovered the coaſt of New Holland, riſing very high between N. E. by N. and W. by S. and diſtant eight leagues, being in latitude 37° 50′ S. and longitude 31° 00′ W. from

103. Magra: the discovery of the East coast of Australia

FIRST EDITION, FIRST ISSUE

103. [FIRST VOYAGE] [MAGRA, James] A Journal of a Voyage round the World in His Majesty's Ship Endeavour, in the years 1768, 1769, 1770 and 1771...

Quarto; a good copy in later half green morocco. London, T. Becket and P. A. De Hondt, 1771.

First edition, first issue: the earliest published account of Cook's first voyage to the Pacific. Published anonymously some two months after the return of the *Endeavour* and nearly two years before Hawkesworth's official account, its author remains unknown, though the great Cook scholar Beaglehole has demonstrated that the sailor James Magra is the likeliest candidate (*Journals*, I, pp. cclvi-cclxiv).

Cook's great first voyage in the *Endeavour* ranks with the voyages of the great explorers of all periods, from Columbus on. As the first description of the voyage to be published, this is the most important of all voyage books for the history of the east coast of Australia. It contains the first printed account of the east coast, and its publication predates the naming of Botany Bay, here called Sting-ray Bay as it was originally christened by Captain Cook.

This is an example of the rare first issue with the leaf of dedication to Banks and Solander, quickly suppressed by the publishers after legal action against them. Most copies seen of the book are of the second issue, with the dedication leaf removed.

Beddie, 693; Holmes, 3. **$38,500**

104. Magra

DUBLIN PIRACY

104. [FIRST VOYAGE][MAGRA, James] A Journal of a Voyage round the World, in His Majesty's Ship Endeavour...

Duodecimo; a nice copy in contemporary calf, slightly rubbed. Dublin, Printed for J. Exshaw, [etc.], 1772.

The earliest printed account of Cook's first voyage and the discovery of the east coast of Australia, in its very scarce Dublin printing. Irish publishers of the eighteenth century were swift and efficient in their piracy of new books; their versions of likely best-sellers were often on the market within a matter of days after London publication.

The original London edition appeared in two issues: the first had an unauthorised dedication to Banks and Solander, soon withdrawn. This Irish version was prepared from one of the earliest copies issued as the dedication appears here in full.

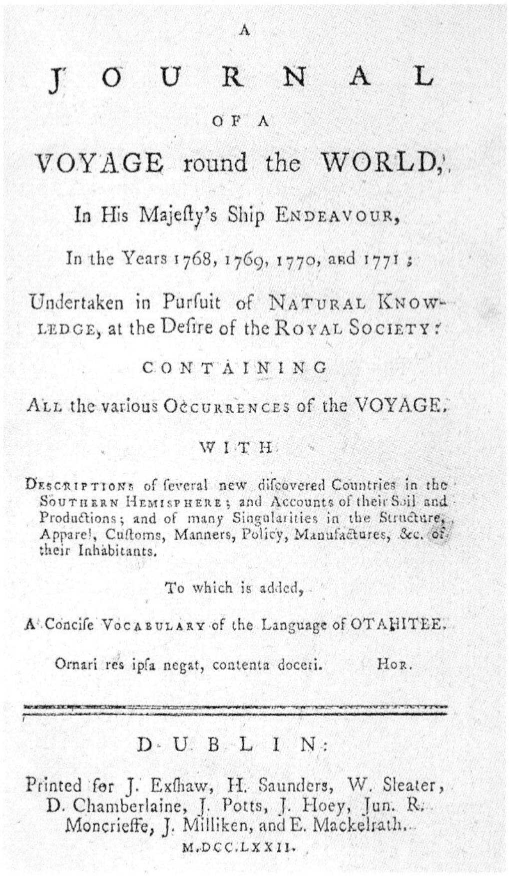
A
JOURNAL
OF A
VOYAGE round the WORLD,
In His Majeſty's Ship ENDEAVOUR,
In the Years 1768, 1769, 1770, and 1771;
Undertaken in Purſuit of NATURAL KNOWLEDGE, at the Deſire of the ROYAL SOCIETY:
CONTAINING
ALL the various OCCURRENCES of the VOYAGE.
WITH
DESCRIPTIONS of ſeveral new diſcovered Countries in the SOUTHERN HEMISPHERE; and Accounts of their Soil and Productions; and of many Singularities in the Structure, Apparel, Cuſtoms, Manners, Policy, Manufactures, &c. of their Inhabitants.
To which is added,
A Conciſe VOCABULARY of the Language of OTAHITEE.
Ornari res ipſa negat, contenta doceri. HOR.
DUBLIN:
Printed for J. Exſhaw, H. Saunders, W. Sleater, D. Chamberlaine, J. Potts, J. Hoey, Jun. R. Moncrieffe, J. Milliken, and E. Mackelrath.
M.DCC.LXXII.

104. Magra: the Dublin edition

Beddie, 695; Kroepelien, 216. **$5850**

FIRST FRENCH EDITION

105. [FIRST VOYAGE] [MAGRA, James] Journal d'un Voyage Autour du Monde... Traduit de l'Anglois, par M. de Fréville.

Octavo; contemporary French marbled calf, flat spine gilt in compartments. Paris, Saillant & Nyon, 1772.

A fine copy of the first French account of Cook's *Endeavour* voyage, and the first French book on the east coast of Australia. Fréville's translation was also issued in 1772 by the same publishers as a supplementary third volume to the octavo edition of Bougainville, but that issue is clearly identified by the wording of the title-page.

As Rolf du Rietz points out, "it is evident that the French, before the appearance of Hawkesworth's work in 1773, believed the anonymous account of 1771 to be the official narrative of the expedition, the announced forthcoming work by Hawkesworth being supposed to deal with the scientific observations only...".

Beddie, 696; Kroepelien, 218. **$5750**

THE FULL OFFICIAL ACCOUNT

106. [FIRST VOYAGE] HAWKESWORTH, John. An Account of the Voyages… in the Southern Hemisphere.

Three volumes, quarto, with 52 finely engraved maps and plates; in contemporary calf skilfully rebacked; slight staining to edges of a few pages in the first volume, otherwise a very good set. London, W. Strahan, 1773.

First edition: the official narrative of Cook's great first voyage into the Pacific during the course of which he discovered and charted the entire east coast of Australia, naming it New South Wales. This is the official account, edited from Cook's journals by the professional writer Hawkesworth. Cook's voyage occupies volumes 2 and 3; the first volume contains the official accounts of the voyages of Byron, Wallis and Carteret, with the result that Hawkesworth's compendium actually contains the cream of eighteenth-century English exploration.

Beddie, 650; Hill, pp. 139-140; Holmes, 5 (note); Kroepelien, 535. **$12,500**

107. Parkinson

FIRST EDITION

107. [FIRST VOYAGE] PARKINSON, Sydney. A Journal of a Voyage to the South Seas, in His Majesty's Ship, the Endeavour… Embellished with Views and Designs, delineated by the Author, and engraved by capital Artists.

Quarto, with frontispiece portrait, map and 26 plates; occasional light marking and offsetting, but a very nice copy, large as usual, in contemporary tree calf, rebacked. London, Printed for Stanfield Parkinson, 1773.

An excellent, large copy of the first edition: the most handsome of the unofficial accounts of Cook's first voyage. Engaged by Banks as botanical artist on the *Endeavour*, Parkinson produced an enormous number of magnificent botanical and natural history drawings of Tahiti, New Zealand and Australia. At the end of the voyage, en route from Batavia to the Cape of Good Hope, he died of a fever.

His manuscripts and drawings became a matter of dispute: Banks considered that they were his, while Parkinson's brother Stanfield claimed them under the provisions of his brother's will. When Hawkesworth learned of the impending publication of this work, he got an injunction to delay its appearance until some

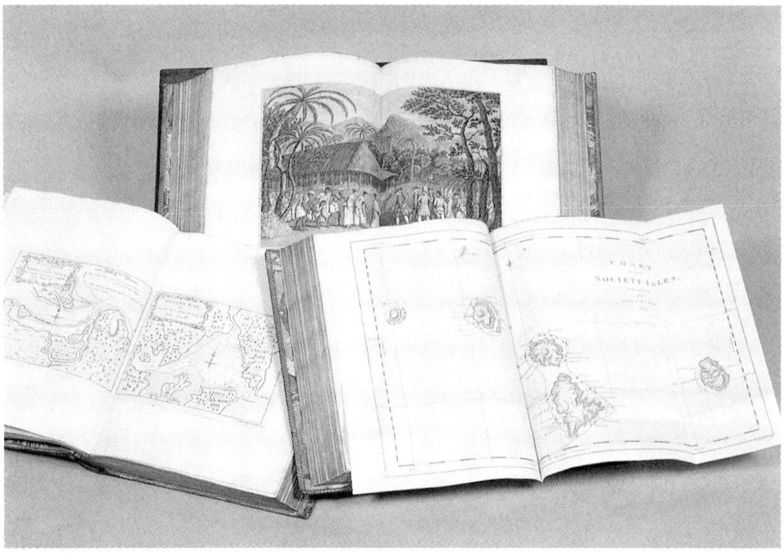

106. Hawkesworth

time after his official account (see catalogue number 106), and retaliated by deliberately omitting Parkinson's name from the narrative: even the botanical illustrations in the official account have no credit to the artist.

Parkinson himself was responsible for the original drawings for twenty-three of the twenty-seven plates here (he did not draw the map of New Zealand, nor plate 2, which was done by Alexander Buchan, nor plates 13 and 26, done by S.H. Grimm). His original artwork and these engravings made from it are one of the chief visual sources for Cook's first voyage, and one of the first views European observers had of such South Pacific scenes. Parkinson's journal also has some of the earliest natural history observations on the region, and contains the first published use of the word kangaroo (as 'kangooroo', p. 149).

Beaglehole, I, pp. ccliii-cclv; Beddie, 712; Hill, pp. 223-4; Holmes, 7; Kroepelien, 944; O'Reilly-Reitman, 371. **$7000**

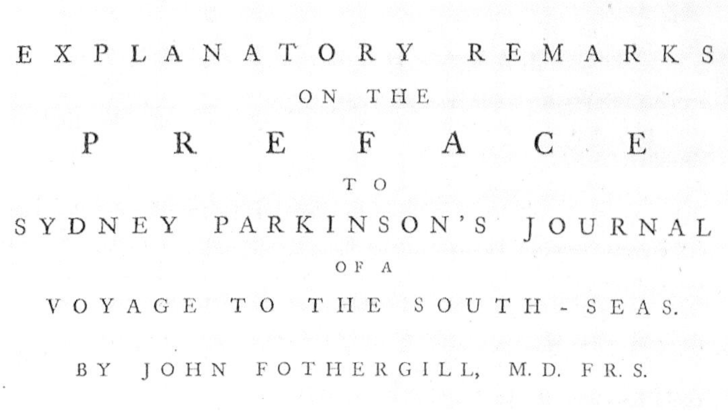

EXPLANATORY REMARKS
ON THE
PREFACE
TO
SYDNEY PARKINSON'S JOURNAL
OF A
VOYAGE TO THE SOUTH-SEAS.
BY JOHN FOTHERGILL, M.D. F.R.S.

108. Parkinson

INTERIM EDITION

108. [FIRST VOYAGE] PARKINSON, Sydney. A Journal of a Voyage to the South Seas, in His Majesty's Ship, the Endeavour…

Quarto, with frontispiece portrait, a map and 26 plates; a fine, large copy in modern half calf. London, Printed for Stanfield Parkinson, the Editor, 1773.

First edition, with the addition of Fothergill's additional 22-page *Explanatory Remarks on the*

Preface, prepared in the interim period between publication of the first edition in 1773 and the 1784 reissue (see below).

After Stanfield Parkinson's death, Dr. John Fothergill, like Parkinson a Quaker and also a friend of Banks, stepped in and bought the 400 or so remaining copies. His act, while saving the Parkinson family's financial bacon, may not have been entirely altruistic: he wished as well to defend himself and Banks against remarks made in the original preface, and prepared a reply which he had printed, at some time before 1777 (see Holmes for details). This 22-page addition would later appear, along with other new material, in the second edition of the book published in 1784; at the time of its original printing it was to be made available to those who had bought the first issue, but this does not seem to have been an offer that was taken up at all widely, as very few copies indeed of the first issue with the extra printing have been identified.

The extra material is also seen in rare copies of the book presented by Fothergill himself before publication of the second issue. As this copy has a modern binding, it is not possible to say whether it is an augmented first edition or one of the copies issued privately by Fothergill during the interim period between editions.

Beaglehole, I, pp. ccliii-cclv; Beddie, 712; Hill, pp. 223-4; Holmes, 7; Kroepelien, 944; O'Reilly-Reitman, 371. **$11,000**

109. Parkinson

SECOND EXPANDED EDITION

109. [FIRST VOYAGE] PARKINSON, Sydney. A Journal of a Voyage to the South Seas, in His Majesty's Ship, the Endeavour...

Large quarto, with a portrait, two maps and 27 numbered plates; a few leaves lightly spotted, light offset from some plates; a very good, clean and large copy in half calf antique, spine panelled in gilt, crimson label; originally in the library of the Inner Temple, London, with an attractive title-page stamp and a few unobtrusive stamps elsewhere. London, Printed for Charles Dilly, 1784.

The rare second and much expanded issue of the first edition; less than 400 copies were published. This is the best edition of Parkinson's journal, made up partially of the reissued sheets of the first edition and partially of added material published here for the first time.

The first edition had appeared in 1773. Some years later, Stanfield Parkinson went insane and the unsold sheets of the book were sold to Dr. John Fothergill. Fothergill added some "explanatory remarks" to Stanfield's ill-tempered preface, a double-sheet map of the world, and a history of Cook's explorations which almost doubled the size of the book, binding this new material with the sheets of the first section and supplying a new title-page. The second edition is considered the best, combining as it actually does both editions in one.

Material new in this edition also includes a two-page preface by John Lettsom, a lengthy (about 140 pp.) extension of the text, a new double-hemisphere world map, an additional resumé of the voyages of Byron, Wallis, Carteret and Bougainville, as well as the summary of Cook's second and third voyages. This second edition of Parkinson is thus one of the earliest volumes to collect information on all three of Cook's voyages.

Beddie, 714; Davidson, pp. 54-56; Hill, pp. 224-5; Holmes, 49; Kroepelien, 945. **$12,500**

HMS ENDEAVOUR

110. [FIRST VOYAGE] BAYLDON, Francis J. Dimensional Drawing of HMS Endeavour.

Colour lithograph, 400 x 550 mm.; framed. Sydney, Angus & Robertson, n.d. [probably 1968].

Scarce on the market: although this was published by Angus and Robertson, it was not generally distributed. Bayldon's original drawing was done in 1923 and this print was evidently issued for the bicentenary of the *Endeavour* voyage. **$425**

HE SHOOK COOK'S HAND

111. [FIRST VOYAGE] [PATUONE] DAVIS, Charles Oliver. The Life and Times of Patuone, the Celebrated Ngapuhi Chief.

Duodecimo, with an actual photograph of Patuone mounted as frontispiece; front joint just rubbed, but an excellent copy in the original gilt-lettered green cloth. Auckland, J.H. Field, 1876.

Biography of the great Maori figure, "chief of the Ngatihao at Hokianga... a Christian convert and powerful friend of the pakeha... who took part in some of the great inter-tribal campaigns, although Davis is at pains to emphasise his role as peace-maker later..." (NZNB). Patuone lived until the age of 108, and it is this accident of longevity that accounts for something evidently unique in this book: a photograph of someone who had met Cook. Patuone was a child when Cook visited the Bay of Islands in 1769, but lived well into the age of photography. The portrait-frontispiece here is an actual photograph, by G. Redfern; the owner of this copy has pasted two other photographs onto the endpapers, one of them another portrait of Patuone.

NZNB, 1542; Hocken, p. 300. **$785**

BENEATH YON BREAD-TREE

112. [FIRST VOYAGE] SCOTT-WARING, John. A Second Letter from Oberea, Queen of Otaheite, to Joseph Banks.

Quarto, engraved vignette on title; a few stains but good in quarter morocco. London, T.J. Carnegy, 1774.

Rare joke at the expense of Joseph Banks. Illustrated with an engraving that actually depicts a New Zealand Maori, this poem masquerading as a letter from the Tahitian Queen pokes fun at stories of Banks's amorous activities whilst in Tahiti on the first Cook voyage. A first *Letter* from the Queen had been published the year before.

The Queen's envoy, Otaipairoo, is supposed to have

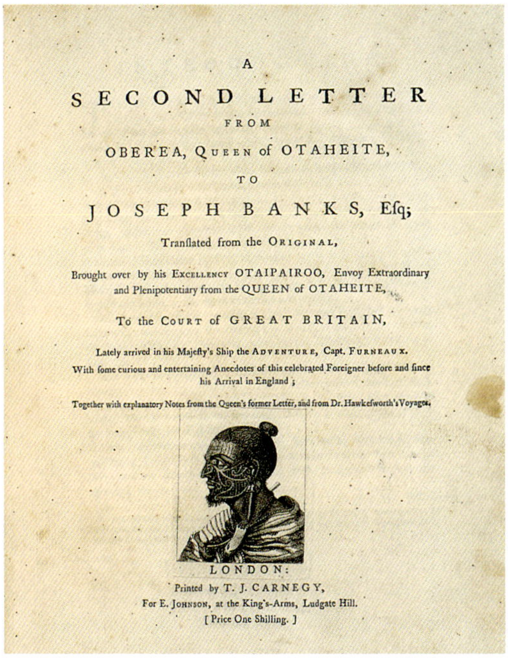
A SECOND LETTER FROM OBEREA, QUEEN of OTAHEITE, TO JOSEPH BANKS, Esq;

Translated from the ORIGINAL,

Brought over by his EXCELLENCY OTAIPAIROO, Envoy Extraordinary and Plenipotentiary from the QUEEN of OTAHEITE,

To the COURT of GREAT BRITAIN,

Lately arrived in his Majesty's Ship the ADVENTURE, Capt. FURNEAUX.
With some curious and entertaining Anecdotes of this celebrated Foreigner before and since his Arrival in England;

Together with explanatory Notes from the Queen's former Letter, and from Dr. Hawkesworth's Voyages.

LONDON:
Printed by T. J. CARNEGY,
For E. JOHNSON, at the King's-Arms, Ludgate Hill.
[Price One Shilling.]

112. Scott-Waring

travelled back to London on the *Adventure*, taking the "love serenades" with him. "Though now some European maid you woo,/ of waist more taper, and of whiter hue;/ Yet oft with me you deign'd the night to pass,/ Beneath yon bread-tree, on the bending grass".

Beddie, 3919; Holmes, 13; Kroepelien, 1171; O'Reilly-Reitman, 9797. **$1800**

COOK IN SWEDISH

113. [FIRST VOYAGE] WHITE, John. John Whites Resa till Nya Holland aren 1787 och 1788...

Duodecimo, four plates; neat repair to title, fine in original boards. Uppsala, J. Edmans, 1793.

A rare abridged version of White's *Voyage to New South Wales* in Swedish. The last quarter of the book is devoted to Cook's first voyage and the discovery of the east coast of Australia.

Beddie, 675; Ferguson, 172. **$1100**

OFFICIAL ACCOUNT

114. COOK, Captain James. A Voyage towards the South Pole, and Round the World, Performed in His Majesty's Ships the Resolution and Adventure...

Two volumes, quarto, with frontispiece portrait and 63 engraved charts, portraits and plates; a few slight spots or stains; an excellent set in contemporary lightly sprinkled calf, flat spines panelled in gilt; joints skilfully renewed. London, W. Strahan and T. Cadell, 1777.

First edition of the official account of Cook's great second voyage, prepared for publication by the navigator himself, dissatisfied with Hawkesworth's treatment of the first voyage. The superb engravings are mostly the work of William Hodges whose presence on the voyage resulted also in a superb series of oil-paintings.

This was historically the most important of Cook's three voyages. For the first time the Antarctic circle was crossed when, at the beginning of the voyage, Cook cruised as far south as possible, round the edge of the Antarctic ice. In the Pacific, he visited New Zealand again, and either discovered or revisited many of the islands, including New Caledonia, Palmerston and Norfolk Islands, Easter Island, the Marquesas, New Hebrides, Tonga, the South Sandwich Islands and South Georgia.

Beddie, 1216; Hill, p. 61; Holmes, 24; 'Printing and the Mind of Man', 223; O'Reilly-Reitman, 390. **$10,500**

ALTERNATIVE ACCOUNT

115. [SECOND VOYAGE] FORSTER, Georg (?and Johann Reinhold). A Voyage round the World in His Brittannic Majesty's Sloop Resolution, commanded by Capt. James Cook during the years 1772, 3, 4, and 5.

Two volumes, quarto, with a large folding map; some spotting; old marbled calf, well rebacked at some time, flat spines gilt in compartments with double labels, rubbed. London, B. White, 1777.

First edition of this alternative account of Cook's second voyage, "an important work and a necessary supplement to the official account" (Davidson). When the Admiralty decided to allow the elder Forster to take no part in the official account of the voyage, the Forsters set to work to forestall it with an account of their own, and succeeded in doing so by about six weeks. Georg Forster in the preface states that he wrote the account himself but consulted his father's journal. Wales, the astronomer, roundly accused him of being author in name only, the real author being

114. Cook

his father. This Georg Forster vehemently denied.

"But whatever the respective shares of father and son in the composition of this book, it is pertinent to observe that no acknowledgement is made of the assistance derived from Cook's journal, the proof sheets of which had in accordance with the agreement signed at the Admiralty on 13 April 1776, been placed at the elder Forster's disposal" (Holmes).

Beddie, 1247; Hill, p. 108; Holmes, 23; Kroepelien, 450; O'Reilly-Reitman, 382. **$5850**

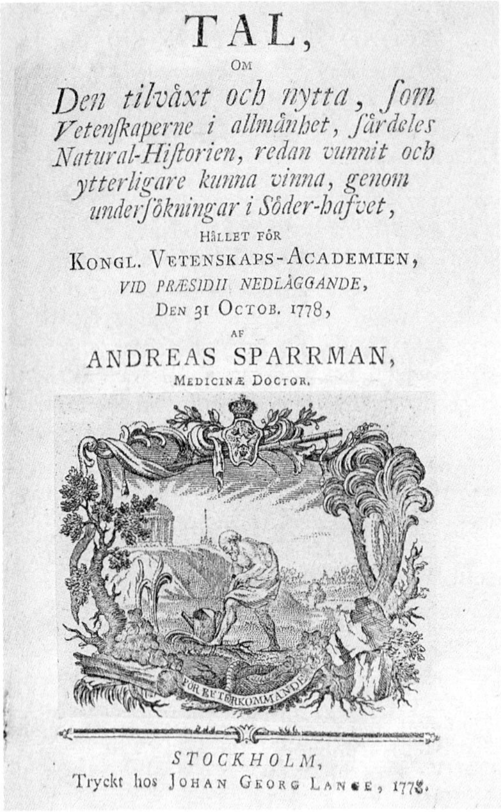

TAL,
OM
Den tilväxt och nytta, som Vetenskaperne i allmänhet, särdeles Natural-Historien, redan vunnit och ytterligare kunna vinna, genom undersökningar i Söder-hafvet,
HÅLLET FÖR
KONGL. VETENSKAPS-ACADEMIEN,
VID PRÆSIDII NEDLÄGGANDE,
DEN 31 OCTOB. 1778,
AF
ANDREAS SPARRMAN,
MEDICINÆ DOCTOR.

STOCKHOLM,
Tryckt hos JOHAN GEORG LANGE, 1778.

116. Sparrman

SCIENCE IN THE SOUTH SEAS

116. [SECOND VOYAGE] SPARRMAN, Anders. Tal, om den tilväxt och nytta, som vetenskaperne i allmänhet, särdeles natural-historien, redan vunnit och ytterligare kunna vinna, genom undersökningar i Söder-hafvet...

Octavo, 40 pp., engraved vignette pasted to title as issued; a fine copy, uncut, in a fine modern limp vellum binding. Stockholm, Johan Georg Lange, 1778.

Very scarce indeed. Sparrman, botanist on Cook's second voyage, reports to the Swedish Academy of Science on his recent voyage and on the present state of scientific knowledge of the South Seas. He discusses the benefits to science and natural history already gained and likely to be gained in the future, by discoveries and investigations in the Pacific.

This rare paper was translated into English as "What the South Seas have to offer" in *The Captain Cook Bicentenary* (ed. B. Danielsson, Ethnos, 1969).

Not recorded by Beddie; not in Hill; Kroepelien, 1217; O'Reilly-Reitman, 8405. **$2100**

ASTRONOMICAL OBSERVATIONS

117. [SECOND VOYAGE] WALES, William and William BAYLY. The Original Astronomical Observations made in the course of a Voyage towards the South Pole and Round the World, in His Majesty's Ships the Resolution and Adventure...

Quarto, with a double-page plate, two folding plates and a folding map; neat stamp and release of the Royal Library, Berlin on verso of title-page; some slight stains, tear in title-page restored on verso; contemporary calf, rubbed, spine wearing, old repair to top and tail. London, W. and A. Strahan, 1777.

First edition and rare: the official astronomical results of Cook's second voyage, including a fine double-page engraving of meteorological phenomena after the expedition's official artist, William Hodges.

The long introduction is by Wales and in it he attacks Dalrymple and the other proponents of the Great Southern Continent theory – "the notion which some persons have got concerning the necessity of a counterpoise, is so very unphilosophical, that I am much surprised how so many ingenious gentlemen have happened to adopt it...".

Publication of Wales's and Bayly's book was underwritten by the Commissioners of Longitude; it was probably produced in relatively small numbers because of its scientific nature. It is "as important as it is rare... I have not noted a copy offered for sale in recent years and should a copy become available it would certainly arouse great competition between collectors..." (Davidson).

Beddie, 1287; not in the Hill collection; Holmes, 26; Kroepelien, 1336; O'Reilly-Reitman, 3999. **$11,250**

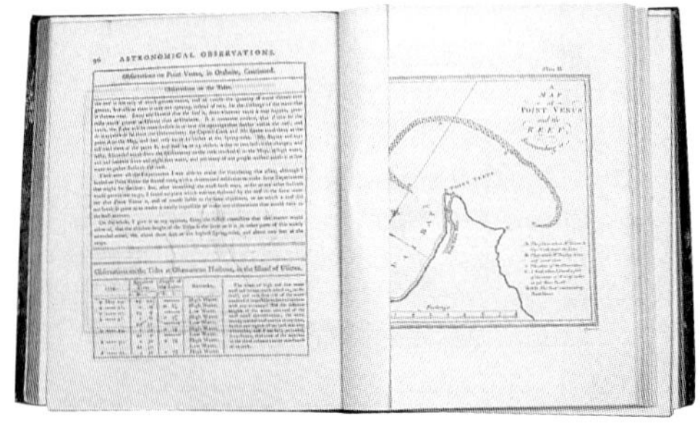

117. Wales & Bayly

WALES'S REMARKS

118. [SECOND VOYAGE] WALES, William. Remarks on Mr. Forster's Account of Captain Cook's last Voyage round the World, in the years 1772, 1773, 1774, and 1775.

Octavo; some staining, marginal defects to a few leaves at the end made good with several words on both sides of the last leaf supplied in facsimile; a large untrimmed copy, in recent quarter calf. London, J. Nourse, 1778.

One of the scarcest of all the Cook publications and nowadays only very rarely seen on the market, this is a particularly interesting product of the second voyage. The account of the voyage by George Forster published in 1777 had been a controversial affair: as Beaglehole describes the reaction, "The Forster

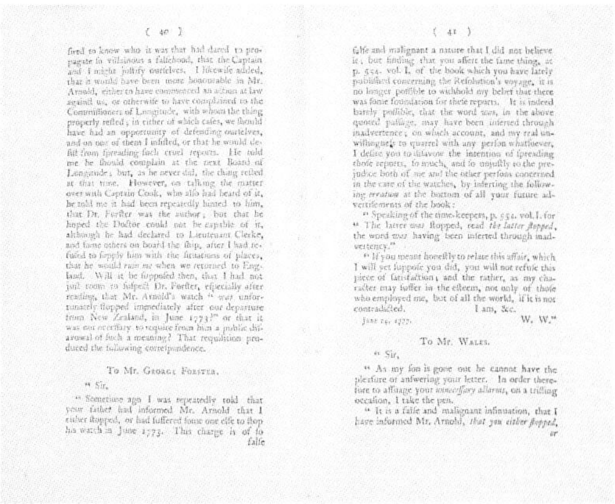

(40)

fired to know who it was that had dared to propagate so villainous a falsehood, that the Captain and I might justify ourselves. I likewise added, that it would have been more honourable in Mr. Arnold, either to have commenced an action at law against us, or otherwise to have complained to the Commissioners of Longitude, with whom the thing properly rested; in either of which cases, we should have had an opportunity of defending ourselves, and on one of them I insisted, or that he would desist from spreading such cruel reports. He told me he should complain at the next Board of Longitude; but, as he never did, the thing rested at that time. However, on talking the matter over with Captain Cook, who also had heard of it, he told me it had been repeatedly hinted to him, that Dr. Forster was the author; but that he hoped the Doctor could not be capable of it, although he had declared to Lieutenant Clerke, and some others on board the ship, after I had refused to supply him with the situations of places, that he would *ruin me* when we returned to England. Will it be supposed then, that I had not just room to suspect Dr. Forster, especially after reading, that Mr. Arnold's watch "*was* unfortunately stopped immediately after our departure from New Zealand, in June 1773?" or that it was not necessary to require from him a public disavowal of such a meaning? That requisition produced the following correspondence.

To Mr. George Forster.

"Sir,

"Sometime ago I was repeatedly told that your father had informed Mr. Arnold that I either stopped, or had suffered some one else to stop his watch in June 1773. This charge is of so false

(41)

false and malignant a nature that I did not believe it; but finding that you assert the same thing, at p. 554. vol. I. of the book which you have lately published concerning the Resolution's voyage, it is no longer possible to withhold my belief that there was some foundation for these reports. It is indeed barely possible, that the word *was*, in the above quoted passage, may have been inserted through inadvertence; on which account, and my real unwillingness to quarrel with any person whatsoever, I desire you to disavow the intention of spreading those reports, so much, and so unjustly to the prejudice both of me and the other persons concerned in the care of the watches, by inserting the following *erratum* at the bottom of all your future advertisements of the book:

"Speaking of the time-keepers, p. 554. vol. I. for "The latter *was* stopped, read *the latter stopped*, the word *was* having been inserted through inadvertency."

"*If you meant honestly to relate this affair*, which I will yet suppose you did, you will not refuse this piece of satisfaction; and the rather, as my character may suffer in the esteem, not only of those who employed me, but of all the world, if it is not contradicted. I am, &c.

June 14, 1777. W. W."

To Mr. Wales.

"Sir,

"As my son is gone out he cannot have the pleasure of answering your letter. In order therefore to assuage your *unnecessary alarms*, on a trifling occasion, I take the pen.

"It is a false and malignant insinuation, that I have informed Mr. Arnold, *that you either stopped*, or

118. Wales

temper and the Forster recklessness being so prominently displayed in certain passages, it could hardly escape criticism. This was applied by Wales, 'our accurate and indefatigable astronomer', in an octavo pamphlet of 110 indignant pages…

"Wales frankly disbelieved that George had written the book, its prejudices were so much and so transparently those of Reinhold – in which his acumen as a literary critic certainly failed him. But he made some damaging points. Not only did he spring to the defence of the 'poor seamen' but he took a series of specific statements that outraged him and rebutted them with a vigour and heat that are still alive, far removed from the genial fun he had poked in the private pages of his journal… Apart from the controversy, he adds to our knowledge a little that is valuable in the annotation of Cook. George could not let this attack go unnoticed, and produced a quarto "Reply"…" (Beaglehole, *Journals*, II, pp. cli-clii).

Beddie, 1292; Holmes, 30; Kroepelien, 1335; O'Reilly-Reitman, 388. **$27,500**

MOYENS EMPLOYÉS PAR M. COOK

119. [SECOND VOYAGE] CONSEIL DE MARINE, Toulon. Original manuscript report: Décisions. Précaution contre le Scorbut… observation sur les moyens employés par M. Cook pour conserver la santé des gens de mer…

Folio, 3 pp., manuscript in ink on paper; in fine condition. Toulon, 9 April 1777.

An unusual example of speedy continental reaction to one of Cook's more famous achievements, the conquering of the scurvy threat. This original manuscript extract of proceedings at a meeting of the Conseil de Marine in Toulon, signed by the Conseil's secretary, Molliere, as a true record of the decisions of the meeting, shows that of five matters considered by far the most important was the success of Captain Cook in combating scurvy. A letter from the Minister was read to the meeting inviting them to consider the question, as well as a memoir "sur les moyens employés par M. Cook pour conserver la Santé de l'équipage du Vau. Anglois la Résolution pendant un Voyage autour du monde…". Medical reports were given, and in view of the high expense associated with appropriately provisioning a ship, the members of the Council were asked to consider the matter at length. **$4200**

GIFTS FROM BANKS & LINNÆUS

120. [SECOND VOYAGE] ROSCOE, William. An Address, delivered before the Proprietors of the Botanic Garden, in Liverpool, previous to Opening the Garden…

Octavo; original pink boards with printed paper label, cloth spine. Liverpool, J. M'Creery, 1802.

Rare. The Liverpool Botanic Garden was first proposed in 1800 as a private enclave for the 300 original Proprietors (listed here) and their guests, but it opened to the public in 1803. It boasted a substantial library with many of the books donated by Sir Joseph Banks; its Herbarium was of particular importance, the core of which was the unrivalled collection of the naturalist Reinhold Forster and his son Georg, collected during Cook's second voyage on the *Resolution*. Other donors included Linnæus and Sir James Smith. **$650**

OFFICIAL ACCOUNT

121. COOK, James and James KING. A Voyage to the Pacific Ocean… for making Discoveries in the Northern Hemisphere…

Three volumes, quarto, with the normal suite of 87 engraved plates and maps plus three extra engravings; a little spotting, some plates cut close at margins; contemporary marbled calf, sides bordered with floral ornament in gilt; edges worn, inner joints strengthened, spines well renewed. London, Nicol and Cadell, 1785.

121. Cook & King

The official account of Cook's last voyage: second edition, with the engraved vignettes of the Royal Society's Cook medallion that did not appear in the first edition. This copy has as added frontispieces to the first and second volumes the stipple-engraved portraits of Cook and King by Bartolozzi after Webber and the engraving "Death of Captain Cook" by Bartolozzi after Byrne and Webber.

The full story of Cook's great third voyage, and his murder while revisiting the Hawaiian Islands that he had discovered earlier during the voyage, was so eagerly awaited that the entire first edition sold out, at the then huge price of four pounds fourteen shillings and sixpence, within three days and copies were soon changing hands at up to ten guineas.

The official artist on the voyage was John Webber, and his romantic views of the islands of the Pacific published here remain the most evocative portrayals of the islands.

This copy, with contemporary bookplates of Sir John Eliot, Bart., has the engraved plates bound into their appropriate places in the text volumes rather than into a separate atlas volume. The "Directions for placing the plates" shows that a separate atlas was anticipated but by no means universal: "Many of the Purchasers of this Work may choose to preserve the larger-sized Plates in a separate volume in folio… Booksellers are cautioned not to have them bound up, with the rest of the Plates, in the places of these [text] volumes pointed out by the respective References, unless they receive particular directions for that purpose…".

Beddie, 1552; Hill, pp. 61-62; Holmes, 47. **$10,500**

123. The full set of Cleveley's aquatint views of the Pacific

ARMORIAL CALF

122. COOK, James and James KING. Troisième voyage de Cook, ou Voyage à l'Océan Pacifique…

Five volumes, quarto, with 88 engravings in the fifth (atlas) volume; a fine set in contemporary French calf, gilt with arms. Paris, Hôtel de Thou, 1785.

Handsome set in armorial bindings of the scarce first French edition of Cook's third voyage. The coat-of-arms stamped heavily on each cover has the name Ruolz Montchal and the motto "Toujours pret".

This French edition of the voyage is illustrated in the manner of the English publication, and includes a version of the famous "Death of Cook" plate based on the original drawing by John Webber.

Beddie, 1556. **$12,500**

FULL SET OF THE CLEVELEY VIEWS

123. [THIRD VOYAGE] CLEVELEY, James. The series of four separately issued coloured aquatint views of the Pacific: "View of Owhyhee, one of the Sandwich Islands"; "View of Morea, or Eimo, one of the Society Islands"; "View of Huaheine, one of the Society Islands, showing the Resolution and Discovery at anchor in the magnificent harbour of O Wharre"; "View of Charlotte Sound in New Zealand" [sic; actually Matavai Bay, Tahiti].

The complete set of four aquatints, handcoloured, each 595 x 430 mm.; some ageing and browning, the margins trimmed; a good set, mounted and preserved in a cloth folding case. London, Drawn on the spot by James Cleveley. Painted by John Cleveley. Aquatints by Francis Jukes, 1787-1788.

These fine aquatints, prepared from watercolours by John Cleveley after original drawings made by his brother James in the South Seas, are probably the most magnificent Pacific marine prints issued. They show *Resolution* and *Discovery* at anchor in both the Society and Sandwich Islands. The settings are ideals of Pacific beauty.

James Cleveley was ship's carpenter on the expedition; although not formally trained, he was a proficient artist. After the voyage returned to England in October 1780 his brother John, a professional draughtsman, took the sketches made on the voyage ("drawn on the spot", according to the original prospectus issued for the prints), and worked them up into finished watercolours.

It was from his finished watercolours that these aquatints were prepared in 1787-8. In the process of preparing them for European eyes, the landscape has been tamed and the islanders portrayed in a manner both dignified and classical. The finished works portray the Pacific islands and their inhabitants in a picturesque and idealised manner, which perhaps has more to do with conforming to European notions of a Pacific Arcadia than specific representations of the islands as they were.

Beaglehole, III, pp. ccxiv; Beddie, 1752-1783; Nan Kivell and Spence, p. 68; Joppien and Smith, III, pp. 216-221, plates 205-209. **$34,500**

OMAI'S HOUSE ON HUAHEINE

124. [THIRD VOYAGE] CLEVELEY, James. View of Huaheine.

Coloured aquatint, 490 x 690 mm., mounted and framed; some general ageing but very good. London, 1787.

An attractive example of the famous aquatint by John Cleveley, one of the four remarkable views of the Pacific done after the Cook voyages. Huaheine, one of the Tahitian Islands, was of special interest to Cook, as it was here that he decided to leave Omai. "Cook negotiated with the chieftains of the island for a suitable piece of land… a garden was laid out, on which Cook tells us "Shaddocks, Vines, Pineapples, Millons and several other articles" were planted "all of which were in a flourishing state before we left the island" (Joppien & Smith).

Omai's house can be seen in the foreground of this superb aquatint, under construction. The *Resolution* and *Discovery* are shown riding at anchor, and the crew can be seen setting out in long boats accompanied by traditional Tahitian craft – the artist depicts a moment of great serenity and beauty.

Nan Kivell and Spence, p. 68; Joppien and Smith, III, pp. 216-221, plate 205. **$7500**

RARE UNCOLOURED VERSION

125. [THIRD VOYAGE] CLEVELEY, James. A view of Moorea one of the Friendly Islands in the South Seas…

Uncoloured aquatint, 490 x 690 mm.; mounted. London, 1787.

An unusual uncoloured version of the famous aquatint by John Cleveley, one of the four remarkable views of the Pacific done after the Cook voyages. The Cleveley aquatints were issued in various different states; one version, for example, had the printed lines appearing as faintly as possible, so that when the prints were coloured by hand they had the appearance of original watercolours. This uncoloured version is interesting for the amount of detail that appears in the actual printing of the plate, and has no subsequent hand-finishing.

This view of Moorea is incorrectly titled "one of the Friendly Islands", when it is in fact part of the Society Islands, with Morea just to the north-west of Tahiti.

Nan Kivell and Spence, p. 68; Joppien and Smith, III, pp. 216-221, plate 206. **$2500**

COOK & THE RUSSIANS

126. [THIRD VOYAGE] COXE, William. Account of the Russian Discoveries between Asia and America. To which are added, the Conquest of Siberia, and the History of the Transactions and Commerce between Russia and China.

Octavo, four folding charts and a plate; contemporary calf, spine worn, rejointed. London, J. Nichols, 1787.

The scarce third edition, the first to contain the important supplementary piece "A Comparative View of the Russian Discoveries with those of Captains Cook and Clerke", a 46-page section with separate title-page.

The book is an important source on Russian exploration and expansion into the Northern Pacific, and the attempt to open trade with Alaska and the Aleutians. Coxe, travelling in Russia, collected, compared and translated Russian journals and maps, consulted with authorities and included several journals "never before given to the public", directing the "curious and inquisitive reader" to compare them to Cook's then unpublished journals.

Hill, p. 385; Holmes, 107; Lada-Mocarski, 29. **$2650**

THE ARTIST'S ACCOUNT

127. [THIRD VOYAGE] ELLIS, William. An Authentic Narrative of a Voyage performed by Captain Cook and Captain Clerke, in His Majesty's Ships Resolution and Discovery... including a faithful Account of all their Discoveries, and the unfortunate Death of Captain Cook...

Two volumes, octavo, with a folding chart and 21 engraved plates; a fine copy in a splendid recent binding (incorporating original tree calf boards) of marbled polished calf gilt, spines ornately gilt in compartments with double crimson labels, yellow edges, by Aquarius. London, G. Robinson, J. Sewell and J. Debrett, 1782.

First edition. Ellis, surgeon's mate and talented amateur artist, sailed first on the *Discovery* and later on the *Resolution*. Despite the prohibition by the Admiralty of the publication of any unauthorised account of the voyage, Ellis, in financial straits, sold his narrative to a London publisher for 50 guineas, thus earning the condemnation of Sir Joseph Banks, who wrote to him in January 1782 that "I fear it will not in future be in my power to do what it might have been, had you asked and followed my advice".

Ellis's narrative contains much valuable information on Alaska, the Northwest Coast, and Hawaii, and the attractive engraved plates, after the author's drawings, include eight of Hawaii, two of Alaska, and three of the Northwest Coast, among the earliest published of these areas; Choris's famous views did not appear until almost forty years later.

Beddie, 1599; Hill, p. 95; Holmes, 42; Judd, 59; Kroepelien, 399; Lada-Mocarski, 35. **$11,500**

SECOND EDITION

128. [THIRD VOYAGE] ELLIS, William. An Authentic Narrative of a Voyage performed by Captain Cook and Captain Clerke, in His Majesty's Ships Resolution and Discovery...

Two volumes, octavo, folding chart and 21 engraved plates; a good copy, very light stain at start of second volume but generally clean and large; contemporary sprinkled calf, sides a little worn, rebacked. London, G. Robinson, 1783.

The second, revised edition. "Containing a worthwhile history of the expedition, and being an early account of the voyage, the work was popular and a second edition was published...Today either edition is rare..." (Davidson).

Beddie, 1600; Judd, 59; Kroepelien, 400. **$3850**

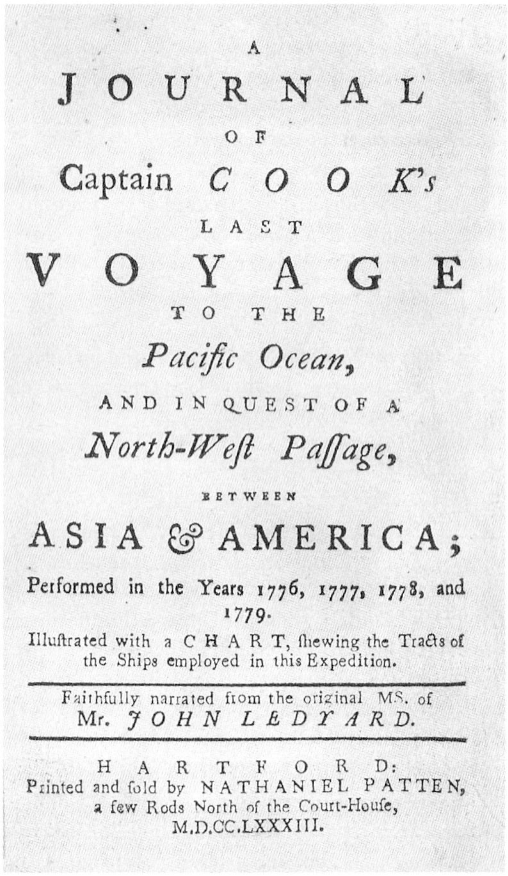
A
JOURNAL
OF
Captain COOK's
LAST
VOYAGE
TO THE
Pacific Ocean,
AND IN QUEST OF A
North-West Passage,
BETWEEN
ASIA & AMERICA;
Performed in the Years 1776, 1777, 1778, and 1779.
Illustrated with a CHART, shewing the Tracts of the Ships employed in this Expedition.
Faithfully narrated from the original MS. of
Mr. JOHN LEDYARD.
HARTFORD:
Printed and sold by NATHANIEL PATTEN, a few Rods North of the Court-House,
M,D,CC,LXXXIII.

129. Ledyard

THE ONLY AMERICAN ACCOUNT

129. [THIRD VOYAGE] LEDYARD, John. A Journal of Captain Cook's last Voyage to the Pacific Ocean, and in quest of a North-West Passage, between Asia and America; performed in the years 1776, 1777, 1778, and 1779... Faithfully narrated from the original ms. of Mr. John Ledyard.

Small octavo, lacking the map (as do virtually all copies known; it is supplied here in expert facsimile on old laid paper); the paper slightly discoloured and stained as always, in the publisher's binding of plain sheep, rebacked in period style, dark red leather label; quarter morocco case. Hartford, Nathaniel Patten, 1783.

An extremely scarce surreptitious narrative of Cook's third voyage. Ledyard, one of several Americans on the voyage bur the only one to publish, sailed as a corporal of marines. His book "is now a very scarce item, and is so seldom catalogued that I have been unable to locate a current price..." (Davidson).

It is now generally acknowledged that Ledyard obtained a copy of Rickman's narrative from which he compiled this account after his return to America. He himself describes the sealing of all diaries, drawings, memorandums, and charts of all officers and crew aboard ship at Canton. However, the narrative contains considerable information not available elsewhere, including the first published description of the Russian presence on Unalaska, the first permanent Russian settlement in Northwest America. The chart supposed to have been issued with the book, which seems to have been an almost direct copy of the Rickman map, is missing in almost every copy, and in view of the erratic nature of American eighteenth-century publishing it is nowadays accepted that it was not generally issued.

Ledyard's description of his stay at Hawaii, his expedition inland and the death of Cook occupies 64 pages of the text. The account of New Zealand and Australia is short but includes his observation that "the island of New-Holland (for its boundaries are now ascertained) is by much the largest known, and most eligably situated on the map of nature... even the Empress of Russia might be gratified with such a portion".

Beddie, 1603; Hill, p. 176; Holmes, 45; Judd, 108; Kroepelien, 717; Lada-Mocarski, 36; O'Reilly-Reitman, 432. **$18,500**

LIFE OF LEDYARD

130. [THIRD VOYAGE] [LEDYARD] SPARKS, Jared. Memoirs of the Life and Travels of John Ledyard.

Octavo; a good copy in later quarter calf. London, Henry Colburn, 1828.

Soon after Ledyard's death in Egypt in 1789, his cousin Dr. Isaac Ledyard of New York began collecting materials for a biography. The project was abandoned, but the material was preserved and ultimately turned over to Sparks. A substantial portion of the text relates to Cook's third voyage, and contains critical comments on Ledyard's narrative found nowhere else.

Sparks, noted American historian, owner and editor of the *North American Review*, and later President of Harvard (1849-1853) added "important original letters" and his compilation remains one of the great American biographies.

Beddie, 1659; Holmes, 86. **$1250**

NOT KNOWN TO BEDDIE, FERGUSON OR HOLMES

131. [THIRD VOYAGE] [LEDYARD] ST. JOHN, James Augustus. The Lives of Celebrated Travellers.

Three volumes, duodecimo; a few spots but generally fine in contemporary half green calf. New York, Harper, 1832.

A nice copy of this popular set, published as part of "Harper's Family Library", with the biographies of some thirty notable explorers ranging from Marco Polo to various nineteenth-century travellers. Of particular interest is the biography of John Ledyard, which gives a full account of the writing of his controversial narrative.

Not recorded by Beddie, Ferguson or Holmes. **$375**

SAMWELL ON COOK'S DEATH

132. [THIRD VOYAGE] SAMWELL, David. Détails nouveaux et circonstanciés sur la mort du Capitaine Cook, traduits de l'Anglois.

Octavo, 54, (ii) pp.; neo-classical vignette on title, woodcut head- and tail-pieces; a fine uncut copy in original plain wrappers; quarter morocco box. "A Londres, et se trouve à Paris, chez Née de la Rochelle, Libraire...", 1786.

First French edition and extremely rare: Samwell's account – in its English printing or in this French form – is one of the most difficult of all Cook-related pieces to acquire. We know of only two other copies of this edition offered for sale in recent years (the Kahn copy, now in Hawaiian State Archives, and another sold by ourselves in 1993).

Samwell's account of Cook's death is one of the earliest books on Hawaii, preceded only by the official account and the handful of unofficial accounts of Cook's third voyage. Of all the early books on Hawaii, it ranks with Zimmermann's narrative of the voyage and Shaw's collection of tapa cloth as among the rarest and most significant.

"Apart from its rarity, this pamphlet is of the greatest importance, since it fills in gaps, e.g., as to the responsibility for Cook's death, which are suppressed in the official account. Samwell's estimate of Cook's character, coming as it did from an educated man who knew him well, needs to be read alongside that of King in the official account to get a true picture of Cook as he appeared to those under his command..." (Holmes).

Samwell sailed as surgeon's mate on the *Resolution*, but became surgeon on the *Discovery* in mid-1778. His eye-witness account of the events at Kealakekua Bay forms the basis of our knowledge of the details of the event, particularly since the visual record is so muddled by myth-making aspects. As a surgeon he was also well placed to make the observations that appear here about venereal disease in the islands.

Only two editions (this and the English one) of Samwell's narrative were published, unlike most of the other third-voyage accounts which appeared in numerous European, Russian or Scandinavian editions as well as Irish, and even American, piracies.

Apart from Kippis's use of the material in his later *Life of Cook*, the book was not reprinted until David Magee republished the Samwell text in 1957 (*Captain Cook and Hawaii*), with an introduction by the Cook bibliographer Sir Maurice Holmes who noted that "The fullest, most detailed and most objective [account of Cook's death] is that by David Samwell, which is here reprinted. Such a reprint is certainly called for. In its original form it is of great rarity and correspondingly expensive...". The Hill catalogue notes of the English edition (acquired at the Streeter sale in 1969) that "this exceedingly rare work may perhaps be considered the highlight of a Cook collection".

In Australian libraries, Beddie records only the Mitchell Library's copy of this version, whilst in American libraries NUC records only two copies.

Barbier, 'Dictionnaire des ouvrages anonymes', I, 916 (without identifying the translator); Beddie, 2446; Hill, p. 265 (not this edition); Holmes, 62; Kahn, 2/34; Kroepelien, 1144; O'Reilly-Reitman, 453; Quérard, VIII, 434. **$34,500**

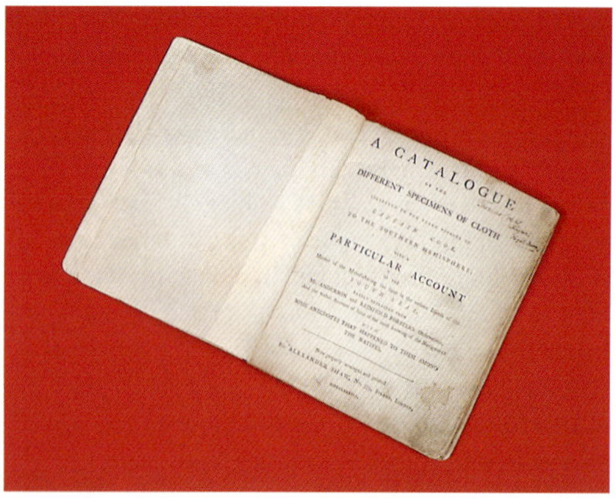

133. Shaw, A Catalogue; with a typical page

THE CLOTH BOOK

133. [THIRD VOYAGE] SHAW, Alexander. A Catalogue of the Different Specimens of Cloth collected in the three voyages of Captain Cook, to the Southern Hemisphere; with a Particular Account of the Manner of the Manufacturing the same in the various Islands of the South Seas...

Quarto, 222 x 175 mm.; eight pages printed text, followed by actual cloth specimens arranged between blank leaves, amounting to considerably more than the normal full complement of 39 pieces; all leaves uncut; original marbled boards, later but old quarter calf spine; in generally fine condition. London, Alexander Shaw, 1787.

One of the rarest and most exotic of all the Cook productions: an extraordinary book, which assembles actual specimens of *tapa* cloth collected by the expedition in the Pacific islands, particularly Hawaii and Tahiti.

Published only a few years after the return of the *Resolution* it is not known how many of the cloth-books were prepared; the limited supplies of the actual cloth must have dictated a very small edition. In modern times the book has become one of the great rarities of eighteenth-century Pacific exploration.

A total of 39 specimens is listed in Shaw's preface (although a proportion of surviving copies of the book in fact have less than the full set), the last of which was from Jamaica and had nothing to do with Cook; all are present in this copy, which actually contains a total of 51 pieces – presumably the extra pieces were added by the book's original owner, perhaps the Royal Navy surgeon Thomas Cocks whose ownership inscription is in ink on the title-page. This copy also has early pencil notes identifying many of the specimens.

Each island group used designs unique to its culture and the interest of Europeans in this material equalled the passion aroused by the extraordinary wood-carvings and exotic shells brought back by Cook. It was evidently a late-eighteenth-century habit to construct what have become known as 'snippet books' of the Cook *tapa* cloths, by cutting pieces from copies of Shaw's volume and then compiling them as separate books. As a result of eighteenth-century depredations, copies of the original Shaw book survive in various states of completeness, sometimes with only small fragments of the once full-page specimens remaining. In this copy, some of the samples have had cuttings taken from them, but most of the samples remain substantial, many of them full-page.

The preface contains detailed descriptions from various officers on board Cook's ships of the importance in the Pacific of the manufacture and use of the cloth. *Tapa* is made from the inner bark of the paper mulberry or the breadfruit trees, sometimes used together. The dyes for the decoration were made from roots, berries, leaves, bark and flowers and the predominant colours were black, brown, yellow and red. In the eighteenth century the cloth was used to denote rank and the status of different social positions and was an essential part of island life. Some cloths stretched to over one hundred feet.

Each of the 39 cloth examples is described in more or less detail, the most interesting being "No 34 From Otaheite, worn as garments by the ladies. – A number of the natives being on board of the Resolution, one of the chiefs took a particular liking to an old blunt iron and taking hold of a boy about nine years of age, offered him in exchange pointing to the iron. The gentleman although he knew he could not keep the youth, yet willing to see if he would willingly stay; or if any of the rest would claim him, took the child and gave the savage the iron; upon which a woman, who appeared rather young for the mother, sprung from the other side of the ship... upon which, after putting her hands twice upon her head, she unbound the roll of cloth which was round her body, and from which this specimen was cut, and having spread it before him, seized the boy, and jumping into the sea both swam ashore...".

Beddie, 3640 (calling for only 38 specimens); Hocken, p. 26; Holmes, 67; not in Hill, Lada-Mocarski, or Kroepelien; Kaeppler, pp. 117-118. **$87,500**

134. Webber & Prestel. The set of four trial sepia aquatints.

THE RARE ISSUE IN SEPIA AQUATINT

134. [THIRD VOYAGE] WEBBER, John and Marie Catherina PRESTEL. Complete set of the four Pacific aquatints drawn and etched by Webber and coloured by Prestel.

Set of four aquatints, each 380 x 530 mm.; some slight foxing mainly to margins, otherwise in very good condition; mounted. London, J. Webber, No. 312 Oxford Street, 1 February 1787 (two) or 1788 (two).

These four rare sepia aquatint engravings are the very first trials done by John Webber, the official artist on Cook's third voyage, for his celebrated *Views in the South Seas*. In 1786, perhaps encouraged by the response to his South Seas paintings exhibited at the Royal Academy, Webber collaborated with Marie Catherina Prestel, "an aquatint artist of some note, who had recently come to London from Frankfurt". (Joppien & Smith, p. 192). In 1786 and 1787 he prepared four aquatints with her – "A View of Otaheite", "View in Ulietea", "A View in Annamooka, one of the Friendly Islands" and "A View in Pulo Condore". All are offered here.

Webber's old friend and teacher, the Parisian J.-G. Willie, mentions in his diary Webber's visit to him and the work in progress on these aquatints. "He brought me several prints… which the artist had etched himself after his drawings brought back from his voyage around the world with Captain Cook. He was on the spot when this famous navigator was killed in the Sandwich Islands. Those etchings have been washed (*lavées*) in bistre by Madame Prestel, who is actually now in London but her husband remains in Frankfurt".

Perhaps it was at the suggestion of Willie that Webber abandoned these trials in aquatint engraving, and also his relationship with Prestel, to investigate the new technique of soft-ground etching. We know that by August 1788 he had issued the first three of the sixteen soft-ground etchings that were to form his famous publication, *Views in the South Seas*, perhaps the most beautiful of all view books on the Pacific. Such was its success that later issues of these plates were still being sold well into the nineteenth century.

This fine set of atmospheric and beautiful views represent not only the entire body of work produced by Webber in his collaboration with Madame Prestel, but also the embryonic stage of his determination to put the beauties of the newly-discovered Pacific before an eager, public audience.

Beddie, 1869 (examples of these issues in an album in the Dixson Library); Joppien & Smith, 3.120Aa; 3.157Ab; 3.44Aa,b; 3.397Ac. **$19,500**

COLOUR VIEW OF PULO CONDORE

135. [THIRD VOYAGE] WEBBER, John. View in the Island of Pulo Condore.

Handcoloured etching, 420 x 540 mm.; small repaired tear to lower margin; mounted. London, J. Boydell, 1809.

The coloured etching, an example of Webber's South Sea views at a much later stage. This view, first issued as an aquatint in 1787 (see above), was still being published long after the artist's death as part of his *Views in The South Seas*. **$750**

136. *Webber*

DRYING THE FISH

136. [THIRD VOYAGE] WEBBER, John. Balaguns or Summer Habitations…

Handcoloured etching, 420 x 540 mm.; mounted. London, J Boydell, 1809.

This view of the Russian province of Kamchatka near the Bering Sea, shows the method of drying fish used by the natives. This handcoloured aquatint is also from the 1809 publication *Views in the South Seas*, and like other views by John Webber it is based on an original drawing by the artist made during Cook's third voyage to the Pacific. **$750**

DEATH OF COOK

137. [THIRD VOYAGE] WEBBER, John and William BYRNE, after. BARTOLOZZI, F., Engraver. Death of Captain Cook.

Engraving, 490 x 610 mm.; repairs to margins; mounted. London, J. Webber, No. 312 Oxford Street and W. Byrne, No. 79 Titchfield Street, 1 January 1784.

The rare, separately issued engraving of Cook's death. John Webber was the official artist on the third voyage, and although not an eye-witness to Cook's death was at Kealakekua Bay on 14 February 1779. On his return to London he painted the most celebrated version of "the death"; in 1784 this engraving was published, based on the famous oil painting. Webber was responsible for the figures in the engraving whilst William Byrne produced the landscape. Both were the publishers of this view which is the most famous of all eighteenth-century depictions of the massacre.

Beddie, 2603; Nan Kivell & Spence, p. 80. **$5750**

FRENCH VERSION OF THE DEATH

138. [THIRD VOYAGE] WEBBER, John (after). Mort de Cook.

Engraving , 250 x 380 mm.; mounted; in good condition. France, n.d. [1790].

The French version of Webber's death, prepared by Benard. **$550**

FRENCH EDITION

139. [THIRD VOYAGE] RICKMAN, John. Troisième voyage de Cook, ou Journal d'un expédition faite dans la Mer Pacifique du Sud & du Nord en 1776… 1780. Traduit de l'Anglois…

Octavo, with a folding engraved frontispiece; a good copy in contemporary quarter French calf. Paris, chez Belin, 1783.

The scarce French version of Rickman's surreptitious account of Cook's third voyage: second edition of J.N.

137. *Webber, Death of Cook*

Demounier's translation. Rickman's book had first appeared in English in 1781; even in this French version it was still reaching the public a long time before the official narrative was published.

As one of the first descriptions of the voyage to appear in print, this was one of the earliest descriptions of Hawaii. Rickman's narrative differs from the official account in many respects – particularly regarding the death of Cook, for which his book is a prime source. The image of Cook's death in the various editions of Rickman – here as the frontispiece, entitled "Mort du capitaine Cook à Owhy-hée, Fevrier 1779" – was one of the earliest representations of Hawaii.

The Cook bibliographer Beddie calls in error for two engravings: the Mitchell Library copy that was the basis for this collation proves on examination to have been extra-illustrated by an owner.

Beddie, 1613. **$1850**

COOK'S CHARACTER LEADS TO HIS DEATH

140. [THIRD VOYAGE] [RICKMAN, John] FORSTER, Johann Reinhold. Tagebuch einer Entdedekkungs Reise nach der Sudsee in den Jahren 1776 bis 1780 unter Anfuhrung der Capitains Cook, Clerke, Gore and King…

Octavo, with a frontispiece and a folding map (small tear at a fold); original plain grey boards, handwritten label; bookplate and small ownership stamp on title; boxed. Berlin, Haude und Spener, 1781.

A rare, important and early third voyage piece: the first publication of Forster's German edition of Rickman's *Journal of Captain Cook's last voyage*, the surreptitious account of the voyage which had appeared earlier the same year. Rickman's English and Zimmermann's German accounts both appeared in 1781 and it is uncertain which can claim to have been the first description of the voyage.

Forster's discussion of Cook's character and how it led to his death in Hawaii is quite different to the almost universal heroic adulation. Forster's more questioning perspective is all the more pertinent since he had sailed as senior naturalist on Cook's second voyage and wrote with the benefit of personal knowledge of Cook as well as first-hand experience of Pacific voyaging.

Forster "was vitally interested in the results of Cook's third voyage and particularly in communicating some of these results to an eager German public… Early in 1781 John Rickman, a lieutenant on the third voyage, anonymously rushed out his *Journal*… and one of Forster's well placed correspondents immediately sent off a copy to Halle, together with any supporting gossip he could glean… In publishing *Tagebuch*… Forster took the opportunity to announce for the first time in Germany his own plans for further papers on the natural history of the voyages [and attacked] some of his English 'enemies'. Banks, Solander, Forster and his son, we are informed, had really been civilising influences upon James Cook during the first two voyages; their presence had prevented Cook from being too harsh with the South Sea islanders. The absence of educated, scientific gentlemen on the third voyage had,

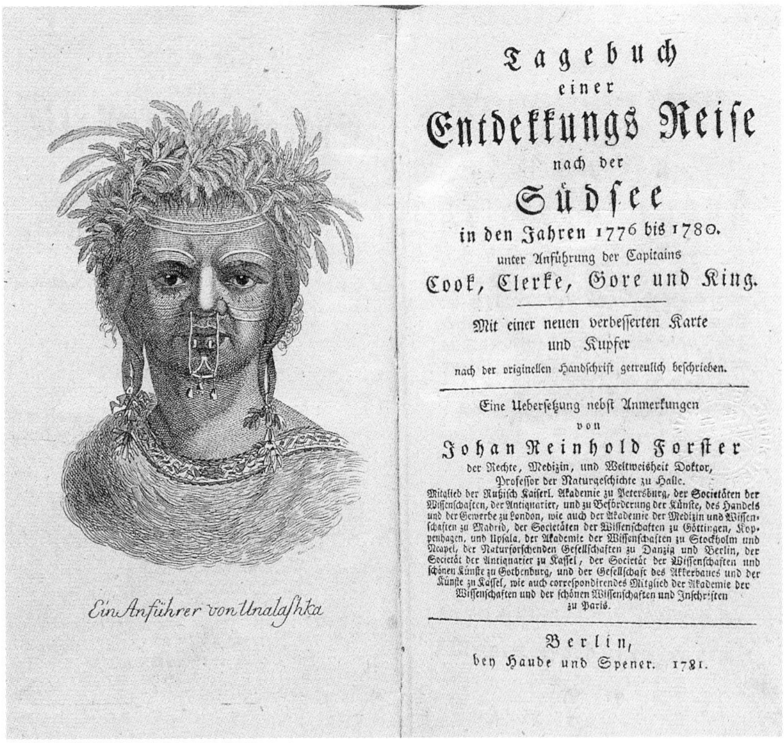

140. Forster's Rickman

in Forster's opinion, led to the tragedy of Cook's death on Hawaii. Even before the third voyage Cook, "a cross-grained fellow who sometimes showed a mean disposition and was carried away by a hasty temper", had developed an "overbearing attitude… , the result of having his head turned by Lord Sandwich"…" (Hoare, *The tactless philosopher: Johann Reinhold Forster*, p. 237).

Besides his introduction, Forster included a revised version of Rickman's map and many footnotes to the text – "a quite valuable… contribution to the ethnology, geography and natural history of the Pacific" (Hoare, p. 238).

Beddie, 1609; Hoare, 64; Kroepelien, 1079. **$14,500**

SET OF COOK'S VOYAGES

141. [COLLECTED VOYAGES] COOK, Captain James. A set of the voyage accounts, comprising:

FIRST VOYAGE. HAWKESWORTH, John. An Account of the Voyages undertaken by the order of His present Majesty for making Discoveries in the Southern Hemisphere… First edition, first issue, before printing of the erratum, the chart of the Streights of Magellan and the "Directions for Placing the Cuts". *Three volumes, quarto, 52 engraved plates and maps, many folding. London, Strahan and Cadell, 1773.*

SECOND VOYAGE. COOK, James. A Voyage towards the South Pole, and Round the World… Fourth edition. *Two volumes, quarto, 64 engraved plates and maps, many folding. London, Strahan and Cadell, 1784.*

THIRD VOYAGE. COOK, James and James KING. A Voyage to the Pacific Ocean… First edition. *Three volumes, quarto, 24 engraved maps and coastal profiles; with the separate folio atlas, containing two charts and 62 engraved plates. London, Strahan and Cadell, 1784.*

141. *Set of Cook's voyages*

Together eight volumes, quarto, and a folio atlas; a handsome set, internally in excellent condition, the eight text volumes in contemporary polished marbled calf gilt bindings, spines expertly renewed to match, double crimson and green labels; the atlas in half calf to match, marbled sides. London, 1773-1784.

The full series of official narratives of Cook's voyages – the cornerstone of any collection of books relating to Australia or the Pacific. Each of the three narratives is illustrated with engravings based on the work of the official artists on the voyages, including Parkinson, Hodges, and Webber, and the series stands as the great monument to Cook's achievements.

These were the best-sellers of the second half of the eighteenth century; very expensive when published, the first editions were sold out within a few days of publication. Their popularity meant that many copies were almost literally read to pieces; as a result, good sets of the voyages have become relatively scarce.

This set comprises the first edition of the first voyage; the fourth edition of the second voyage (the edition currently in print at the time of publication of the third voyage); and the first edition of the third voyage. Sets of the voyages are seen in many combinations of editions: this particular combination, generally regarded as a good way to have the set, is one of those seen with some regularity.

Beddie, 648, 1229, 1543; Hill, pp. 139–40, 60–62; Holmes, 5, 24, 47. **$45,000**

COOK FOR CHILDREN

142. [COOK] CAMPE, Joachim Heinrich. Bibliothèque Géographique et Instructive des Jeunes Gens…

Ten volumes in five, 16mo, with nine frontispieces (one folding), a folding map, and a large folding table of the entire series; an excellent set in contemporary half calf. Paris, J.E. Gabriel Dufour, 1803-1807.

A very rare set of Cook's three voyages for children. Translated from Campe's original German, this French version was part of the series *Bibliothèque Géographique et Instructive des Jeunes Gens*, a finely produced series for a young, and evidently sophisticated and literate, audience. Campe's narrative of Cook follows closely the official account and most of the plates are fairly faithful reproductions in miniature of the originals by Hodges and Parkinson. There also is a fine folding "Mort du Capitaine Cook", after Webber.

Joachim Campe (1746-1818) was a prolific writer for children, producing a vast array of works mainly concerned with disseminating information about voyages of discovery. His work was translated not only into French but also English and a number of Scandinavian languages. The full run of the series *Bibliothèque Géographique* – 72 volumes bound in 18 – was published between 1802 and 1807, and covered all the major voyages of exploration. This set of five bound volumes includes the full set of seven volumes pertaining to Cook, and also includes volumes relating to Vancouver, Labat and others.

Not in Beddie; not in Gumuchian; see Muir 1295 (other titles in this series) & 1297 (unseen by Muir). **$2250**

142. *Campe*

NORTHERN WORTHIES

143. [COOK] COLERIDGE, Hartley and Samuel Taylor. Lives of northern worthies. A new edition, with the corrections of the author, and the marginal observations of S.T. Coleridge.

Three volumes, small octavo; a very fine set in the original green ribbed cloth. London, Edward Moxon, 1852.

Scarce: the first complete and best edition, including a long biography of James Cook.

Hartley Coleridge began this series of provincial biographies in 1833 under the general title *Biographia Borealis*. Portions were issued at Leeds at two-month intervals, but only three parts had appeared when the project foundered. This edition, edited by the poet (Hartley's brother), brings together the thirteen long biographies. The 171-page biography of Captain Cook is its chief interest.

Beddie, 2190. $1250

LIFE OF COOK

144. [COOK] KIPPIS, Andrew. The Life of Captain James Cook.

Quarto, with engraved portrait of Cook; a good large copy in contemporary quarter calf, base of spine just worn. London, G. Nicol and G.G.J. and J. Robinson, 1788.

First edition of the first biography of Cook, the standard life of the navigator for the next century. The proposed settlement of Botany Bay is mentioned as a result of the voyages: "If it be wisely and prudently begun and constructed, who can tell what beneficial consequences may spring from it, in future ages?"

This copy has tipped in at the end a large French map by Jaillot showing the tracks of all three of Cook's voyages, a double-hemisphere world map in fine contemporary colouring.

Beddie, 32; Hill, p. 163; Holmes, 69; Kroepelien, 647; Lada-Mocarski, 40; O'Reilly-Reitman, 455. $4850

FIRST SPANISH EDITION

145. [COOK] KIPPIS, Andrew. Historia de la vida y viages del Captain Jaime Cook... traducida al castellano por Don Cesareo de Nava Palacio.

Two volumes, small quarto; in fine condition in contemporary vellum, spines lettered in ink. Madrid, en la Imprenta Real, 1795.

A fine copy of the very scarce first Spanish edition of the first biography of Cook, the standard life of the navigator for the next century.

Beddie, 56. $3400

WHITBY COAL-CATS

146. [COOK] TAYLOR, Henry. Memoirs of the Principal Events in the Life of Henry Taylor...

Octavo; a fine copy, uncut in the original printed boards, spine repaired; boxed. North Shields, T. Appleby for the author, 1811.

The memoirs of a merchant-seaman, of great interest for the light that it casts on Cook and the North Sea coal trade – that "great nursery for seamen".

From 1758, Taylor was master of the Whitby coal-cat *The Mary*. Eight years earlier, one of the ordinary crewmen whose name appeared on the Muster Roll of the same ship had been one "James Cook, seaman". Cook's biographer Beaglehole felt that Taylor's "experience, coal trade and North Sea, was much like Cook's and his picture of the life is the best we have" (*Life of Cook*, p. 8). Beaglehole dwells on the importance to these sailors of their practical skills, learned under hard conditions: "Henry Taylor, whose memoirs are so instructive about the conditions of the coal trade in which both he and Cook served, nevertheless found it worth while to point out that 'Another necessary duty, especially on this coast, is that of frequently casting the lead... Even in crossing the North Sea it is necessary to have a recourse to the lead'; and adds a cautionary tale about a master who scorned it..." (*op. cit.*, p. 13).

Taylor and Cook shared a patron in John Walker, the Whitby ship-owner and coal-trader, and must have known each other. Taylor refers to Cook as "the famous circumnavigator" who had also "served his apprenticeship in the coal and coast trade and acquired almost all his knowledge of books after he went to sea".

This copy has an appropriate provenance: it belonged originally to the naval officer Henry Parkyns Hoppner, who accompanied Lord Amherst's embassy to China and subsequently sailed on the first three of William Parry's Arctic Expeditions.

Not recorded by Beddie; NUC locates three copies. $2400

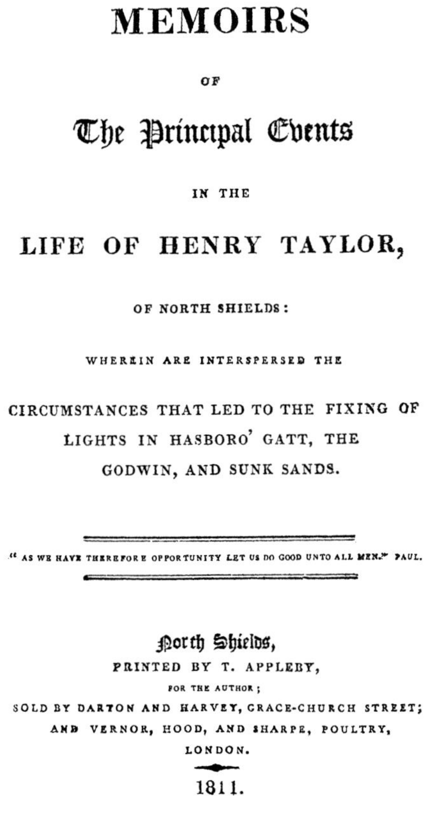
MEMOIRS

OF

The Principal Events

IN THE

LIFE OF HENRY TAYLOR,

OF NORTH SHIELDS:

WHEREIN ARE INTERSPERSED THE

CIRCUMSTANCES THAT LED TO THE FIXING OF LIGHTS IN HASBORO' GATT, THE GODWIN, AND SUNK SANDS.

"AS WE HAVE THEREFORE OPPORTUNITY LET US DO GOOD UNTO ALL MEN." PAUL.

North Shields,
PRINTED BY T. APPLEBY,
FOR THE AUTHOR;
SOLD BY DARTON AND HARVEY, GRACE-CHURCH STREET;
AND VERNOR, HOOD, AND SHARPE, POULTRY,
LONDON.

1811.

146. Taylor

5: AUSTRALIAN SETTLEMENT

BARRINGTON'S VOYAGE

147. BARRINGTON, George. A Voyage to Botany Bay with a description of the Country, manners, Customs, religion, &c. of the Natives... [*and*] A Sequel... comprising an Interesting Narrative of the Transactions and Behaviour of the Convicts; the Progress of the Colony...

Octavo, two works bound together, the first with an engraved frontispiece and vignette on title, the second bound without half-title; both works in fine condition in a neat contemporary French quarter calf binding, a little chipped. London, C. Lowndes and H.D. Symonds, c. 1800 [and] 1801.

An early pairing of the two initial 'Barrington' Botany Bay works; a number of editions were published in this form, all of which are scarce. The first work is dated 1795 by Ferguson, but actually has 1800 watermarks. The vignette on the first title-page showing convicts in chains being disembarked is one of very few illustrations of convicts in the early period, a subject surprisingly consistently avoided by contemporary illustrators.

Bound with it is the very scarce second edition of the *Sequel*. As with most Barrington material this is fraudulently attributed to him but is still significant on two counts: firstly, because it collects and summarises the information making its way to England not only from the obvious sources like First Fleeter narratives but also from more minor and personal or anecdotal sources; and secondly, because it represents precisely the generally perceived view of the new colony and its early history.

Ferguson, 206 (dated c. 1795) and 328. **$3500**

WITH COLOURED PLATES

148. BARRINGTON, George. An Account of a Voyage to New South Wales.

Octavo, portrait frontispiece and nine handcoloured plates, folding map (laid down); some browning and staining but an adequate copy in contemporary half calf. London, Jones, 1810.

The second edition (first published 1803) of this substantial account of the voyage to New South Wales via the Cape of Good Hope, with incidents partly derived from Barrington's experience on the Third Fleet.

Abbey, 606 (note); Ferguson, 486. **$1550**

BARRINGTON'S HISTORY

149. BARRINGTON, George. The History of New South Wales, including Botany Bay, Port Jackson, Pamaratta [sic], Sydney, and all its Dependancies from the Original Discovery of the Island: with the Customs and Manners of the Natives...

Octavo, frontispiece, with handcoloured title-page and 16 handcoloured plates as usual, with an additional four handcoloured plates and folding map; a fine copy in recent period style half calf. London, Printed for M. Jones, 1802.

The first coloured views of Sydney: this is the first edition of Barrington's *History*, a substantial and important early account of the colony that transmitted much information to the English reader about life in the colony at the beginning of the nineteenth century.

This copy contains, as well as its normal suite of sixteen plates, a further four plates ("South View of Sydney", "Cape of Good Hope", "East View of Sydney", "A Native Family"). The frontispiece and the two extra views of Sydney are the first colourplate Australian views, and are also among the earliest engravings to show proper settlement.

Abbey, 565; Ferguson, 345. **$4000**

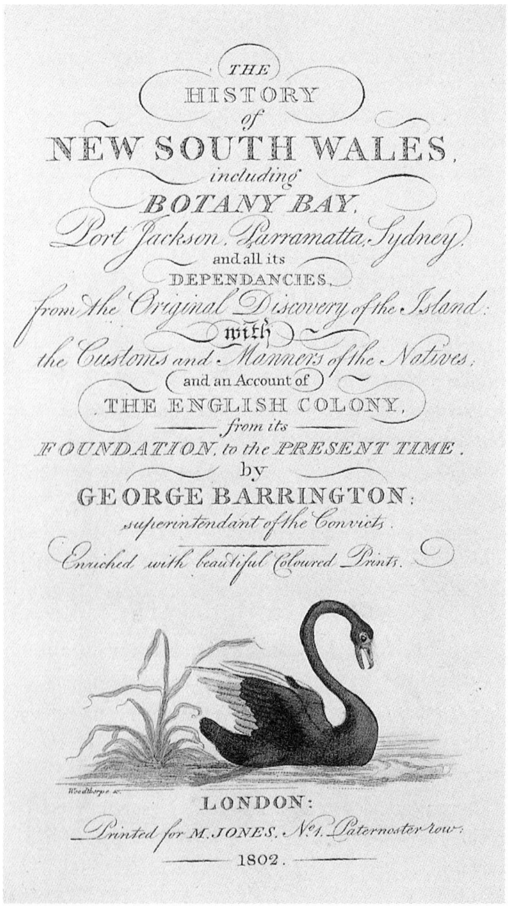

THE HISTORY of NEW SOUTH WALES, including BOTANY BAY, Port Jackson, Parramatta, Sydney, and all its DEPENDANCIES, from the Original Discovery of the Island: with the Customs and Manners of the Natives; and an Account of THE ENGLISH COLONY, from its FOUNDATION, to the PRESENT TIME. by GEORGE BARRINGTON; superintendant of the Convicts. Enriched with beautiful Coloured Prints.

LONDON: Printed for M. JONES, No. 1, Paternoster-row. 1802.

149. Barrington

EARLY SETTLEMENT

150. COLLINS, David. An Account of the English Colony in New South Wales, from its first Settlement in January 1788, to August 1801: with Remarks on the Dispositions, Customs, Manners, &c. of the Native Inhabitants of that Country. To Which are Added, Some Particulars of New Zealand; Compiled, By Permission, from the Mss. Of Lieutenant-Governor King.

Quarto, with 23 engravings and a folding map; some very slight ageing throughout; handsome later binding of half calf, gilt. London, T. Cadell and W. Davies, 1798.

First edition: a very good copy. Collins arrived with the First Fleet as Judge-Advocate and was Secretary to Governor Phillip. His handsome book is a valuable account of the early settlement by an educated and observant resident of ten years, illustrated with full-page engravings prepared from sketches done in the colony by the convict artist, Thomas Watling. These are the first views to have been published of British settlements at Sydney and Parramatta.

A second, supplementary volume was published in 1802 (see below). This first volume, published in 1798, is scarce today.

Ferguson, 263. **$5000**

THE FULL FIRST EDITION

151. COLLINS, David. An Account of the English Colony in New South Wales…

Two volumes, quarto, bound in three, with two maps, 23 full-page engraved plates (three coloured) and eight half-page engravings including two coloured vignettes; contemporary diced russia leather, marbled endpapers and edges, spines recently renewed. London, 1798-1802.

The full first edition of David Collins's classic account of the early settlement of Australia, that is the main work published in 1798 and the supplementary volume published four years later.

This is an unusual and attractive set, bound by its first owner into three volumes rather than the usual two – quite a sensible arrangement as the first volume is normally bulky. The binding was evidently arranged by Thomas Hutton whose contemporary bookplate appears in each volume; he has supplied half-title and title-pages for the extra volume in handsome manuscript.

Ferguson, 263 & 350. **$12,500**

SECOND EDITION

152. COLLINS, David. An Account of the English Colony in New South Wales…

Quarto, with a portrait, two maps, 23 engraved plates and eight half-page engravings; a few very light marginal stains, otherwise a fine copy in modern polished calf. London, T. Cadell and W. Davies, 1804.

The second edition, considerably changed from the original two-volume edition.

Collins's wife, Maria, continued her husband's revision and edited the book, completing it with newinformation available to the end of 1803. Collins's earnings had been severely affected by his long service in New South Wales, and he had originally published his history of the colony in an attempt to eke out the family income. He and Maria decided to issue this cheaper edition to bring it up to date, and to reach new readers who could not afford the expensive two-volume edition.

This is the preferred issue of the book, with the natural history plates in original handcolouring.

Ferguson, 390. **$3850**

153. Gillray

THE HOUSE IS ADJOURNED TO BOTANY BAY

153. GILLRAY, James. Consequences of a Successfull French Invasion…

Engraving, 370 x 270 mm., handcoloured as issued; framed. London, n.d., but mid-nineteenth century.

A famous, satirical look at the decision of the English parliament to create a penal settlement at Botany Bay, in contrast to the atmosphere of reform and concern with civil liberties in the newly created republic of France and in the American colonies. This cartoon was first published in 1798; this attractive reissue evidently appeared about fifty years later.

Gillray turns the tables on the politicians of the House of Commons responsible for the Botany Bay decision: a party of French revolutionaries led by Napoleon storms the parliament, gagging the Speaker and placing all the Members in chains and dressing them in the garb of convicts. All the symbols of the House are destroyed: the mace is smashed, and the statutes are discarded on the floor; inscribed below the Speaker's chair are the words, "The House is adjourned to Botany Bay".

Reproduced by Jonathan King, 'The Other Side of the Coin: A Cartoon History of Australia', p. 18. **$1500**

154. Gosse

HEROIC FOUNDATION

154. GOSSE, Thomas. Founding of the Settlement of Port-Jackson at Botany Bay In New South Wales.

Mezzotint engraving, 535 x 610 mm.; with generous margins, in very fine and fresh condition; in an early rosewood frame. By the engraver, 1799.

A beautiful example of this extremely rare and important image: the single ambitious myth-making attempt at providing a noble or heroic foundation for the Australian settlement, and the only "high art" treatment of the subject in the early period. Gosse's work is unique in trying to lend to the settlement of New South Wales something of the dignity of images of other new nations, most notably America.

This splendid mezzotint, separately issued, was drawn and engraved as well as published by Thomas Gosse (1765-1844), an account of whom appears in *The Life of Philip Henry Gosse, F.R.S.* by the well-known literary figure Edmund Gosse: he emerges as an interesting and eccentric figure.

The first historical representation of white settlement in Australia, it actually stands for much more: it is, although an engraving, Australia's first and for a long time only "history painting". It was not issued as a copy of a milestone painting: it appeared only in this form, not as an original in oils but as an engraving. It must have been published in a very small number, as only very few examples survive today. To our knowledge there has not been another example offered publicly in several decades.

The settlement of the new colony at Port Jackson does not seem to have interested other European artists of the period, perhaps because of the generally perceived ignoble beginnings of the "thief-colony". Nor was the subject attempted by colonial artists, who concentrated rather on natural wonders, landscape, flora and fauna, sometimes the Aborigines, almost always topography – the images that could ultimately show people in the "old country" what the new world looked like.

In this sense, Gosse's work fits into the familiar pattern: colonial art shows the exotica of the new country, while the old world provides the high art. This same notion would provide the beginnings, in the form of copies and imitations of old world masterpieces, of many if not most of our major public collections of art.

Sir William Dixson's copy of the engraving, in the State Library of New South Wales, was recently shown in the Library's exhibition *A Rage For Curiosity*. The curator Richard Neville pondered in the catalogue: "Was it because of this surfeit of criminality, and the moral disgrace it visited upon Australia that the country was never the subject of history painting? Or was this deficiency the result of a number of things: the colony's unusual history, its lack of a picturesque indigenous culture and the absence of sophisticated artists who could turn the country into art? Only English artist Thomas Gosse attempted to elevate the settlement into high art with his 1799 mezzotint...The obvious fantasy packs curious flora and fauna, a negroid Aboriginal, the military, a demure young girl (a 'good' girl, though presumably a convict) into a landscape that already appears to be settled...".

The figures are assembled in a posed tableau style. The naval officer at the centre is Captain Phillip, the first governor of the colony. The other main features of the composition are the British seamen with a marine and workmen establishing buildings and fencing; an Aborigine appears at the right, and on the left a man is shooting an exotic bird, while another is turning turtles onto their backs. Cattle appear in the image, as well as a pig.

Also illustrated as frontispiece to this catalogue. For Gosse, see also catalogue number 222. **$48,500**

BOTTOMY BAY

155. HEATH, William. New Panorama... A Startling Interrogation.

Handcoloured engraving, 300 x 410 mm.; in fine condition, framed. London, T. McLean, 1829.

A fine impression of the coloured issue of this rare cartoon, satirising the colony which at this time was desperately trying to attract free immigration. The cashier at the Panorama (probably the Colosseum in Regents Park: see catalogue number 316) asks "Do you wish to go to Hell or Botany Bay? Sir." To which the prospective immigrant replies: "I wonts to go to Bottomy Bay"; a young girl exclaims to her enormously fat mother "La Mama I should like to see the Naughty Place better than any thing".

William Heath is noted by Bénézit: "on cite aussi de lui des caricatures pleines d'humour sur les moeurs anglaises, vers 1825".

Detail illustrated opposite catalogue number 102.

Reproduced (although not attributed, and misdated) by Jonathan King, 'The Other Side of the Coin: A Cartoon History of Australia', p. 23. **$4500**

HUNTER'S JOURNAL

156. HUNTER, John. An Historical Journal of the Transactions at Port Jackson and Norfolk Island, with the Discoveries which have been made in New South Wales and in the Southern Ocean, since the publication of Phillip's Voyage, compiled from the Official Papers; Including the Journals of Governors Phillip and King, and of Lieut. Ball; and the Voyages from the First Sailing of the Sirius in 1787,

to the Return of that Ship's Company to England in 1792.

Quarto, with 17 engraved plates, folding maps and charts; a good copy, in later quarter calf and marbled boards (the often cropped date on the title-page intact). London, John Stockdale, 1793.

First edition of Hunter's *Journal*, equal in importance to Phillip's and White's as an account of the first years of settlement at Sydney. Captain John Hunter was appointed second captain of HMS *Sirius* under Phillip for the voyage to Botany Bay. Once in New South Wales he was actively engaged in surveying and exploration, and only left for England in late 1791 after the loss of the *Sirius* at Norfolk Island while under his command. He spent the next few years in England, where he wrote the *Journal*.

Hunter gives an excellent account of many activities, particularly exploration and the settlement at Norfolk Island, which are treated more cursorily by the other First Fleet chroniclers. The engraved plates and maps, many of the latter from original cartography by Hunter, Dawes and Bradley, are very fine.

Ferguson, 152. **$6000**

THE RARE OCTAVO EDITION

157. HUNTER, John. An Historical Journal of the Transactions at Port Jackson and Norfolk Island…

Octavo, portrait, engraved title-page (shaved at outer margin), large folding map and a folding plate; a very good copy in contemporary half calf, spine gilt. London, John Stockdale, 1793.

The condensed octavo edition of Hunter's journal, published in the same year as, but scarcer than, the quarto first edition; this version was intended "sufficiently [to] gratify the curiosity of those who may not have leisure to peruse, or are not desirous of purchasing the quarto edition…".

Ferguson, 153. **$2450**

SWEDISH EDITION

158. HUNTER, John. Johan Hunters Resa til Nya södra Wallis, üren 1787, följande; jämte Nyaste underrättelser om Engelska Nybygget i Port Jackson, Nya Holland och Norfoks-On, af Capit. Tench och King samt Cap. Edvards Resa omkring Jorden…

Small octavo; a fine copy in contemporary Swedish quarter calf. Stockholm, Johann Pfeiffer, 1797.

Very scarce: the first Swedish edition of Hunter's journal of events at Port Jackson and Philip Gidley King's Norfolk Island journal, abridged and joined with material from both Tench's books and Hamilton's account of the *Pandora* voyage.

Ferguson, 250 (Kroepelien collection; Mitchell and National Libraries); but not in the Kroepelien catalogue. **$1250**

FIRST ACCOUNT BY A WOMAN

159. PARKER, Mary Ann. A Voyage round the World, in the Gorgon Man of War: Captain John Parker… Performed and written by his widow; for the advantage of a numerous family. Dedicated, by permission, to Her Royal Highness the Princess of Wales.

Octavo; the title-page worn, and with marginal repairs; otherwise a good copy of the book, in old half morocco. London, Printed by John Nichols, 1795.

First edition. In this, the first account of the colony by a woman, Mary Ann Parker paints a vivid picture of life in New South Wales in the late eighteenth century. She had sailed with her husband, captain of the *Gorgon*, to the struggling colony with much-needed provisions. Philip Gidley King, who was travelling to take up his post as governor of Norfolk Island, was a fellow passenger.

This account provides us with detailed descriptions of Port Jackson, ships in the harbour, disease on the transports, a discussion of "the propriety of establishing a whale-fishery on the coast of New Holland", breakfast with Governor Phillip aboard the *Gorgon*, shore visits at Sydney Cove and Botany Bay with descriptions of flora and fauna, and an excellent chapter on the Aborigines, including a description of a meeting with Bennelong after her return to London.

Ferguson, 229. **$5850**

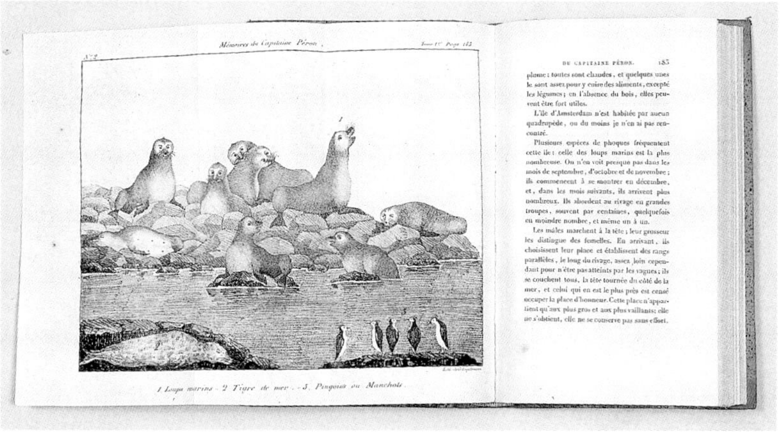

160. Péron

ONE OF FEW FIRST-HAND DESCRIPTIONS

160. PERON, Captain Pierre F. Mémoires du Capitaine Péron, sur ses Voyages…

Two volumes, octavo, with four folding maps and two folding plates lithographed by Engelmann; slight paper loss in blank outer margin of some leaves in the second volume, otherwise a fine copy in contemporary French half calf. Paris, Brissot-Thivars, 1824.

Rare. Péron's description of Van Diemen's Land, his journey up the east coast of Australia, and his visit to Sydney mark this as one of surprisingly few first-hand descriptions of eighteenth-century Australia. This memoir of a life at sea also includes much on the Northwest coast of America. Of particular note is the extensive material on the escaped Scottish martyr Th omas Muir, who sailed alongside Péron on the *Otter*, after a dramatic escape from Port Jackson. Péron describes their time together on the Northwest coast, particularly in Juan de Fuca strait, and various experiences with the Indians.

Ferguson, 980. **$3650**

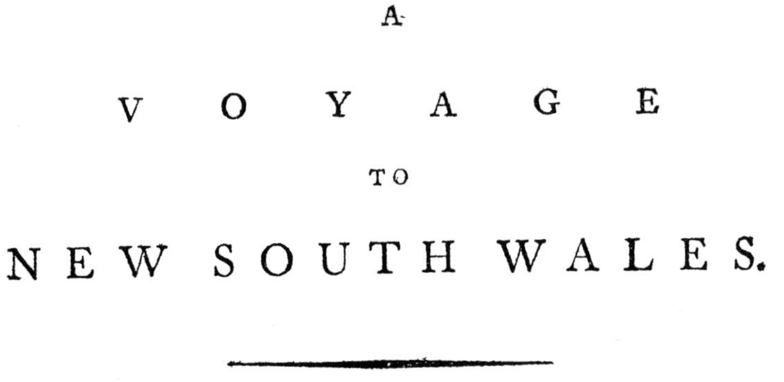
A

VOYAGE

TO

NEW SOUTH WALES.

161. Phillip

THE GOVERNOR

161. PHILLIP, Arthur. The Voyage of Governor Phillip to Botany Bay; with an Account of the Establishment of the Colonies of Port Jackson & Norfolk Island...

Quarto, frontispiece portrait, 55 plates, maps & charts (many folding), uncoloured as usual; a couple of early press clippings inserted at front; some foxing and ageing but generally a good copy in quarter calf. London, John Stockdale, 1789.

First edition of the official account of the first settlement of Australia, the single most important book to describe the journey to Botany Bay and the foundations of settlement.

Based on the governor's journals and despatches, "this volume contains all that is yet known of the settlement at Sydney Cove"(Dedication). Included are the first descriptions and engravings of a large number of Australian birds, animals, fish and plants as well as primary views of Botany Bay and Port Jackson. It contains a list of the names of the 700 convicts brought to Sydney on the six transport vessels that were part of the First Fleet.

Ferguson, 47. **$7000**

THIRD EDITION

162. PHILLIP, Arthur. The Voyage of Governor Phillip to Botany Bay...

Large octavo, engraved title-page and 19 engraved plates and maps, mostly double-page or folding; a very nice copy, with wide margins, in old half brown morocco, top edges gilt, other edges rough-trimmed only. London, Stockdale, 1790.

The scarce third edition of this classic, rare on the market.

Ferguson, 91. **$2750**

FIRST FRENCH EDITION

163. PHILLIP, Arthur. Voyage du gouverneur Phillip à Botany-Bay.

Octavo; a good uncut copy in later quarter calf over old boards. Paris, Buisson, 1791.

The first French edition of the classic account of Australian settlement.

Ferguson, 112. **$1250**

THE HULK

164. ROWLANDSON, Thomas. A Hulk at Sheerness.

Watercolour, with pencil, ink and pen on paper, 215 x 343 mm., laid down on card, signed lower left "Rowlandson". N.d., but c. 1800.

A delightful watercolour by Rowlandson of an unusual subject: one of the notorious "hulks" or decommissioned vessels, some of which were used to house convicts awaiting transportation; others – like this one – had become de facto housing.

Rowlandson was especially respected for his maritime paintings, a genre in which he excelled. This is a typical example of his loose and relaxed style, and his interest in documentary and social history. It was probably painted on the spot, in his characteristically unhesitant style; the people and boats at lower left are sketched in pencil, but not completed in watercolour.

The hulk was a common eighteenth-century sight on the Thames and along the English coast. After their seagoing days were over, the lives of many of "the wooden walls of old England" were considerably extended thanks to their stout oak construction. Many served as training ships or dockyard auxiliaries, although some were used as hospitals, even chapels, and others, notoriously, were used as prisons – such as Duncan Campbell's Thames Hulks, which housed convicts. A good number also served as housing; Brayley's *Beauties of England and Wales* (1808) states that the population of the Fortress and Dockyard at Sheerness was perhaps two thousand, "including those who reside in the houses without the Fortress, and in the old ships of war which have been stationed on the shore as breakwaters". The hulls of these ships "are occupied by about seventy or eighty families, and altogether present a very singular appearance, the chimneys being raised of brick from the lower gun-decks".

Roe suggests that the topsy-turvy "Heath Robinson" construction had an essential humour to it which would have appealed to Rowlandson. However the sketch is also evocative of this period when not only the prisons were appallingly overcrowded – potentially of such importance to Australia's convict history – but so were the living conditions of ordinary folk. This is perhaps equally as important for understanding Australia's history of free settlement.

This was once in the collection of Captain Bruce S. Ingram, a collection noted for its maritime paintings, and for Rowlandson material. It is similar to a work in the Mellon Collection (Baskett and Snelgrove, *Drawings of Thomas Rowlandson in the Paul Mellon Collection*, London, 1977, no. 279). Ours has no inscription, although it was known by its present title whilst in the Ingram Collection, though the title of the Mellon version would be equally suitable for this work, "One of the Old men of War made a Breakwater and fitted for Shipwrights at Sheerness".

Literature: F. Gordon Roe, "Drawings by Rowlandson: Captain Bruce S. Ingram's Collection", pp. 85-91, 'Connoisseur', Dec. 1946 (illustrated).

Provenance: Ingram Collection (with his collector's mark in pencil on original backing paper, presaerved) sold Sothebys, London, February 1965. **$15,500**

164. Rowlandson

SYDNEY COVE ON A BINDING

165. SYDNEY VIEW. Painted view on the fore-edge of a volume of the Bible, Ezekiel–Malachi…

Octavo; fine contemporary binding of dark blue straight-grained morocco, ornately gilt, gilt edges, the fore-edge with a painted view concealed below the gilt. London, John Reeves, 1802 [painting c. 1820–1830].

A delightful early nineteenth-century painting of Sydney Cove, loosely based on the first of John Eyre's four aquatints, *View of Sydney from the East side of the Cove.* Eyre's aquatint views, issued in 1810, are the best documentary evidence for the development of Sydney at the beginning of the nineteenth century – a most suitable choice for this fore-edge painting.

Almost certainly executed in England, the painting does not strictly follow the Eyre view – there are more ships in the foreground and on the left side of the painting there is an imaginary waterway. The buildings depicted show the development of the Rocks area, and include the residence of the Governor, the Military Barracks, storehouses and warehouses, and a government windmill high on the hill.

Painting on the fore-edge was an early book art: there are examples from the Renaissance. The technique acquired novelty when it was realised that a painting done on the slanting surface of a fore-edge with the book splayed open becomes invisible when the book is closed: it only becomes visible again when the edges are spread out. This curious art form was popular throughout the nineteenth century, but few examples are known of Australian subjects.

The identity of the artist who created this rare and interesting work is unknown: fore-edge paintings are usually unattributed, and difficult to date accurately. Our research suggests that this was probably executed before 1830. **$8500**

165. Fore-edge painting

FIRST EYE-WITNESS ACCOUNT

166. TENCH, Captain Watkin. A Narrative of the Expedition to Botany Bay; with an Account of New South Wales, its Productions, Inhabitants, &c., to which is subjoined a list of the Civil and Military Establishments at Port Jackson.

Octavo, lacking the half-title and the final leaf of Debrett's advertisements; a good copy in old green half calf, expertly rebacked, retaining original spine. London, J. Debrett, 1789.

The first eye-witness account of Australia's settlement. This important book was the very first authentic account of settled Australia to appear in print, the first copies of the first edition appearing for sale on 24 April 1789. It was extremely popular, not surprisingly in view of the large public that would have been curious for news of the colony, and three editions in English, a Dublin piracy and French, German and Dutch translations all appeared quickly.

Tench spent altogether four years in the colony (in 1793 producing a second book on the settlement). He was a lively and cultured member of the new society, and these qualities come through in his book which gives a vivid picture of the voyage out, and the establishment of the town at Sydney Cove. Apart from its importance as the first genuine description of the new colony, Tench's *Narrative* provides us with the clearest of the surviving images of the crucial first months of settlement.

Ferguson, 49; Wantrup, pp. 54-57. **$7500**

DUBLIN PRINTING

167. TENCH, Captain Watkin. A Narrative of the Expedition to Botany Bay.

Octavo; some browning as usual, but very good in modern half calf and marbled boards. Dublin, H. Chamberlaine, 1789.

The very scarce Dublin printing of the first authentic account of settled Australia to appear in print. The book was published quickly, the first copies of the first edition appearing for sale on 24 April 1789. It proved extremely popular, not surprisingly in view of the large public that would have been curious for news of the colony, and this Dublin piracy, as well as three English editions, and French, German and Dutch translations were soon called for.

Ferguson, 51. **$2400**

FRENCH EDITION

168. TENCH, Watkin. Voyage à la Baie Botanique.

Octavo, folding map; contemporary German glazed boards, some uneven sunning but a fine copy. Paris, chez Letellier, 1789.

One of two French editions of Tench's *Narrative* published in 1789, of uncertain priority. This is the more extended version with a long, 85-page additional section entitled "Récit historique de la découverte de la Nouvelle Hollande et du nouveau Pays de Galles Méridional", which the economic historian K.M. Dallas sees as significant evidence in the continuing debate about the reasons for the decision to settle New South Wales. This is the issue with the folding map noted by Ferguson.

Ferguson, 53. **$2100**

TENCH, PATTERSON & LA PEROUSE

169. TENCH, Watkin and George PATTERSON. Voyages dans le pays des Hottentots... à la Baye Botanique.

Octavo; contemporary French marbled sheep, front cover marked but a very good copy. Paris, chez Letellier, 1790.

The second Letellier edition of Tench, along with Patterson's narrative of South Africa and an account of La Pérouse at Botany Bay.

Ferguson, 96. **$1200**

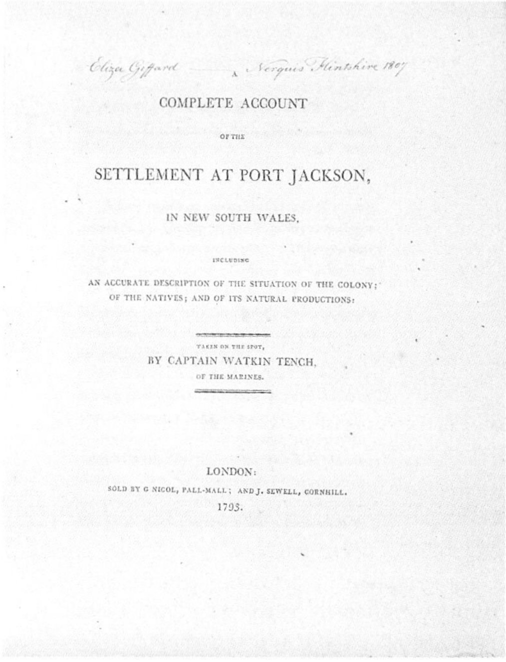

A

COMPLETE ACCOUNT

OF THE

SETTLEMENT AT PORT JACKSON,

IN NEW SOUTH WALES,

INCLUDING

AN ACCURATE DESCRIPTION OF THE SITUATION OF THE COLONY;
OF THE NATIVES; AND OF ITS NATURAL PRODUCTIONS:

TAKEN ON THE SPOT,

BY CAPTAIN WATKIN TENCH,

OF THE MARINES.

LONDON:

SOLD BY G. NICOL, PALL-MALL; AND J. SEWELL, CORNHILL.

1793.

170. Tench

SUBSCRIBER'S COPY

170. TENCH, Watkin. A Complete Account of the Settlement at Port Jackson, in New South Wales...

Quarto, with folding map; a splendid copy, completely uncut in the original blue-grey boards, paper spine and label renewed; quarter morocco bookform box. London, Nicol and Sewell, 1793.

A lovely copy of the first edition. One of the scarcest First Fleet accounts, this is the second of Tench's two books and brings his account of settlement in New South Wales down to the end of the first four years. Tench explored the country around Sydney, discovered the Nepean River, and tried to cross the still impregnable Blue Mountains; his book is thus also considered the first journal of an inland explorer.

This is a subscriber's copy of this important book, inscribed "Eliza Giffard – Nerquis Flintshire 1807". Tench's patron, and the dedicatee of the book, was Sir Watkin Williams Wynn of Flintshire, whose near-neighbour Mrs Giffard was.

Ferguson, 171; not in Hill. **$13,850**

AN
ACCOUNT
OF A
VOYAGE
TO ESTABLISH A COLONY AT
PORT PHILIP IN BASS's STRAIT,
ON THE SOUTH COAST OF
NEW SOUTH WALES,
IN HIS MAJESTY's SHIP CALCUTTA,
IN THE YEARS 1802-3-4.

BY J. H. TUCKEY, ESQ.
FIRST LIEUTENANT OF THE CALCUTTA.

" Bear Britain's thunder, and her cross display,
" To the bright regions of the rising day;
" Tempt icy seas, where scarce the waters roll,
" Where clearer flames glow round the frozen pole,
" Or under southern skies, exalt their sails,
" Led by new stars, and borne by spicy gales."
POPE's WINDSOR FOREST.

171. Tuckey

A FIRST FLEET FOR VICTORIA & TASMANIA

171. TUCKEY, James Hingston. An Account of the Voyage to Establish a Colony at Port Philip in Bass's Strait, on the South Coast of New South Wales, in His Majesty's Ship Calcutta...

Octavo, bound without the half-title, as often; a very good copy in later half calf. London, Printed for Longman, Hurst, Rees, and Orme, 1805.

First edition: the only contemporary publication describing Victoria's abortive and Tasmania's successful First Fleet. The *Calcutta* sailed from England in 1802, with the intention of establishing a settlement at Port Phillip under the command of Lieutenant-Governor Collins. The hostility of the natives, the lack of water, and other problems of environment caused the abandonment of the settlement after some months, and part of the expedition continued to found Hobart.

Tuckey, first lieutenant on the *Calcutta*, was one of those who continued from Port Phillip to Sydney and then back to England. He gives a lucid narrative description of the events of the voyage and of life on the short-lived Port Phillip settlement. Port Phillip was not settled for more than twenty years as a result of the expedition's failure, and was used only occasionally as a harbour by the whalers and sealers of Bass Strait, until the military settlement of Western Port – once again abortive – in 1826.

Ferguson, 418. **$3850**

THE CHIEF SURGEON'S JOURNAL

172. WHITE, John. Journal of a Voyage to New South Wales...

Quarto, with 65 engraved plates; a clean and large copy in old half morocco, joints and extremities rubbed but an attractive copy. London, J. Debrett, 1790.

First edition of this important journal of the first Australian settlement, by the chief surgeon of the First Fleet. White was also a keen amateur naturalist and after arriving at Port Jackson found time to accompany Phillip on two journeys of exploration. On joining the First Fleet he had begun to keep a journal in which he later made notes about birds in the new colony. This manuscript journal formed the nucleus of his book.

The natural history content makes White's particularly noteworthy amongst the First Fleet journals. Many of the plates were drawn in England by leading natural history artists of the day, such as Sarah Stone, from original sketches done in the colony. White's interest in natural history continued until he left New South Wales in December 1794. When the convict artist Thomas Watling arrived in the colony in October 1792 he was assigned to White and in the next two years made many drawings of birds for him. It is possible that White himself had some skill as an artist and that he was responsible for the original sketches of some of the engravings here.

White's journal also contains a good description of the voyage from London, with long, detailed accounts of the stops at Rio de Janeiro, Cape Town and of the colonial voyages to Norfolk Island.

Davidson, pp. 81-6; Ferguson, 97; Ford, 2495. **$6850**

172. White

173. White

DELUXE COLOUR ISSUE

173. WHITE, John. Journal of a Voyage to New South Wales…

Quarto, with the 65 engraved plates in fine original colouring; small tear to title-page which is also slightly soiled; later half calf. London, J. Debrett, 1790.

First edition: the rare deluxe issue of the famous First Fleet book in which the plates were coloured by hand. In this form, White's *Journal* is one of the most beautiful of Australian colour-plate books, and one of the most attractive, as well as one of the earliest, Australian bird books.

Abbey, 605; Ferguson, 97. **$14,000**

FIRST FRENCH EDITION

174. WHITE, John. Voyage à la Nouvelle Galles du Sud, à Botany-Bay…

Octavo, with two plates; contemporary calf, some worming and wear to joints. Paris, chez Pougin, An 3 de la République (1795, vieux style).

First French edition of White's journal: a rare example with two engraved plates. This first edition is normally seen without plates, although Ferguson described a copy which apparently, like the present example, had the engravings (opposite pp. 10 & 140) normally found in the 1798 second edition.

Ferguson, 231. **$1200**

6: AUSTRALIAN EXPLORATION

OVER THE MOUNTAINS

175. BLAXLAND, Gregory. A Journal of a Tour of Discovery across the Blue Mountains, New South Wales in the year 1813.

Small octavo; some foxing but a very good copy in original fawn paper wrappers; quarter morocco case. Sydney, Gibbs Shallard & Co., [1870].

Second edition and scarce; the first edition is one of the notorious rarities among Australian exploration books.

Blaxland's was the first great feat of Australian inland exploration. A series of previous attempts had failed to find a path into the hinterland and for twenty-five years the formidable barrier of the Blue Mountains had locked the settlers out of the interior. A landholder in the Parramatta district, Blaxland set out in 1813 to conquer the Blue Mountain barrier and discover new pastures to the west of the ranges.

With his companions Lawson and Wentworth, he finally accomplished his mission on 31 May 1813, sighting rich lands from the summit of the mountain later called Mount Blaxland. From the top they could see forests and grasslands capable of supporting the colony's herds for the next thirty years. Surveyor George Evans, on direct instructions from Macquarie, followed the path blazed by Blaxland and broke through to the magnificent pastures of the Bathurst Plains; within a year William Cox had completed his road to the new territories. By 1815, a settlement had been made at Bathurst.

Ferguson, 7133; Wantrup, 103c. **$7000**

EARLY ISSUE

176. [BURKE AND WILLS] "ARGUS". The Burke and Wills Exploring Expedition: an account of the crossing the continent of Australia, from Cooper's Creek to Carpentaria. Reprinted from "The Argus".

Octavo, pp. iv, 32 (double column); in good condition, bound without wrappers in modern morocco gilt. Melbourne, Wilson and Mackinnon, 1861.

Very scarce: one of the early impressions or editions of this important pamphlet account of the Burke and Wills expedition drawn from the original Melbourne *Argus* reports. It was reprinted many times between November 1861 and January 1862.

This is identified as an early issue, and probably the first, by the two misprints in the second column on page 25: "Fred. Mueller" for "Ferd."; "explanaions" for "explanations".

Ferguson, 7702; Maria, 19; Wantrup, 163. **$1800**

BURKE'S OWN JOURNALS

177. [BURKE & WILLS] JACKSON, Andrew. Robert O'Hara Burke and the Exploring Expedition of 1860.

Octavo, with title-page vignette, a folding map and advertisements; a few marks but a fine, largely unopened, copy in the original green cloth; MHC shelfmark in pencil. London, Smith, Elder and Co., 1862.

Only edition of the one substantial contemporary publication of Burke's own journals. Jackson knew Burke, was an officer in the same regiment as his father, and published this account in memory of Burke and dedicated it to Burke's sister Hester "whose name was the last utterance from the lips of her loved and gallant brother".

This copy was once in the famous library of Sir Thomas Phillipps, the most omnivorous of book-collectors. His nineteenth-century books, largely unread and even sometimes as here literally unopened, are often still in remarkable condition.

Ferguson, 10857; Maria, 98; Wantrup, 173. **$2250**

ROYAL COMMISSION REPORT

178. [BURKE AND WILLS COMMISSION] Report of the Commissioners Appointed to Enquire into and Report Upon the Circumstances Connected with the Sufferings and Death of Robert O'Hara Burke and William John Wills, the Victorian Explorers.

Folio, printed on blue paper; a good copy in a modern binding of textured cloth. Melbourne, John Ferres, 1862.

The scarce and important official report of the Royal Commission appointed to enquire into the Burke and Wills expedition. The report apportions blame between Burke himself as the expedition's incompetent leader, the careless overseer William Wright, and the indecisive Exploration Committee of the Royal Society of Victoria.

Maria, 35; Wantrup, 167. **$1750**

OFFICIAL REPORTS OF THE EXPLORATION COMMITTEE

179. [BURKE AND WILLS] ROYAL SOCIETY OF VICTORIA. Progress Reports and Final Report of the Exploration Committee of the Royal Society of Victoria...

Foolscap folio, printed on blue paper; in fine condition, in modern quarter blue calf. Melbourne, 1863.

A scarce volume, in fine condition. This is the collective issue of the official reports of the Exploration Committee, giving the full history of the Burke and Wills expedition from its inception, preparation, and departure to its tragic end.

The original Royal Society papers represented here are: First report, 1857; Second report, 1858; Third report, 1858; Fourth report, 1860; Special report, 1859; Fifth progress report, 1860; Sixth progress report, 1861; Seventh and final report, 1863; Supplementary final report, 1872.

Copies of this collection of papers can vary, some evidently having been assembled from separate papers and others issued by the Royal Society of Victoria in volume form, as noted in the final page of the report itself, suggesting a maximum of 200 copies published. In any event it is very scarce today, and was probably in practice issued in a limited way to institutions, corresponding societies and important individuals.

Ferguson, 15185; Maria, 31. **$6000**

180. *Field*

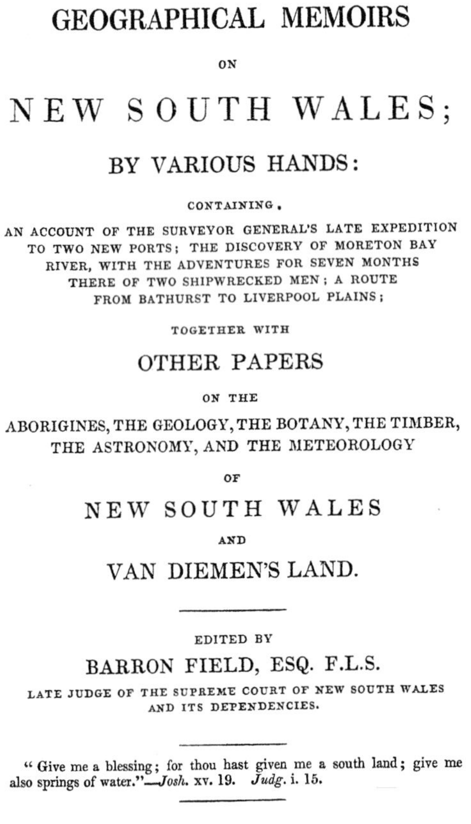

GEOGRAPHICAL MEMOIRS

ON

NEW SOUTH WALES;

BY VARIOUS HANDS:

CONTAINING,

AN ACCOUNT OF THE SURVEYOR GENERAL'S LATE EXPEDITION TO TWO NEW PORTS; THE DISCOVERY OF MORETON BAY RIVER, WITH THE ADVENTURES FOR SEVEN MONTHS THERE OF TWO SHIPWRECKED MEN; A ROUTE FROM BATHURST TO LIVERPOOL PLAINS;

TOGETHER WITH

OTHER PAPERS

ON THE

ABORIGINES, THE GEOLOGY, THE BOTANY, THE TIMBER, THE ASTRONOMY, AND THE METEOROLOGY

OF

NEW SOUTH WALES

AND

VAN DIEMEN'S LAND.

EDITED BY

BARRON FIELD, ESQ. F.L.S.

LATE JUDGE OF THE SUPREME COURT OF NEW SOUTH WALES AND ITS DEPENDENCIES.

" Give me a blessing; for thou hast given me a south land; give me also springs of water."—*Josh.* xv. 19. *Judg.* i. 15.

BY VARIOUS HANDS

180. FIELD, Judge Barron. Geographical Memoirs on New South Wales; by various hands...

Octavo, aquatint view of the Campbell River by Lewin, two botanical plates and four maps; usual light offsetting from plates but an excellent copy in old polished calf, spine renewed in style retaining original label. London, John Murray, 1825.

First edition: a significant publication, which collects various papers that sprang from members of the literary and cultural coterie that flourished in Macquarie's New South Wales. It is of high exploration interest and includes the only edition of John Oxley's journal of the exploration of Moreton Bay and the discovery of the Brisbane River together with John Uniacke's account of the same expedition; as well there is the only published account of Allan Cunningham's exploration to the Liverpool Plains, Currie's excursion to the southward of Lake George, the journal of Field's own trip across the Blue Mountains, and numerous other papers. The volume also includes the third – and first realistically available – edition of Field's own *First Fruits of Australian Poetry*, the first volume of verse published in New South Wales (in 1819 and, with additions, in 1823).

This copy once belonged to the great woman collector, Frances Mary Richardson Currer.

Ferguson, 1009; Wantrup, 108. **$2850**

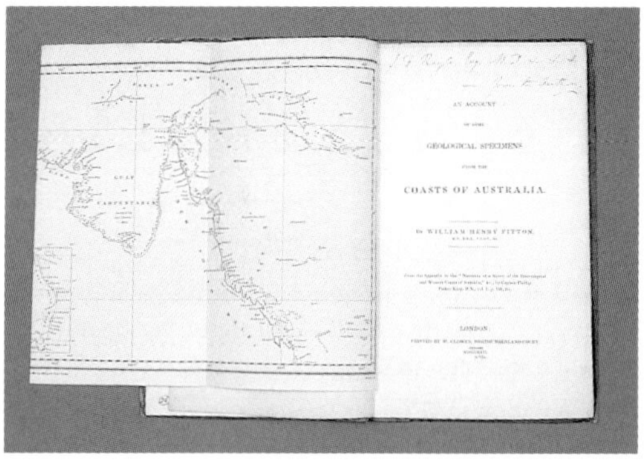

181. Fitton

PRESENTATION COPY OF THE SEPARATE PRINTING

181. FITTON, William. An Account of some geological specimens from the coast of Australia...

Octavo, folding frontispiece map and three lithograph plates; a very good copy in original grey boards, later spine. London, W. Clowes, 1826.

First edition, and particularly rare: the geological results of an important coastal exploration. Fitton's *Account*, which describes the geological discoveries made during King's coastal voyage, also appears partly reset as the appendix to King's official account of the voyage (*Narrative of a Survey*: catalogue number 189). This separate printing was evidently for very limited private distribution, and most of the few copies that have appeared on the market, including this one, have been presentation copies.

Ferguson, 1071. **$3850**

FIRST PUBLICATION OF THE FRANCIS NARRATIVE

182. FLINDERS, Matthew. Matthew Flinders' Narrative of his Voyage in the Schooner Francis: 1798, preceded and followed by notes on Flinders, Bass, the wreck of the Sidney Cove, &c. by Geoffrey Rawson...

Folio, with a map and wood-engravings by John Buckland Wright; owner's presentation inscription in ink on front flyleaf; an excellent copy, uncut in the original green cloth, gilt ship ornament on the upper cover, top edge gilt. Golden Cockerel Press, 1946.

One of 750 numbered copies of the important first publication of the manuscript in the State Library of Victoria. This gives Flinders's account of his 1788 voyage in the *Francis* that led to his exploration of Bass Strait, and to the discovery of the Illawarra coal seams by Bass. **$750**

ONE OF 100 COPIES IN MOROCCO

183. FLINDERS, Matthew. Matthew Flinders' Narrative of his Voyage in the Schooner Francis ...

Folio, map and wood-engravings; a fine copy, uncut in the original full green morocco gilt by Sangorski, slipcase. Golden Cockerel Press, 1946.

The scarce de luxe issue, one of only one hundred numbered copies in full morocco. **$4250**

SOUTH AUSTRALIAN VIEW

184. [FLINDERS] WESTALL, William. Port Lincoln (taken from the South).

Engraving, 250 x 340 mm.; mounted. London, Ackermann & Co., 1835.

Fine, rare image of the South Australian coast by the artist on Flinders's circumnavigation of Australia. Some thirty years after the voyage, Westall became involved in the plans to form a new colony in South Australia. A series of six views by Westall was planned, but only this one was actually issued, to accompany the *First Report of the Directors of the South Australian Company* published in 1836. $1750

185. Grant

THE LADY NELSON

185. GRANT, James. The Narrative of a Voyage of Discovery, performed in His Majesty's Vessel The Lady Nelson, of sixty tons burthen, with sliding keels, in the years 1800, 1801 and 1802, to New South Wales...

Quarto, large folding plan (carefully laid down on linen), chart, coloured plate of a cockatoo and five other engravings including a portrait of Bennelong; complete with the rare "List of the Encouragers"; a handsome copy in old half calf, spine ornately gilt with double labels. London, C. Raworth, 1803.

First edition: one of the most important of the early Australian coastal voyages, and "a rare and most desirable item" (Davidson).

Lady Nelson was the first ship to be built with sliding keels, to facilitate the exploration of shallow waters; Grant sailed her through Bass Strait, the first to do so from the west, arrived at Sydney to discover that he had missed Flinders – to whom he was to deliver the ship – and was sent back to survey the south-western coast of the continent. He concentrated his efforts on Bass Strait: the chart that he published here was the first to appear of the newly-discovered Victorian coast.

On his return to Sydney, Grant was sent to examine the Hunter River, and it was as a result of his report that Newcastle was established. Two engravings in this book illustrate *Lady Nelson's* exploration of the Hunter River: they were probably engraved after sketches by the colony's first professional artist, John William Lewin, who accompanied the expedition.

Ferguson, 375; Hill, p. 127; Davidson, pp. 125-6; Wantrup, 75. $11,000

CLASSIC VOLUMES OF AUSTRALIAN LITERATURE

186. GREY, Captain George. Journals of Two Expeditions of Discovery in North-West and Western Australia...

Two volumes, octavo, with 22 plates and two large maps, illustrations in the text; a good copy, recased in the original cloth and joints of one volume repaired, slightly faded. London, T. and W. Boone, 1841.

First edition of this famous account of Grey's important expeditions to examine the otherwise neglected north-west interior of the continent. The book includes scientific appendices on birds by John Gould, mammals, reptiles, and amphibians by John Edward Gray, and insects by Adam White; some of the plates illustrate the extraordinary Aboriginal rock paintings found on the first expedition.

"Grey's honest enthusiasm and the wealth of information these volumes contain on the countryside he crossed make these classic volumes of Australian literature, recording the first impressions of the Australian landscape of a man who was to be of considerable importance in the history of Australia..." (*People, Print and Paper*).

Ferguson, 3228; Wantrup, 131. $3850

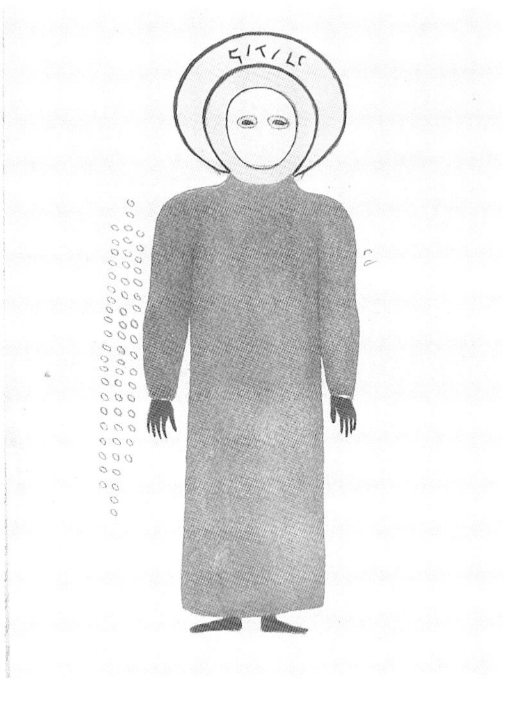

186. Grey

THIRD & STRONGEST

187. HUME, Hamilton. A Brief Statement of Facts in Connection with an Overland Expedition from Lake George to Port Phillip, in 1824... Third Edition: with Addenda.

Octavo, with a frontispiece portrait of Hume; a very good copy in the original limp cloth with printed label. Yass, J.J. Brown ,Courier Office, 1874.

Hamilton Hume's own account of the discovery of the rich Port Phillip district in 1824, an expedition that was previously only generally publicised in the 1837 edition of his companion Hovell's notebook. Hume gave his first public account in 1855, when newspaper reports gave him the impression that Hovell was claiming all the credit for their discoveries. Both men spent their final years in a pamphlet war.

This third edition was published shortly after Hume's death in 1873, with a new addenda containing several letters to confirm Hume's statement "of facts" including one by a survivor of this famous overland journey from Lake George to Port Phillip.

Ferguson, 10665; Wantrup, 113a. **$3250**

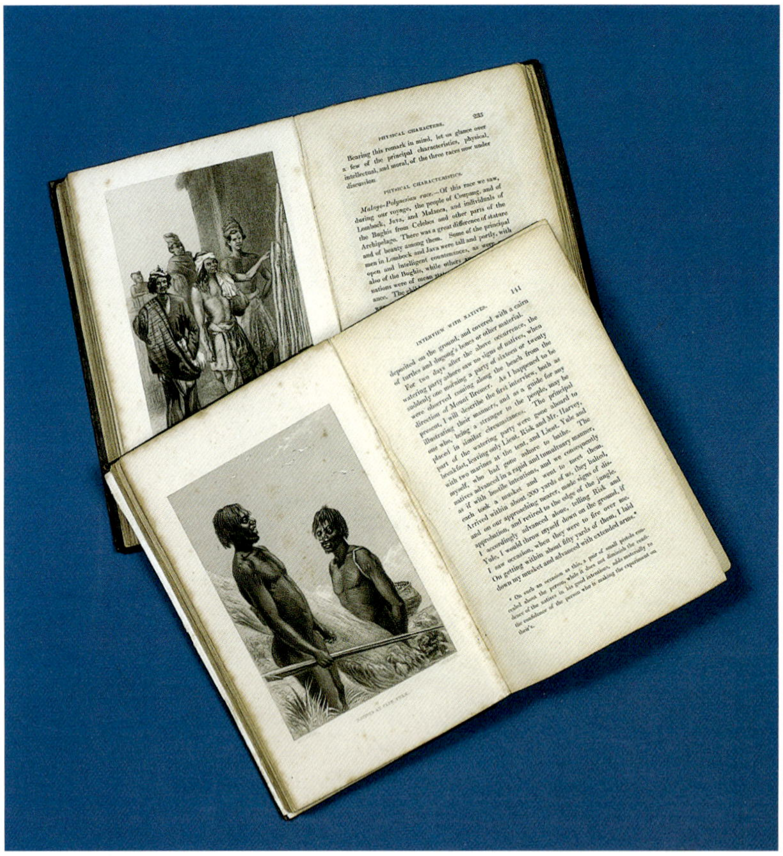

188. Fly

THE FLY VOYAGE

188. JUKES, J. Beete. Narrative of the Surveying Voyage of HMS Fly, commanded by Captain F.P. Blackwood, R.N. in Torres Strait, New Guinea, and other islands of the Eastern Archipelago...

Two volumes, octavo, with many engravings, folding map at the end of each volume; some light spotting but a good copy, with the advertisements, in the original blue-grey blindstamped cloth, spines faded a little but uniformly. London, T. & W. Boone, 1847.

First edition of this important surveying voyage, the last of the series of early coastal explorations. Joseph Beete Jukes was naturalist to the *Fly* expedition and wrote this official narrative. The survey of Torres Strait and of the Great Barrier Reef, as well as the various New Guinea explorations, were all of great importance. The proper scientific understanding of the Barrier Reef could not begin until the completion of the survey, which Jukes charted for the first time in detail. As well as charting, his close examination of the reef was significant, and his chapter on the subject "is an invaluable record. His observations strongly supported Darwin's theory of the formation of coral reefs..." (Davidson).

Davidson, pp. 129–30; Ferguson, 4549; Hill, p. 159; Wantrup, 92a. **$7200**

189. King

THE RARE 1826 ISSUE

189. KING, Phillip Parker. Narrative of a Survey of the Intertropical and Western Coasts of Australia. Performed between the years 1818 and 1822... With an Appendix, containing various subjects relating to Hydrography and Natural History...

Two volumes, octavo, with a folding chart, 13 plates and a folding engraving; a couple of leaves at the end of vol. 1 lightly stained; bound without the colophon leaf required by Ferguson at the end of vol. 2, but complete with the final leaf of errata (intended to be cut to form two errata slips); contemporary calf gilt (surface of leather somewhat cracked), recently well rebacked. London, John Murray, 1826.

First edition, first issue: the very rare form of this important book with the publication date of 1826 on both title-pages. The book was properly published in 1827, with the title-pages reset to show that date. The only complete copy of this variant known to Ferguson was in the State Library of Victoria; a number of copies have appeared since that time, but this remains one of the more difficult books to find.

Admiral Phillip Parker King, Australian-born son of the third governor Philip Gidley King, became the Navy's leading hydrographer. His coastal voyages and Oxley's expeditions inland were the great expansionary undertakings of the Macquarie era. Despatched to complete Flinders's interrupted survey and firmly to establish Great Britain's claim to the north coast, King charted the greater part of the west, north and north-east coasts and also carried out important surveys in the area of the Barrier Reef. His hydrographical work is still the basis of many of the modern charts for the areas he surveyed.

This copy has the book plate "College, Regent's Park" and the binder's label of Henington, Ingham Court, Fenchurch St.

Not in Abbey; Davidson, pp. 127-8; Ferguson, 1084; Wantrup, 84a. **$12,500**

QUEENSLAND RELIEF EXPEDITION

190. LANDSBOROUGH, William. Journal of Landsborough's Expedition from Carpentaria, in search of Burke and Wills. With a Map showing his Route.

Octavo, frontispiece and large folding map, handcoloured in outline; original yellow printed boards with linen spine, a very good copy. Melbourne, F.F. Baillière, 1862.

The first public printing of Landsborough's narrative, based on official papers and reports. Landsborough led the Queensland relief expedition, one of four relief expeditions in search of Burke and Wills; the discoveries he made contributed greatly to opening up pastoral lands in outback Queensland.

This is the quite rare superior form of the Baillière issue of the book, with the map handcoloured in outline and with the 16-page botanical appendix by von Mueller.

This issue not in Ferguson; Maria, 109; Wantrup, 174d. **$3250**

PRESENTATION COPY

191. [LANDSBOROUGH] BOURNE, George. Bourne's Journal of Landsborough's Expedition from Carpentaria, in search of Burke and Wills.

Octavo; original yellow printed wrappers, the title repeated on the upper wrapper; wrappers a bit stained, some wear around the spine, a few spots in the text, but still a good unsophisticated copy. Melbourne, H.T. Dwight, 1862.

First edition: a very scarce book, especially in its original wrappers. This is especially appealing as a presentation copy, inscribed on the front wrapper to "John MacGregor Esq. MLA, with the Author's Compts".

This is the only independent account of the Landsborough expedition, by its second-in-command; Bourne is quite critical of his leader. Quite as important to the history of the expedition as Landsborough's own narrative, this is in fact much harder to find.

Ferguson, 7303; Maria, 15; Wantrup, 176a. **$4500**

ORIGINAL WRAPPERS

192. [LANDSBOROUGH] LAURIE, James Stuart, editor. Landsborough's Exploration of Australia from Carpentaria to Melbourne…

Octavo, with a double-page map; a good copy in the original pink wrappers, neatly repaired. London, Thomas, Murby, [1866].

First edition, the issue in wrappers of the scarce and attractive 1866 London edition of Landsborough's official account of his expedition. "The volume was chiefly dictated with the help of materials recorded in the log-book of the journey, by Mr. Landsborough himself" (Ferguson 11330).

Ferguson, 11330; Maria, 112; Wantrup, 175. **$1250**

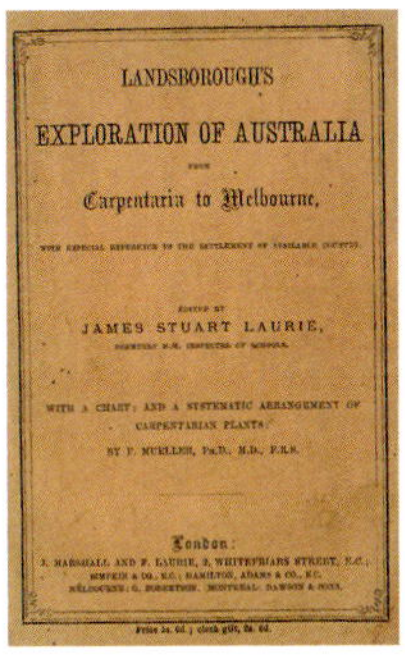

192. Laurie

LEICHHARDT'S FIRST EXPEDITION

193. [LEICHHARDT] HODGSON, Christopher Pemberton. Reminiscences of Australia, with Hints on the Squatter's Life.

Duodecimo, frontispiece and folding map (foxed); neatly repaired original cloth, light external wear, a very nice copy. London, Simpkin and Marshall, 1846.

A scarce and important account relating to Leichhardt's first expedition and the search party sent out for him, and a good account of the hard life of a squatter on the north-east frontier.

Ferguson, 4310; Wantrup, 140. **$1250**

HANDSOME SET, WITH THE MAPS

194. LEICHHARDT, Ludwig. Journal of an Overland Expedition in Australia, from Moreton Bay to Port Essington, a distance of upwards of 3000 miles, during the years 1844-1845.

Two volumes, octavo, with a frontispiece and six aquatint plates (one folding) and many woodcut illustrations; the three sheet maps laid down on fine old linen in a separate volume with matching binding; some scattered foxing, a very good set finely bound in nineteenth-century half red morocco. London, T. & W. Boone, 1847.

A most handsome set of this classic of Australian land exploration, with the nineteenth-century ownership stamp of Arthur Milson (a member of one of the first families to settle Sydney's North Shore; he lived at Walaya in Milson's Point, the area named after his family).

Leichhardt's meandering journey from Brisbane to Port Essington, the military base on the far north coast, ranks among Australia's most successful

194. Leichhardt

exploring expeditions. His scientific and geographic work were greatly admired during his lifetime, but his reputation as an expeditionary leader has been criticised since. His life and unknown death have become part of Australian mythology.

In the course of their harrowing fourteen-month journey – one of the longest inland expeditions undertaken – Leichhardt and his party travelled over three thousand miles of completely unknown country. The expedition opened up a vast and valuable area of the interior which became among the most valuable pasture on the continent. It was generally believed that Leichhardt and his party had all perished, so when they appeared at Port Essington they were welcomed as if they had returned from the dead. Leichhardt returned to Sydney in March 1846, as a national hero.

This issue with the separate maps is very scarce, and not often found in such a handsome and early binding.

Abbey, 579 (text only); Ferguson, 4571 (text only); Wantrup, 138a and 139b. **$16,000**

THE MCKINLAY EXPEDITION

195. McKINLAY, John. McKinlay's Journal of Exploration in the Interior of Australia...

Octavo, with the three loose folding maps; original brown cloth, repaired. Melbourne, F.F. Bailliere, 1862.

First edition of the first public account of McKinlay's expedition. John McKinlay led the South Australian expedition to search for Burke and Wills in 1861. They reached Cooper's Creek after a two-month journey, and found the grave of one of Burke's party, then explored north of Lake Eyre where they discovered much pastoral and grazing land. This description of the expedition is from McKinlay's journal.

Ferguson, 12057; Maria, 122; Wantrup, 177. **$1850**

THE 'FORMAL' ACCOUNT

196. [McKINLAY] DAVIS, John. Tracks of McKinlay and Party Across Australia... Edited from Mr. Davis's manuscript journal... by William Westgarth...

Octavo, with a tinted lithograph frontispiece and 13 tinted lithograph plates, large folding map loose in the endpocket; a very good copy, neatly recased in the original green cloth. London, Sampson Low, 1863.

First edition of this substantial London publication based on the papers of John Davis, a member of McKinlay's 1861 expedition in search of Burke and Wills. This London edition is generally seen as the 'formal' account of the expedition, and was edited by William Westgarth, who also gives an introductory section on recent Australian inland exploration. The tinted lithograph plates show in vivid detail the trials the party encountered in very remote territory.

Ferguson, 9005; Maria, 57; Wantrup, 180. **$3250**

ORIGINAL CLOTH

197. MITCHELL, Sir Thomas Livingston. Three Expeditions into the Interior of Eastern Australia.

Two volumes, octavo, with lithographic title-pages, large folding map, and 51 plates, a few folding or coloured, and small text illustrations; a particularly good copy in original dark green cloth. London, T. & W. Boone, 1839.

Second, revised and extended edition of Mitchell's narrative of his important first three expeditions. The first expedition (1831-2) surveyed the area between the Castlereagh and the Gwydir Rivers; the second (1835) set out to trace and chart sections of the Murray and Darling Rivers; the third (1836) was intended to complete the survey of the Darling, and this proved to be of the greatest significance when Mitchell, having established that the Murray and the Darling were confluent, decided to explore country to the south that looked more promising. This detour culminated in the discovery of the rich lands of Australia Felix in 1836.

Ferguson, 2811 (misdescribed); Wantrup, 125 (misdescribed). **$3500**

MITCHELL'S LAST EXPEDITION

198. MITCHELL, Thomas Livingstone. Journal of an Expedition into the Interior of Tropical Australia...

Octavo, 12 lithographed plates and seven maps (four folding); a fine copy handsomely bound in contemporary polished calf gilt. London, Longmans, 1848.

First edition of Mitchell's account of his last expedition, with a fine series of tinted lithograph views after his own drawings. Mitchell's work as a topographical artist can be seen at its best here. He started out in December 1845 on this expedition to search for the north-west flowing river and for an overland route to Port Essington and, although the expedition was not entirely successful, he did explore large tracts of previously unknown territory.

Ferguson, 4828; Wantrup, 129. **$2850**

FIRST DETAILED DESCRIPTION OF THE INTERIOR

199. OXLEY, John. Journals of Two Expeditions into the Interior of New South Wales...

Quarto, six aquatint engravings, two of which are coloured, three folding maps and two folding tables; an excellent copy in contemporary half calf (joints neatly repaired). London, John Murray, 1820.

An attractive copy of the first edition of Oxley's great narrative of his two major expeditions, the first detailed description of the Australian interior and the earliest book devoted to Australian inland exploration.

This is physically the most handsome of all Australian exploration journals, a finely produced quarto volume whose appearance recalls the earlier quartos of the First Fleet chroniclers and was clearly designed to rank on the shelf with the books by his illustrious predecessors like Phillip, Hunter, Tench, Collins, White, Grant and Flinders. It "is undoubtedly the chief book-making achievement of the Macquarie period..." (*People, Print and Paper*).

Abbey, 568; Ferguson, 796; Wantrup, 107. **$9850**

THE RATTLESNAKE VOYAGE

200. [RATTLESNAKE] MACGILLIVRAY, John. Narrative of the Voyage of HMS Rattlesnake, commanded by the late Owen Stanley, R.N., F.R.S. &c. during the years 1846-1850...

Two volumes, octavo, 13 plates and a folding chart; a little foxing but a nice copy in period style half calf. London, T. & W. Boone, 1852.

First edition: the last and most difficult to find of the great exploration journals published by Boone during the heroic age of Australian exploration.

The *Rattlesnake* was part of the important series of hydrographical voyages undertaken by the Admiralty in the late 1840s to chart accurately the Great Barrier Reef and north coast and the southern coast of New Guinea. The *Rattlesnake* continued the work that began with Lort Stokes on the *Beagle* and was continued by Blackwood on the *Fly* and *Bramble*. The ship's complement was distinguished by the presence of the naturalist, T.H. Huxley, and the marine artist, Sir Oswald Brierly.

Describing a voyage important in its own right, Macgillivray's account is of even greater value for its reprinting of William Carron's extremely rare account of the tragic Kennedy expedition. It was the *Rattlesnake* which transported Kennedy and his large party to their point of departure on the Queensland coast.

Ferguson, 11972; Wantrup, 145 (14 plates in error). **$8250**

200. Macgillivray

201. Brierly

ANCHORED OFF THE LOUISIADE ARCHIPELAGO

201. [RATTLESNAKE] BRIERLY, Sir Oswald Walter. HMS Rattlesnake in the Louisiade Archipelago.

Pencil and grey wash drawing, 180 x 260 mm.; in fine condition, framed. Inscribed "Morning July 18th 1849 Louisiade".

A fine portrait of the surveying ship *Rattlesnake* in northern Australian waters, on the voyage to chart the Great Barrier Reef and parts of New Guinea, continuing work started by Lort Stokes on the *Beagle*, and continued by Blackwood on the *Fly* and *Bramble*.

On board the *Rattlesnake* during the expedition of 1846-1850 under the command of Owen Stanley were the naturalist T.H. Huxley and the distinguished artist Sir Oswald Brierly. Brierly's marine paintings are justly famous: see also catalogue numbers 247 for his images of the *Mæander* and 257 for the *Galatea*.

This poignant drawing depicts the ship at a particular moment on 18 July 1849 when, as the published account tells us, "we hauled out south and anchored in 22 fathoms, about eight miles from the nearest of the Calvados... from this anchorage we enjoyed an extensive view of the south-eastern portion of the Louisiade Archipelago".

Brierly's original drawings from this voyage are exceedingly scarce. **$5750**

7: LA PEROUSE & THE SEARCH

FULL OFFICIAL ACCOUNT OF LA PEROUSE

202. LA PEROUSE, Jean François Galaup de. Voyage de la Pérouse autour du Monde, publié conformément au décret du 22 Avril 1791, et rédigé par M.L.A. Milet-Mureau.

Four volumes, quarto, and an atlas folio, frontispiece portrait of La Pérouse, atlas with engraved title and 69 engraved maps and plates (21 folding); a fine large copy of the text, completely uncut, the atlas also large and fine with the engraved plates all in strong and fresh impressions; quarter calf over old marbled boards with vellum tips, double labels, an excellent set. Paris, Imprimerie de la République, An V (1797).

First edition of one of the finest narratives of maritime exploration ever written. The official description of the tragic La Pérouse voyage ranks with the official narratives of the Cook voyages in aesthetic terms. The beautiful engravings here are in fact even more splendid than those published to illustrate the Cook accounts.

The tragedy of the disappearance of the entire La Pérouse expedition overshadows what was in reality a great Pacific voyage; its failure certainly curtailed French expansion into the Pacific and Asia.

The most significant results of the voyage are the charts of the imperfectly known Asiatic side of the Pacific. En route to Kamchatka, La Pérouse was the first to safely navigate and chart the Japan Sea and the strait between the Island of Sakahlin and the northernmost island of Japan, which bears his name. At Kamchatka he received instructions to proceed to Australia to assess the extent of British plans in New South Wales. Travelling via Samoa where he discovered Savaii, Manono, and Apolima Islands in December 1787, and through the Tongan group, he arrived at Botany Bay in January 1788, just hours after Governor Phillip had arrived with the First Fleet.

La Pérouse's habit of forwarding records whenever opportunity offered ensured their survival. The first portion was forwarded by sea from Macao; the second (Macao to Kamchatka) went overland with de Lesseps, and the final despatches by courtesy of Governor Phillip from Botany Bay. The folio Atlas contains magnificent maps of Russian Asia, Japan, the Pacific Northwest Coast, San Francisco, and Monterey.

This is a fine set of this important voyage.

Borba de Moraes, p. 449; Cowan, p. 383; Davidson, p. 100; Ferguson, 251; Hill, p. 173; Judd, 102; Lada-Mocarski, 52. **$26,500**

202. La Pérouse

203. La Pérouse

ENGLISH EDITION

203. LA PEROUSE, Jean Francois Galaup de. A Voyage Round the World... performed in the Years 1785, 1786, 1787 and 1788...

Three volumes, octavo, with a portrait frontispiece and 41 engravings and maps, many of them folding; some foxing but a good copy, with early owners' signatures, in contemporary tree calf, London, J. Johnson, 1798.

One of two competing English editions of the official La Pérouse narrative, published almost together just a year after the first French edition.

Ferguson, 270; Hill, p. 471; McLaren, 4. **$2250**

RARE ABRIDGEMENT

204. LA PEROUSE, J.F.G. de la. Rélation abrégée du voyage de La Pérouse, pendant les années 1785, 1786, 1787 et 1788…

Octavo, with an engraved portrait, five plates and a folding world map; contemporary German half calf, spine rubbed. Liepzig, 1799.

Abridged (but quite long) account of La Pérouse's voyage, published two years after the official edition. The book was aimed at voyage enthusiasts, and advertises itself on the title-page as a supplement to La Harpe's series of voyage accounts. Only the National Library and State Library of Victoria copies are recorded by Ferguson.

"The Leipzig imprint is probably fictitious. According to Engelmann, the anonymous compiler was K. Villers…" (R. du Rietz).

Ferguson, 291; Kroepelien, 851; not in the Hill catalogue; not in McLaren. **$1350**

205. Touanne

THE LA PEROUSE MONUMENT

205. [LA PEROUSE] TOUANNE, Vicomte Edmond de la. Monument élevé à la mémoire de la Pérouse.

Handcoloured lithograph, 190 x 280 mm.; mounted. Paris, 1826.

A fine view of the La Pérouse monument, built in Sydney to honour the great French explorer during the visit of the *Thétis* under Hyacinthe de Bougainville in 1825. The building of the monument showed a suitable piety from the French explorers nearly forty years after La Pérouse's disappearance, and especially from Hyacinthe de Bougainville himself, heir to a great French naval tradition as the son of Louis de Bougainville, the first French circumnavigator.

Edmond de la Touanne, who prepared this fine lithograph view of the monument, was a friend and protégé of de Bougainville and sailed on the expedition as lieutenant de vaisseau. Because of the haste with which the expedition was manned, no official artist was sent; as Bougainville remarks, no pictorial record of the expedition would have survived but for de la Touanne's sketches, which were published as lithographs in the official record of the expedition and in de la Touanne's own album. **$1250**

PETER PARLEY

206. [LA PEROUSE] GOODRICH, Samuel Griswold. Peter Parley's Tales of the Sea.

Small square octavo, numerous wood-engravings in the text; a little stained throughout, and extremities worn but a good copy in publisher's quarter roan. Philadelphia, Thomas Cowperthwait, 1848.

Samuel Goodrich was a prolific American author who, under the pen name Peter Parley, wrote and published some two hundred titles for children, some of them – like this one which includes the story of La Pérouse – based on actual accounts of Pacific exploration. This little juvenile is also of interest for the four chapters on whaling. **$250**

SEARCH FOR LA PEROUSE

207. LABILLARDIERE, Jacques Julien Houton de. Rélation du Voyage à la recherche de La Pérouse…

Two volumes, large quarto, with folio atlas containing 44 engraved plates and maps; a uniform set in original buff glazed boards with black calf labels, edges bumped and rubbed, wear to extremities of spines; binding of the atlas stained along upper joint without affecting the book internally which is fine and clean; a very good set, completely untrimmed. Paris, H.J. Jansen, An VIII de la République Françoise [1800].

First edition: the first published results of this voyage in search of La Pérouse. This narrative by the naturalist on the d'Entrecasteaux expedition was the first to appear in print as its author was politically on the correct side; the alternative account, by the royalist d'Entrecasteaux, did not appear until after the restoration of the monarchy.

207. Labillardière

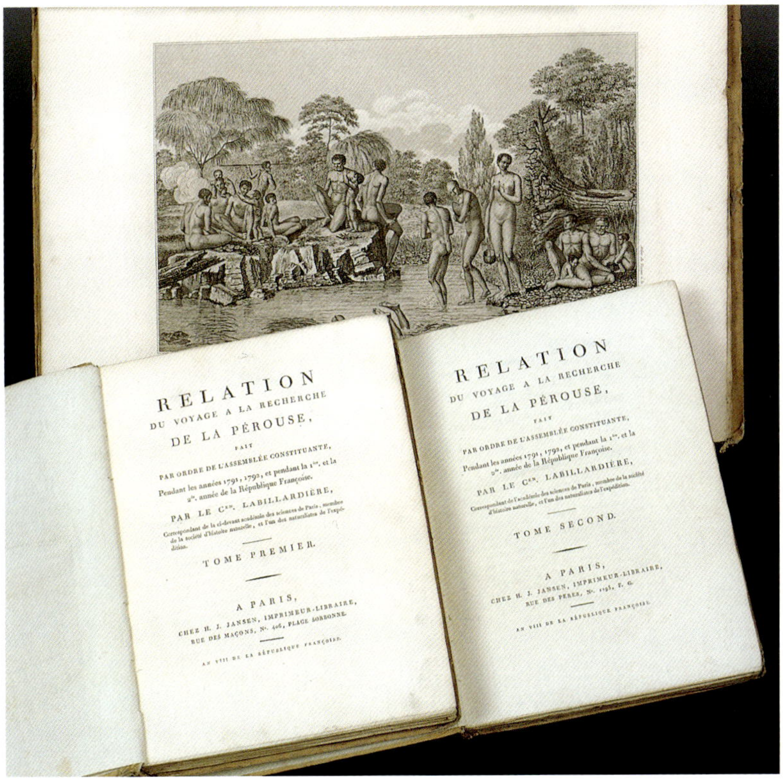

The expedition had no success in its main aim – to discover the fate of La Pérouse – but was important for scientific and geographical discoveries in its own right. Australia was circumnavigated twice, and the islands surrounding investigated for traces of La Pérouse. The work is particularly interesting for its descriptions (and illustrations) of Tasmania, Tonga, New Caledonia and New Guinea, and the Atlas contains outstanding and important views of these areas by the official artist Piron. Included is the first large depiction of the Black Swan. The fourteen botanical plates, all by or produced under the direction of Redouté, the most famous of all botanical artists, include two of Eucalypts and two of Banksias.

Ferguson, 307; Hill, p. 469; Kroepelien, 697; McLaren, 51. **$12,000**

BEST ENGLISH EDITION

208. LABILLARDIERE, Jacques Julien Houton de. Voyage in search of La Pérouse, performed by order of the Constituent Assembly, during the years 1791, 1792, 1793, and 1794.

Quarto, with 46 engraved maps and plates; embrowning to several gatherings including the title and preliminaries, and some foxing, still a good copy in contemporary marbled calf, cleverly rebacked retaining original spine. London, John Stockdale, 1800.

The best and most complete English edition, this was the first English translation to appear, issued simultaneously with a smaller octavo.

Ferguson, 309; this edition not in Hill; Kroepelien, 700; McLaren, 55. **$5500**

210. Labillardière

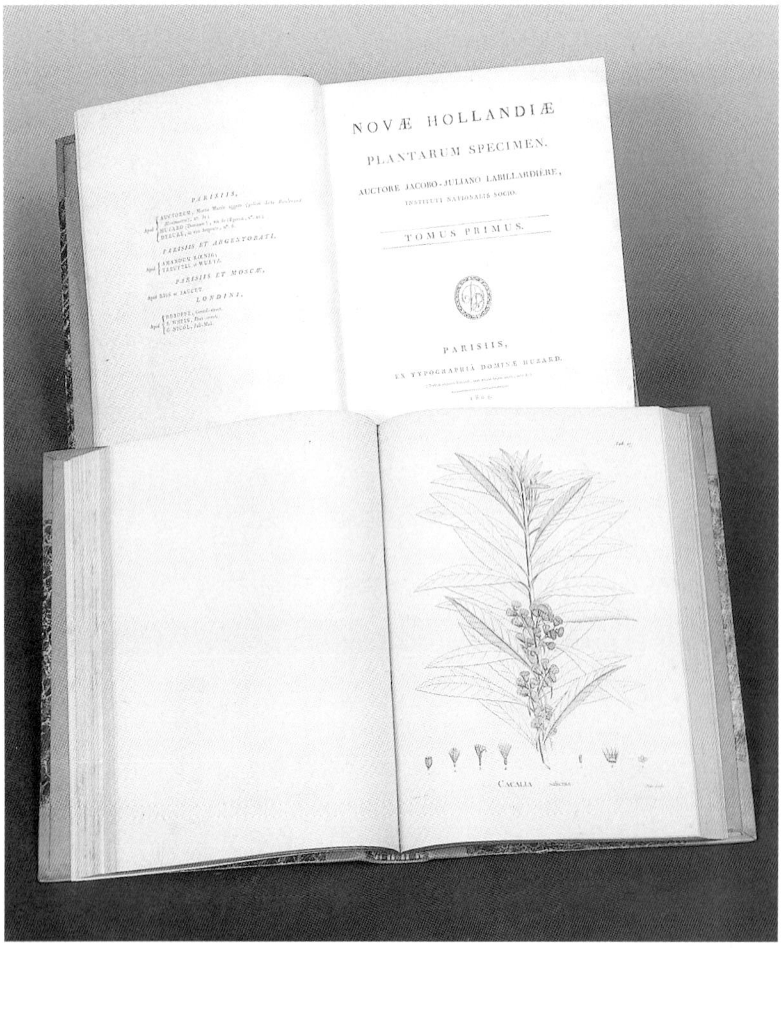

THE ENGLISH OCTAVO

209. LABILLARDIERE, Jacques Julien Houton de. Voyage in search of La Pérouse...

Two volumes, octavo, with 46 plates and maps; sporadic staining to the text; contemporary half calf, rather worn. London, John Stockdale, 1800.

The smaller-sized English edition published by Stockdale alongside the more expensive quarto (see above).

Ferguson, 310; Hill, p. 168; McLaren, 57. **$1500**

FIRST COMPREHENSIVE FLORA

210. LABILLARDIERE, Jacques Julien Houton de. Novæ Hollandiæ Plantarum Specimen.

Two volumes, royal quarto, with 265 engraved plates; some slight spotting, but an excellent copy in later half calf over old marbled boards, Paris, Huzard, 1804.

The most consequential scientific result of the French expeditions up to Baudin: the first comprehensive publication of the Australian flora, with an outstanding series of botanical engravings, this rare publication describes the botanical discoveries made in the course of d'Entrecasteaux's 1791-1795 expedition in search of La Pérouse. Labillardière's was the first scientific and extensively illustrated study of the botany of Australia, so thorough that it remained unsurpassed for decades.

It is remarkable that publication was ever in fact carried out. On the expedition's voyage home, the crew split into two parties, the revolutionary faction led by Labillardière, who then took control of the expedition. Perhaps it was his prominent position that led the Dutch to confiscate his natural history specimens whilst the expedition was in Indonesia in 1794. On their return to Paris, Labillardière pleaded his case with the French Government, who in turn asked for the return of the specimens that had subsequently been sent to England. Sir Joseph Banks became involved; eventually the collection was restored to its owner under a flag of truce.

Publishing work started immediately; the illustrations were prepared in Paris from observations made, and from the "precious" specimens collected during the voyage. The leading French botanical artists working on the project included Redouté himself, the most famous flower-painter of his day, painter to the cabinet of Marie Antoinette and later under the patronage of Josephine Bonaparte. Labillardière was closely involved, as were Turpin and Poiteau, two of Redouté's pupils, considered equal to their master in technical skills. The illustrations are finely stippled engravings, a method not previously used for flower-work that Redouté may have learned in England, where it was developed by Bartolozzi.

This superb and important work, begun under one political régime and completed under another, shows the artistically enriching link that existed between naturalists, scientists and artists of different nationalities, even in spectacularly unsettling times.

Ferguson, 395; 'Great Flower Books', p. 63; McLaren, 65; Nissen, 1116; Pritzel, 4963; Stafleu & Cowan, TL2 4071.

$23,000

THE COMMANDER'S ACCOUNT

211. D'ENTRECASTEAUX, A.-R.-J. de Bruni. Voyage de Dentrecasteaux, envoyé à la Recherche de La Pérouse. Publié par ordre de sa Majesté l'Empereur et Roi... Rédigé par M. de Rossel, ancienne Capitaine de vaisseau.

Two volumes, quarto (with 32 folding engraved plates in volume I), and atlas, imperial folio, containing 39 charts and maps (29 double folding); neat paper repairs to titles (where ownership marks have been excised at an early date) a good set, the text in contemporary half morocco, the atlas in later half calf over original mottled boards. Paris, Imprimerie Impériale, [1807-]1808.

The official commander's account of the search for La Pérouse, published posthumously eight years after Labillardière's narrative of the same expedition.

The twelve magnificent maps of Western Australia and Tasmania in the atlas record much of these coasts accurately for the first time, and are among the most significant ever made. Of prime importance to Tasmania, it was d'Entrecasteaux's explorations of 1792 and 1793 which focussed attention on the Derwent River area (now Hobart) as a suitable place for settlement.

211. D'Entrecasteaux

A very scarce voyage account, of great relevance to Australia and the Pacific.

Davidson, pp. 104-5; Ferguson, 443, 461; Hill, p. 97; McLaren, 49-50; Wantrup, 64a-64b. **$18,500**

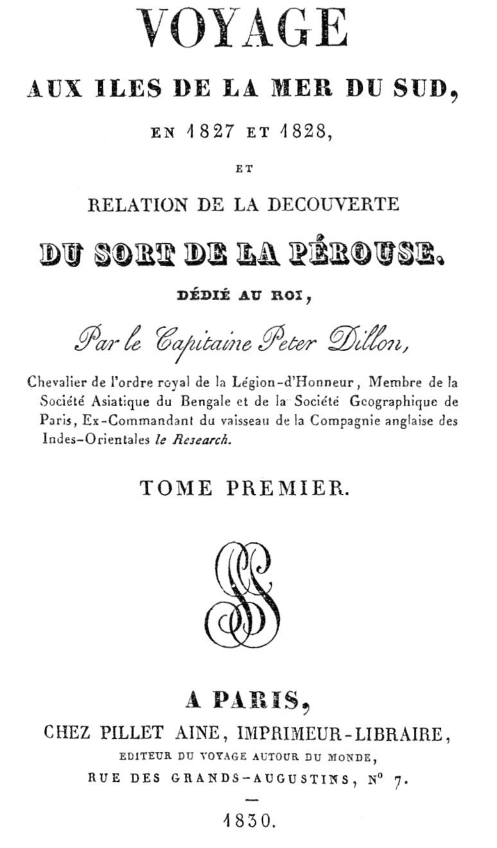
VOYAGE
AUX ILES DE LA MER DU SUD,
EN 1827 ET 1828,
ET
RELATION DE LA DECOUVERTE
DU SORT DE LA PÉROUSE.
DÉDIÉ AU ROI,
Par le Capitaine Peter Dillon,
Chevalier de l'ordre royal de la Légion-d'Honneur, Membre de la Société Asiatique du Bengale et de la Société Géographique de Paris, Ex-Commandant du vaisseau de la Compagnie anglaise des Indes-Orientales *le Research.*
TOME PREMIER.
A PARIS,
CHEZ PILLET AINE, IMPRIMEUR-LIBRAIRE,
EDITEUR DU VOYAGE AUTOUR DU MONDE,
RUE DES GRANDS-AUGUSTINS, N° 7.
1830.

212. Dillon

THE ANSWER AT VANIKORO

212. [LA PEROUSE] DILLON, Peter. Voyage aux Iles de la Mer du Sud, en 1827 et 1828, et Rélation de la Découverte du Sort de La Pérouse.

Two volumes, octavo, three lithographs (two folding) and chart of the island of Mannicolo; an excellent large copy, completely uncut, in contemporary French quarter green roan and glazed boards, rubbed. Paris, Pillet, 1830.

First French edition; with fine plates by Engelmann, the inventor of lithography.

Forty years after the mysterious disappearance of La Pérouse, Peter Dillon, a sandalwood trader, called at the Solomons; when a silver sword guard was shown to him, he suspected that he might have stumbled on the solution. At Vanikoro found many relics including a portion of the stern of the *Boussole*, ships' bells stamped "Bazin m'a fait", monogrammed silver, metal fragments and mill stones known to have been aboard. One native (shown here on a folding plate) had a glass piece from a thermometer in his nose. Dillon was made a Chevalier of the Legion of Honour, his expenses defrayed and granted a pension. His book was commercially successful, appearing in English and French editions promptly and eventually in a number of translations.

Ferguson, 1336; Hill, p. 84; Kroepelien, 297; McLaren, 73.

$3250

8: BLIGH & THE BOUNTY

VOYAGE OF THE BOUNTY

213. BLIGH, William. A Voyage to the South Sea, undertaken by Command of His Majesty, for the purpose of conveying the Bread-fruit Tree to the West Indies, in His Majesty's Ship the Bounty... including an account of the mutiny...

Quarto, engraved frontispiece portrait and seven engraved plates and charts, mostly folding; a fine copy in contemporary calf, expertly respined by Aquarius. London, George Nicol, 1792.

First edition of one of the greatest voyage classics: the official narrative by William Bligh of his doomed voyage in the *Bounty.* This is the full account of the voyage, and includes a slightly altered version of Bligh's description of the mutiny and the open-boat voyage, which had been rushed into print separately two years earlier.

The open-boat voyage, one of the most famous feats of seamanship of all time, was also notable for the coastal discoveries made almost accidentally in the course of the desperate voyage. Bligh's description here is accompanied by his important engraved chart of discoveries made on the coast of present-day Queensland. His achievement in charting large sections of the coast under conditions of terrible hardship partly completed the work of Cook himself on the Australian east coast.

Ferguson, 125; Hill, p. 27; Kroepelien, 93; O'Reilly-Reitman, 550. **$9500**

215. Minutes of the Proceedings

THE OPEN-BOAT VOYAGE

214. BLIGH, William. BACH, Prof. John (editor). The Bligh Notebook... 'Rough account – Lieutenant Wm. Bligh's voyage in the Bounty's Launch...

Two volumes, facsimile and accompanying explanatory text; original calf and quarter calf, silk-lined buckram box. Canberra, National Library of Australia, 1986.

One of only 500 numbered copies published: this well-produced facsimile of Bligh's famous notebook kept during the open-boat voyage gives his original record of the amazing voyage across the Pacific and along the Queensland coast to Timor. **$525**

TRIAL OF THE MUTINEERS

215. [BLIGH, William] Minutes of the Proceedings of the Court-Martial held at Portsmouth, August 12, 1792. On Ten Persons charged with Mutiny on board His Majesty's Ship the Bounty. With an Appendix, containing a full Account of the real Causes and Circumstances of that unhappy Transaction, the most material of which have hitherto been withheld from the Public.

Quarto; an excellent, large copy with generous margins, original stitching-holes visible at inner margins; crimson morocco by Aquarius. London, J. Deighton, 1794.

This is a fine copy of a most important Pacific rarity: the full contemporary account of the trial of the *Bounty* mutineers. It was published in a very small edition for distribution among the interested parties and the ministers of state at the time, and is and always has been very rare. The 1934 catalogue of the remarkable Edge-Partington collection noted that "not more than three or four copies appear to be known, none of which is in the British Museum".

The main work comprises Bligh's deposition to the court and the evidence (with cross-examination) of Fryer, Cole, Peckover, Purcell, Hayward, Hallet, Smith, Captain Edwards, and Lieutenants Larkin and Corner (of the *Pandora*). The appendix is a defence of the conduct of Fletcher Christian by his brother Edward, professor of laws. It also has much information not found elsewhere relating to events preceding and subsequent to the mutiny, none of which was mentioned at the Court Martial. It evoked a reply from Bligh in which he said "...this Appendix is the work of Mr. Edward Christian...written apparently for the purpose of vindicating his brother at my expense...".

This represents a rare opportunity to acquire this legendary Pacific rarity, the most difficult of Bligh-Bounty pieces for any collector.

Ferguson, 175; Hill, pp. 200-201. **$68,500**

BLIGH'S SECOND MUTINY

216. [BLIGH: RUM REBELLION] Proceedings of a General Court-Martial, held at Chelsea Hospital... for the trial of Lieut.-Col. Geo. Johnston...

Octavo, with the errata slip; some browning and staining as usual, lacking some leaves; a decent copy in a near contemporary (c. 1820–30) Sydney binding of half diced calf and marbled boards, the boards somewhat rubbed. London, Sherwood, Neely, and Jones, 1811.

A wonderful association copy of this rare book: the court-martial of the orchestrator of Bligh's second mutiny, the copy once owned by Bligh's star witness. The book sets out the proceedings resulting from the Rum Rebellion in New South Wales, when Bligh was forcibly deposed as governor and kept under house arrest by a group led by Johnston, commander of the New South Wales Corps, for a full year. Bligh himself arranged for this publication of the proceedings of Johnston's subsequent court-martial.

This remarkable copy belonged to John Palmer, a First Fleet officer and Commissary under successive governors, who was a strong supporter of Bligh. He had been dismissed by the rebel administration, and appeared as the leading witness for the prosecution at the court-martial of Johnston – his fellow First Fleeter. This must be the ideal association copy of this important book.

A full history of the copy may be had on request: briefly, Palmer acquired it on publication in England, and it is signed by him in two places; it returned to New South Wales with him, and subsequently belonged to James Laidley, the Assistant Commissary. It was bound in the colony, probably in the 1820s (by that time missing six leaves), with a note almost certainly in Palmer's hand: "The Defence to this Trial is of the most flagititiary [sic] and base character. It is fraught with falshood, perjury and the lowest shifts of meanness and dishonor. An Eye Witness".

Significant association copies of early Australian books, especially when associated with such dramatic figures as Governor Bligh, are now very rarely encountered.

Ferguson, 514; Wantrup, 32. **$9500**

BLIGH IN DUBLIN

217. BLIGH, William, and others. Reports, &c. [on Improvements to Dublin Harbour]...

Folio, 84 pp., with six folding engraved maps; a fine copy in contemporary half calf. Dublin, N.p. [maps with the imprint of Henecy & Fitzpatrick], n.d. [1801].

Very rare: certainly not printed for general distribution. This collection of reports on proposed improvements to the Dublin waterways, submitted to the Directors-General of Inland Navigation in Ireland, contains a significant contribution by William Bligh, commissioned to carry out surveying work in Dublin Bay after the end of his command of the Director in 1799.

Bligh's report deals with the area in general, and has detailed specific comment on the areas proposed for improvement: Dalkey, Bullock, Dunleary, Howth and Ireland's Eye. His work constitutes about a third of the volume, with an accompanying map; other contributors include Joseph Huddart, John Rennie, Daniel Corneille and Richard Broughton as well as Thomas Hyde Page of the Royal Engineers, apparently the overall editor of the work.

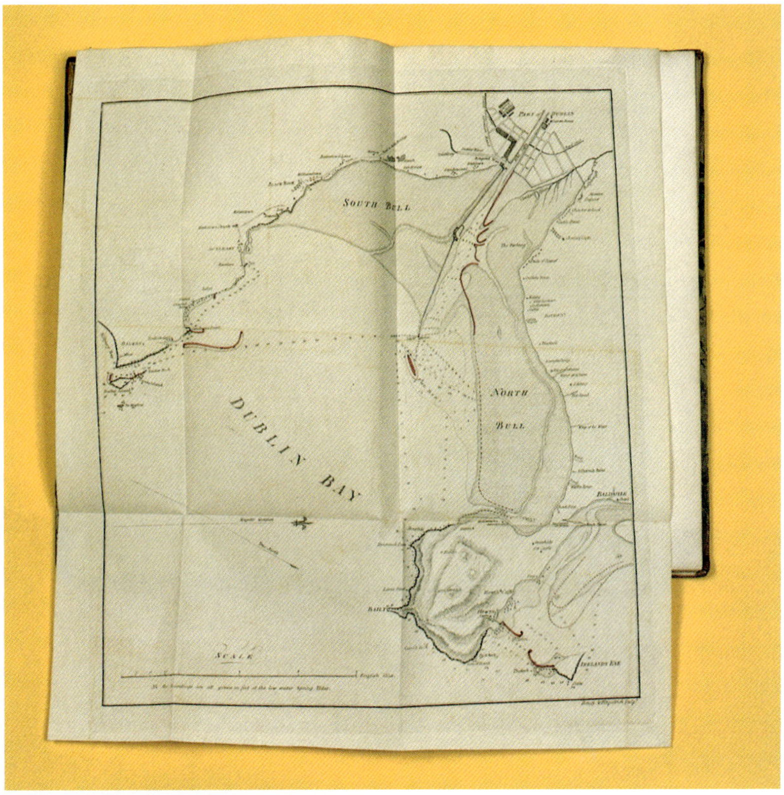

217. Bligh on Dublin Harbour

Bligh wrote to Joseph Banks in October 1800 (the letter is in the Mitchell Library) that he was in Dublin "to make a Survey of Dublin Bay, and give my opinion what can be done for the shelter and security of ships... This survey will keep me for some time here, and I hope will be of considerable benefit...". His work was evidently well received as he wrote to Nepean in February 1801 that "The Board of Navigation in Dublin, having particularly requested by letter that I would publish my survey of Dublin Bay, stating it as their opinion "it would be of infinite importance to their navigation, it being a work so correct and of such authority"... I hope their Lordships will be pleased to permit me to do so, as thereby I will come more correctly before the publick, than by being published in Dublin, and be of some advantage to myself..." (quoted by Mackaness, *Life of William Bligh*, II, p. 61).

Despite Bligh's request, the reports were published in Dublin, and so presumably without financial benefit to him. Bligh's works in print were very few: the *Mutiny* and *Voyage* of 1790 and 1792 (the second actually includes the first), the very rare *Answer to certain assertions* of 1794, and his 4-page *Memoir* of 1803 (known in a single copy – see next item) constitute his entire printed body of work; this significant and rare piece on Dublin harbour is thus of considerable interest. **$11,850**

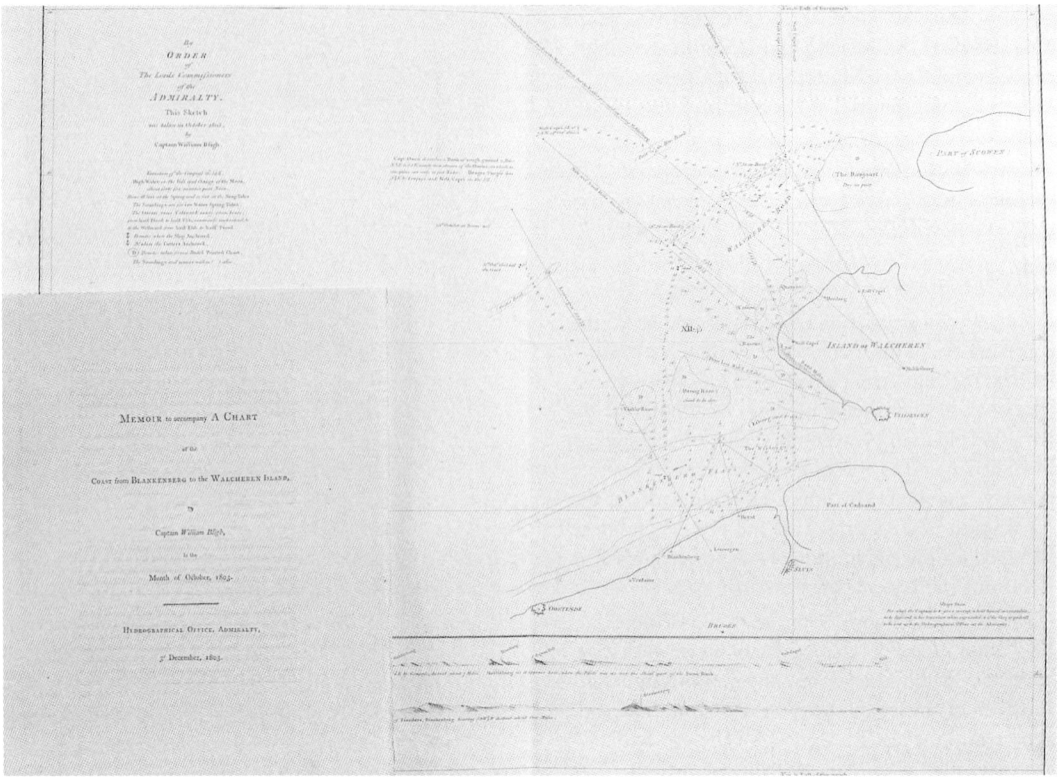

218. Bligh, Memoir

UNDER NAPOLEON'S NOSE

218. BLIGH, William. Memoir to accompany a chart of the coast from Blankenberg to the Walcheren Island... in the Month of October, 1803... [with untitled coastal chart of the approaches to Flushing, captioned:] "By Order of the Lords Commissioners of the Admiralty. This Sketch Was Taken in October 1803, by Captain William Bligh".

Quarto pamphlet, 4 pp., small hole (clear of letters) through the first leaf, unbound and uncut as issued; with the engraved chart, 645 x 615 mm.; the two pieces in fine condition, central fold to the chart. London, Alexander Dalrymple for the Hydrographical Office, 3 December [memoir] and 4 December [chart], 1803.

Exceptionally rare: apparently the only known copy of Bligh's third (and last) printed work, prepared by Alexander Dalrymple in a proof form only, evidently for very restricted issue. We have found no record of another copy of this remarkable piece, which is the result of a clandestine survey, on secret orders, carried out by Bligh of the approaches to the River Scheldt and Flushing on the Dutch coast. In the course of the survey, Bligh approached within miles of the French occupying forces under the partial cover of winter fog.

Flushing, on the southern coast of Walcheren Island, was of great strategic importance during the French occupation of 1795-1814 as the area where Napoleon was engaged in constructing a fleet for the invasion of England. Bligh refers obliquely in the text to the unusual nature of the undertaking: "The Season of the Year, and circumstances attending the undertaking, prevented *this* being a *regular Survey...*". In 1803 "the Admiralty took advantage of the temporary lull in sea fighting to have some important surveys carried out. In September and October Bligh was employed therefore in examining the harbour of Dungeness... No sooner was this work finished than a more important one was entrusted to him, for he was sent in November [sic] to examine and report on the coast of Flushing. This task was completed on the 12th of November..." (Mackaness, *Life of William Bligh*, II, pp. 65-6).

Mackaness's dates are evidently in error; later he discusses how in 1804 "the Admiralty was closely watching the French preparations for the great invasion of England, and as Napoleon had ordered a large number of gunboats and other vessels to be constructed in the various Dutch ports, had elaborated plans for an immense blockade of the entrances of the river Scheldt. To Admiral Keith was committed the carrying-out of this particular duty, and on the 11th of October he was instructed in a "Secret" letter, to despatch Captain Bligh of the *Warrior* with two of the best Flushing pilots to survey the entrance to the Scheldt and to report on the best position for placing the squadron therein... Bligh's skill as a navigator, cartographer and pilot had been so often proved to the satisfaction of the Admiralty, that there can be no reason for doubting that this service was in due course well and truly performed..." (*op. cit.*, pp. 68-9).

Mackaness clearly did not know of the existence of this printing, which actually dates the survey to late 1803. It is indeed of extreme rarity. It was produced by Alexander Dalrymple (see catalogue numbers 64-66) as Admiralty Hydrographer. The present authority on Dalrymple, Andrew Cook of the British Library, has indicated that the Memorial was previously known in the United Kingdom only from photographs of this copy. He has explained that "as Dalrymple had access to (and budget for) only a proofing press for his Admiralty work, all examples of his Admiralty charts on paper with watermark dates before 1808 (the year of his death) were technically proof copies. Authority (and pressure) to print larger quantities for fleet use began only with Dalrymple's successor Thomas Hurd, and by no means all Dalrymple's proof plates were so issued. This quarto pamphlet and engraved chart should be regarded as one of a standard proof run of perhaps twenty or twenty-five copies". It seems that no other copies of this exceptionally interesting printing may have survived. $21,000

LIFE OF BLIGH

219. [BLIGH] MACKANESS, George. The Life of Vice Admiral William Bligh.

Octavo, many plates, two in colour; an excellent copy in the original cloth with dustjacket. Sydney, Angus and Robertson, 1951.

First published in 1931, this is the revised edition of the flawed but standard biography of Bligh. $125

MUTINY ON THE BOUNTY

220. [BOUNTY] [BARROW, Sir John] The Eventful History of the Mutiny and Piratical Seizure of HMS Bounty; its cause and consequences.

Small octavo, with a frontispiece and five etched plates; original beige cloth, title reproduced on front cover, joints repaired. London, John Murray, 1831.

First and apparently the only authorised edition of a popular account of the *Bounty* story, issued as volume XXV of "The Family Library". The plates, after drawings by Batty, include the familiar image "Residence of John Adams, Pitcairns Island".

Ferguson, 1416; Hill, p. 346; Kroepelien, 46; O'Reilly-Reitman, 573. $185

ALECK

221. [BOUNTY] FISKE, Nathan Welby. Aleck, and the Mutineers of the Bounty; or, Thrilling Incidents of Life on the Ocean. Being the history of Pitcairn's Island...

Duodecimo, six full-page wood-engravings and a few vignettes; a fine copy in the original brown cloth, spine slightly rubbed and faded, gilt vignette of ship on front cover. Boston, John P. Jewett & Company, 1854.

An unusual Pitcairn piece: an historical fiction with a distinctly moral tone based on the story of the *Bounty* mutineers and their settlement on Pitcairn Island.

Very scarce: this is one of a number of American printings of this work around mid-century, all of which are uncommon. Ferguson knew only his own copy.

Ferguson, 5822; not in Hill or Kroepelien. $650

TRANSPLANTING THE BREAD-FRUIT

222. GOSSE, Thomas. Transplanting of the Bread-Fruit-Trees from Otaheite.

Mezzotint engraving, 440 x 570 mm., trimmed to margins on three sides; a good impression; some rubbing to plate surface at upper right and lower left, but otherwise in very good condition; mounted. London, T. Gosse, No 212 High Holborn, 1 September 1796.

A rare and much sought-after voyage print by the English engraver Thomas Gosse (see also catalogue number 154): Bligh, after the *Bounty* mutiny, loads bread-fruit trees once again in Tahiti.

In 1791 Bligh was sent back to Tahiti with two ships, the *Providence* and the *Assistant*, to repeat the

222. Gosse

Bounty's task of transplanting bread-fruit trees from Tahiti to the West Indies as a cheap and reliable sustenance for the plantation slaves. Gosse shows the loading of the trees in tubs; the Naval officer to the right is presumably Bligh himself.

Interest in the South Pacific following the voyages of Cook assumed enormous proportions, and heroic prints of this nature became very collectable in the eighteenth century; they have continued in popularity until the present day and most of them, like this fine example, are extremely scarce today.

See also catalogue number 154, and the frontispiece to this catalogue, for Gosse's *Founding of the Settlement.*

Nan Kivell & Spence, p. 33. **$24,500**

223. Delano

DISCOVERY ON PITCAIRN

223. [PITCAIRN] DELANO, Amasa. A Narrative of Voyages and Travels, in the Northern and Southern Hemispheres; comprising three voyages round the world...

Octavo, two engraved plates and a folding map of Pitcairn Island; moderate foxing and staining as in all copies but an excellent copy in original marbled sheep binding. Boston, E.G. House for the Author, 1817.

First edition. Delano describes three important Pacific voyages between 1790 and 1807, including visits to China, New Guinea, the East Indies, India, Chile, Australia, and Hawaii. The most celebrated aspect of the book is the long section on Bligh, the *Bounty*, and Pitcairn's Island – really the first published account of the fate of the mutineers – which includes the first publication of several manuscripts, including one by Captain Edwards of the *Pandora* and others by Folger, who discovered the mutineers on Pitcairn and interviewed Adams himself.

Ferguson, 673; Hill, p. 83; Hocken, p. 37; Judd, 51; Sabin, 19349. **$2100**

MURRAY'S PITCAIRN

224. [PITCAIRN] MURRAY, Rev. Thomas Boyles. Pitcairn: The Island, the People, and the Pastor. To which is added an account of the original settlement and present condition of Norfolk Island...

Duodecimo, frontispiece and 15 engraved illustrations; a fine copy in contemporary black roan gilt. London, SPCK, 1857.

The most famous, and the most often reprinted book on Pitcairn Island. This is the eighth of many editions, of some interest for the two new chapters describing the departure of the Pitcairn islanders for their new settlement on Norfolk Island.

Hill, pp. 208 & 507 (not this edition); Ferguson, 13014; Kroepelien, 882. **$275**

BY A PITCAIRN ISLANDER

225. [PITCAIRN] YOUNG, Rosalind Amelia. Mutiny of the Bounty and Story of Pitcairn Island 1790-1894.

Octavo, photographic plates and numerous illustrations; front inner hinge weak, small defect to final blank leaves; original pictorial blue cloth, a bit rubbed, but a nice copy. Oakland, California, Pacific Press, 1894.

First edition: very scarce. Rosalind Young was Pitcairn's primary-school teacher, and the great-grand-daughter of John Adams, the last survivor of the *Bounty* mutineers. Her book was popular, and there were several later versions.

This copy has a letter from a Canadian naval officer at Esquimault advising a Mr. Jacobus to send magazines to him for onward delivery to Pitcairn, and an autograph letter from Young thanking Jacobus for sending reading material.

Ferguson, 19001; Hill, pp. 625-6 (later editions). **$585**

III: *World Voyages*

9: GENERAL VOYAGES

LIFE IN KAMCHATKA

226. BENYOWSKY, Count Maurice Auguste. Voyages et Mémoires contenant ses Opérations militaires en Pologne, son exil au Kamchatka, son Evasion et son Voyage à travers l'Océan Pacifique, au Japon, à Formose, à Canton en Chine...

Two volumes, octavo, half-title not present in the first volume; a fine copy in old French sprinkled calf, spine gilt with double labels. Paris, F. Buisson, 1791.

First edition in French of a very scarce eighteenth-century Pacific narrative.

Benyowsky's narrative of his travels includes good descriptions of Siberia, the Russian fur trade, and the history, life and customs of Kamchatka, as well as an account of his landing on the Alaskan mainland (although most historians now doubt the validity of this claim) and descriptions of the Kurile and Aleutian Islands.

An Hungarian adventurer, Benyowsky escaped from Kamchatka where he had been exiled by the Russian government with a group of men who sailed to Japan, Formosa and China. As self-proclaimed King of Madagascar he was to inspire an opera by Boiledieu.

Cordier, 453; Hill, pp. 22 & 352 (English and German editions); Lada-Mocarski, 45 (English edition). **$4250**

THE ALEUTIANS

227. [BILLINGS, Joseph] SAUER, Martin. An Account of a Geographical and Astronomical Expedition to the Northern Parts of Russia, for ascertaining the degrees of latitude and longitude of the mouth of the River Kovima; of the whole coast of the Tshutski, to East Cape; and of the islands in the Eastern Ocean, stretching to the American Coast...

Quarto, with a folding map and 14 engraved plates; a few light spots and library stamps, a good copy in old half calf, marbled sides. London, A. Strahan, 1802.

The classic description of the Aleutian Islands.

This is the only edition in English of the Englishman Sauer's full account of the Billings voyage, on which he travelled as secretary to the Captain. This was the last of the major Russian surveys of the eastern coast of Siberia, and accomplished a great deal of geographical research.

The expedition visited Kodiak, Montague Island and Prince William Sound, and saw Mount St. Elias. Billings's naval career had started under Cook: he sailed as an able-bodied seaman on *Discovery*, transferring on Cook's death to *Resolution*.

Sauer's is the "first account in English of the first major exploring expedition sent out by the Russians to the Frozen Ocean and the North Pacific after Bering's second expedition of 1741... The chart was made by Arrowsmith from Sauer's notes and Captain Billings's astronomical observations..." (Streeter). The large folding map is of Bering's Strait and the American and Asian coasts. The appendices include linguistic tables, one of which compares Kamchatkan, Aleutian and Kodiak dialects.

Hill, p. 268; Lada-Mocarski, 58; Sabin, 77152; Streeter, 3499. **$3750**

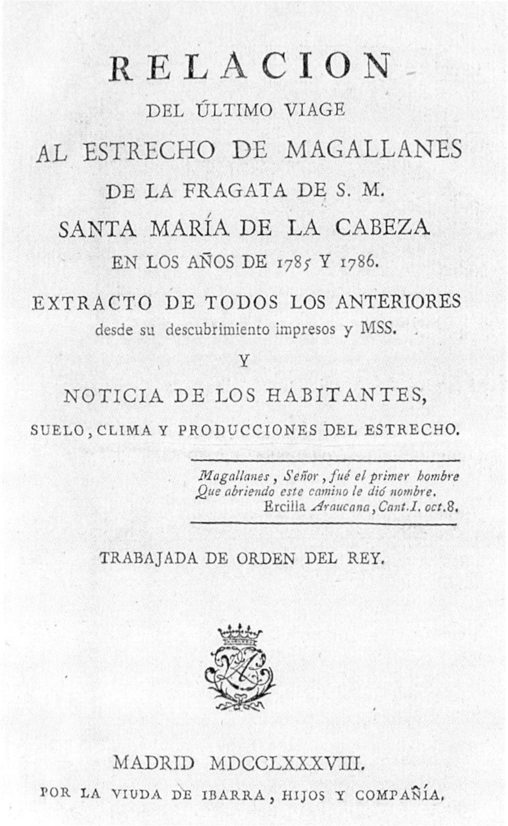
RELACION
DEL ÚLTIMO VIAGE
AL ESTRECHO DE MAGALLANES
DE LA FRAGATA DE S. M.
SANTA MARÍA DE LA CABEZA
EN LOS AÑOS DE 1785 Y 1786.
EXTRACTO DE TODOS LOS ANTERIORES
desde su descubrimiento impresos y MSS.
Y
NOTICIA DE LOS HABITANTES,
SUELO, CLIMA Y PRODUCCIONES DEL ESTRECHO.

Magallanes, Señor, fué el primer hombre
Que abriendo este camino le dió nombre.
Ercilla *Araucana, Cant. I. oct. 8.*

TRABAJADA DE ORDEN DEL REY.

MADRID MDCCLXXXVIII.
POR LA VIUDA DE IBARRA, HIJOS Y COMPAÑÍA.

228. Vargas y Ponce

PATAGONIA & THE STRAITS OF MAGELLAN

228. [CÓRDOBA, Antonio de] VARGAS Y PONCE, José de. Relacion del último viage al Estrecho de Magallanes de la fragata de S.M. Santa María de la Cabeza en los años de 1785 y 1786... [*and*] Apéndice a la Relacion...

Quarto, two volumes bound together, with an engraved portrait of Magellan as frontispiece, five fine large folding maps and five folding tables; contemporary calf-backed marbled boards, some wear but a very good copy. Madrid, Ibarra, 1788-1793.

The complete first edition in two volumes. This is the best description of Patagonia and the Straits of Magellan, then as now an area of strategic and navigational significance. Vargas y Ponce's book gives the official account of the two expeditions made by

Antonio de Córdoba to "ascertain the advisability of Spaniards continuing to use the Strait in voyages – the results indicated the negative. After these expeditions the Spanish trade routes were exclusively to east coast ports and then overland to the Pacific ports..." (Hill). The large folding maps, fine and detailed, are all based on observations made during the Córdoba voyages.

Córdoba's exploration of the Straits of Magellan is accompanied by Vargas y Ponce's summary of the early explorations of the area, mainly from an examination of original manuscripts in the Spanish archives. He discusses various means of navigating the Straits, with detailed sailing instructions, and describes the climate, indigenous peoples and natural history of the region.

Hill, p. 305; Navarrete, II, pp. 127-9; Palau, 352514-5 ("interesante y bien publicada"); Sabin, 16765, 1729. **$4000**

229. Ekelof

SWEDISH LIEUTENANT

229. EKELÖF, Adolf. Ett år i stilla hafvet reseminnen från Patagonien, Chili, Peru, Californien, Britiska Columbia och Oceanien...

Quarto, with 17 images on 12 tinted or chromolithograph plates; small snag to spine but a fine copy in the original cloth. Stockholm, Lith. Art. Anstalt, 1872.

A handsome and finely illustrated account, by a young Swedish lieutenant who sailed on the English frigate *Zealous* to the Pacific in 1867. There are long descriptions of various South American stops, followed by sections on California – particularly San Francisco – and British Columbia, and lengthy descriptions well illustrated on the Pacific Islands, including both Hawaii and Tahiti. Unusually, Ekelof provides both the music and words to an Hawaiian Folk Hymn and a Tahitian National Melody.

Kroepelien, 396; O'Reilly-Reitman, 1260. **$1450**

THE SPECIAL ISSUE WITH EXTRA COLOURING

230. KOTZEBUE, Otto von. Entdeckungs-Reise in die Süd-See und Nach der Berings-Strasse... Unternommen in den Jahren 1815, 1816, 1817 und 1818... auf dem Schiffe Rurick...

Three volumes in one, quarto, with six maps (four large and folding), folding tables, four full-page and four double-page handcoloured aquatints, a plain plate and 11 handcoloured plates of butterflies; a fine copy in contemporary marbled boards, red leather labels, joints rubbed but strong. Weimar, Gebrüdern Hoffmann, 1821.

First edition of Kotzebue's account of his first voyage: the rare special issue on thick wove paper, with extra colouring to the plates. The list of subscribers accounts for 587 copies, only ninety of which were of this issue, including three for the Emperor, and it is very scarce on the market. The plates, many of them beautiful handcoloured aquatints, include two each of Alaska and Hawaii, and four of Micronesia; in this special issue they have exceptionally good colouring, with all the quality of original watercolours. The third volume includes the fine coloured butterfly plates omitted from the English and Russian editions.

230. Kotzebue

This famous voyage narrative is particularly important for its descriptions of Alaska, California, Hawaii and Micronesia; Lada-Mocarski describes the account of Alaska as "rich in early original source material". The *Rurik* sailed along the California coast, stopping at San Francisco; Kotzebue describes the missions, and the work is considered one of the most important early accounts of the state. Kotzebue made a stop in Hawaii in November-December 1816, revisiting again in September-October 1817, anchoring off Hawaii and Oahu. The Hawaiian portion is extensive with important observations on life and customs during the reign of Kamehameha I, whose famous "red vest" portrait by Choris is one of the illustrations.

Cowan, p. 132; Hill, pp. 164-5; Judd, 94; Kroepelien, 670; Lada-Mocarski, 80; Sabin, 38284; Wickersham, 6197; Zamorano Eighty, 48. **$21,000**

CHORIS'S SECOND BOOK

231. [KOTZEBUE] CHORIS, Louis. Vues et Paysages des Regions Equinoxiales, recueillis dans un voyage autour du monde.

Folio, with 24 uncoloured lithographs, half of them stained; the first 14 pp. title and text in facsimile; all the six pairs of original wrappers preserved; a poor copy in half calf. Paris, Paul Renouard, 1826.

Choris's sequel to his *Voyage Pittoresque* with views from the Kotzebue Expedition of 1815-1818. The plates include five of Brazil, three of Chile, six of Micronesia, two each of Hawaii and Kamchatka, and one each of Manila, the Cape of Good Hope, St. Helena and Easter Island. The most famous plate shows the audience of officers of the Kotzebue expedition with King Kamehameha at Kailua, Kona.

This is a poor copy of the uncoloured issue, but a rare title.

Not in Hill; Lada-Mocarski, 90. **$7200**

FIRST ENGLISH EDITION

232. KOTZEBUE, Otto von. A New Voyage Round the World, in the years 1823, 24, 25, and 26.

Two volumes in one, octavo, two engraved plates, one plan and two folding charts; contemporary half russia leather; a fine, clean and attractive copy. London, Henry Colburn & Richard Bentley, 1830.

The first English edition of Kotzebue's second

voyage to the Pacific, in the *Predpriatie*. After calling at Kamchatka, they made a stay of several months on Baranov Island (Sitka), and then continued down the coast of California to the Russian settlement at Fort Ross. Kotzebue gives a good description of San Francisco and the various missions and includes the prophetic comment that "how abundantly and happily might thousands of families subsist here".

The long stay at Hawaii is of special interest. The author describes at length the political and social changes that had occurred since his visit on the *Rurik* in 1816, regretting that the people of Oahu had "lost the simplicity and innocence of character which formerly distinguished them". Other places visited include Rio de Janeiro, Micronesia, and Tahiti, all of which are described in some detail. Although Kotzebue did not visit Pitcairn, he gives a general history including particulars learned in Tahiti from members of Adams's family.

Hill, p. 166; Judd, 96; Kroepelien, 675; Lada-Mocarski, 93 (St. Petersburg edition, 1828); O'Reilly-Reitman, 844; Sabin, 38288. **$3450**

FIRST SWEDISH EDITION

233. KOTZEBUE, Otto von. Resa omkring Jorden aren 1823, 24, 25 och 26…

Octavo, with a frontispiece; contemporary quarter calf. Stockholm, Johann Horberg, 1830.

The rare first edition in Swedish with a lithographic frontispiece of the explorer being welcomed ashore. This is the narrative of Kotzebue's second voyage to the Pacific, in the *Predpriatie*, of great interest for its accounts of Kamchatka, the Northwest coast of America, San Francisco and the missions, and Hawaii.

Kroepelien, 676. **$650**

THE FIRST RUSSIAN CIRCUMNAVIGATION

234. KRUSENSTERN, A.J. von. Reise um die Welt in den Jahren 1803, 1804, 1805 und 1806… auf den Schiffen Nadeshda und Newa…

Two parts in three volumes, small quarto; stipple-engraved portrait of Krusenstern and two other frontispieces both finely coloured by hand, folding engraved map and 12 double-page engraved views; a fine set in contemporary cream and maroon boards, yellow edges. Berlin, Haude und Spener, 1811-1812.

The first edition printed in Germany, published at the same time as the more costly St. Petersburg quarto, of the famous account of the first Russian circumnavigation by Krusenstern, accompanied by a brilliant corps of officers, including Lisiansky, Langsdorff, and Kotzebue. This was the first popular account of this important voyage to be made available to the public. No English edition appeared until 1813.

The voyage was of great importance to Pacific history for the attempt to open Japan to commerce, and for its observations on the Russian-Chinese trade. The *Nadeshda* and the *Neva* were together at the Marquesas and Hawaii; there they separated and from this point on Krusenstern's narrative concerns the western Pacific. The book includes a view of Nagasaki Harbour, one of the earliest views of Japan published in the west, as well as views of Nukahiva, Japan, Macao, Canton, St. Helena and St. Catherine, Brazil.

This is an excellent copy of this scarce edition, in which an extra piece was added by the publisher in the form of the *Nachricht* by Langsdorff on tattooing in Nukahiva and the Washington islands, including a small woodcut showing the instruments. There is also a Prospectus by Langsdorff and his publisher for the German publication of his *Bemerkungen*.

Borba de Moraes, p. 442; Cordier, p. 458; Hill, p. 167; Kroepelien, 691 (but the two plates uncoloured); O'Reilly-Reitman, 728. **$9450**

234. Krusenstern

RARE AMERICAN PRINTING

235. LANGSDORFF, Georg Heinrich von. Voyages and Travels in Various Parts of the World, during the years 1803, 1804, 1805, 1806 and 1807.

Octavo, folding frontispiece; some spotting or browning but an excellent copy, completely uncut, with a 24-pp. Carey catalogue bound at the start; quarter calf. Carlisle, George Philips, and for sale in Philadelphia, By M. Carey & Son, 1817.

First American edition, and – like many early American voyage printings – distinctly scarce. Langsdorff was physician and naturalist aboard the first Russian circumnavigation on the *Neva* under Krusenstern. He went around Cape Horn to Kamschatka, making stops at the Marquesas and Hawaii.

At Kamschatka, Langsdorff left the expedition and travelled on the *Maria* with Rezanov, the Russian official commissioned to study the Russian-American Company in Alaska and to conduct trade negotiations with Japan. Langsdorff states that, though they were politely but firmly refused, their Japanese visit was "the most interesting part of our expedition", devoting nearly 100 pages to their stay at Nagasaki. In their spare time they constructed a Montgolfier-type balloon and made the first aerial ascent in Japan.

The narrative contains a lengthy record of their stay in the Marquesas, and a "fuller account of Sitka and the settlement of San Francisco than any other contemporary account" (Sabin).

Cowan, pp. 382-3; Judd, 101; O'Reilly-Reitman, 736; Sabin, 38896; not in the Hill collection; not noted by Lada-Mocarski. **$2250**

236. Novo y Colson

FIRST FORMAL PUBLICATION OF THE MALASPINA EXPEDITION

236. [MALASPINA, Alejandro] NOVO Y COLSON, Pedro. Viaje político-científico alrededor del mundo por las Corbetas Descubierta y Atrevida al mando de los capitanes de navío D. Alejandro Malaspina y Don José de Bustamante... 1789 a 1794.

Small folio, portrait frontispiece, six engraved views, and a folding lithograph map; contemporary Spanish quarter morocco. Madrid, la viuda e hijos de Abienzo, 1885.

The first formal publication of the Malaspina expedition to America, Australia and the Pacific in the 1790s. This was one of the major eighteenth-century voyages, ranking in importance with those of Bougainville, Cook and La Pérouse; the visit to Port Jackson took place barely five years after the first settlement.

Court intrigue and controversies in Spain meant that no official publication of the voyage ever appeared, despite the original intention to produce a huge published account in the style of the French *grands voyages*. It was a hundred years before this full narrative appeared.

This copy has the initials "J.B.R." on the spine, and the later ink inscription on the title of Joaquin Bustamante de la Rocha, probably a descendant of Malaspina's colleague José de Bustamante.

Hill, pp. 189-90. **$6850**

ONE OF FEW CONTEMPORARY PUBLICATIONS

237. [MALASPINA EXPEDITION] BRAMBILA, Fernando. Vista de la Ciudad de Santiago de Chile desde la falda del Cerro de Santo Domingo.

Copper-engraved landscape view, 304 x 452 mm.; paper watermarked "Ynfantado"; in fine condition, with wide margins; mounted. N.p. [probably Madrid], "Brambila lo delineó y gravó", n.d. [c. 1795].

This fine view of the city of Santiago is one of very few original publications deriving from the Malaspina expedition to America, Australia and the Pacific in the 1790s. The original intention to produce a grand publication (see note to previous item) encouraged the inclusion of a number of artists on the expedition, to record landscape as well as scientific and anthropological observations.

The Milanese artist Fernando Brambila was commissioned, along with his fellow-artist Ravenet, for the latter part of the voyage, joining Malaspina's ships in Acapulco on their return from the Northwest coast. Aboard the *Atrevida* he drew views of Guam, the Philippines, Macao, Vavao, Chile and Argentina. In Spain he prepared some of his drawings for engraving, in anticipation of the full official publication of the voyage – which never eventuated.

Brambila's original drawing for this fine engraving survives in the Bauza collection at the University of Santiago; it is reproduced by Serrano, who describes it as "una espléndida vista de los campos que preceden a la ciudad, cuya silueta se observa al fondo...". Serrano also notes that Brambila's preparatory drawing for the engraving survives in the Museo Naval, Madrid.

Serrano, 'Los Pintores de la Expedición de Alejandro Malaspina', Madrid, 1982, no. 154; see also Bonifacio del Carril, 'La expedición Malaspina en los mares americanos del Sur. La colleción Bauja 1789-1794', Buenos Aires, 1961, plate X; and Julio Guillén y Tato, 'Exposición de dibujos y grabados de los artistas que figuraron en las expediciones cientificas', Las Palmas, 1952, no. 262; and Eugenio Pereira Salas, 'Historia del Arte en el reino del Chile', Santiago, 1965 **$5750**

237. Brambila

238. Michelena y Rojas

"ONE OF THE RAREST AND LEAST-KNOWN"

238. MICHELENA Y ROJAS, Francisco. Viajes científicos en todo el mundo, desde 1822 hasta 1842...

Small folio, with a portrait, two folding maps and 12 lithograph plates; some spotting or browning, light stains in some lower margins, but a very good copy in contemporary quarter green roan. Madrid, I. Boix, 1843.

"One of the rarest and least-known voyages to the Pacific... This work is cited in a few scattered bibliographies with the erroneous annotation that it contains the author's seven voyages. Actually the entire work covers only the Pacific Voyage, the sixth and most important..." (Hill).

This very scarce travel account was printed in Spain for distribution to subscribers in Venezuela (listed at the end of the work). Michelena y Rojas's long description of Australia is illustrated with a folding map of Sydney and two excellent lithographs (the Government Stables and Botanic Gardens, and a group of Aborigines outside their dwelling). Ferguson knew only his own and the National Library copies. The long Hawaiian section is also of great interest, with three excellent lithographs – a portrait of John Adams, a fine view of Honolulu and another romanticised view.

Ferguson, 3697; Hill, p. 498. **$8450**

AUSTRIAN EXPLORING EXPEDITION

239. [NOVARA] SCHERZER, Dr. Karl von. Reise der öesterreichischen Fregatte Novara um die Erde, in den Jahren 1857,1858, 1859...

Three volumes, large octavo, with many maps, plates and illustrations; folding map in a pocket of the binding; original green cloth gilt, a little rubbing to joints but a fine, bright set. Vienna, Kaiserlich-königlithen Hof- und Staatsdruckerei, 1861-62.

First edition: the only major Austrian exploring expedition. Scherzer, author of this narrative, was one of the naturalists on the expedition, in charge of ethnography, linguistics, anthropology and statistics. This is the rare official account of this important voyage to the Pacific, which was commanded by

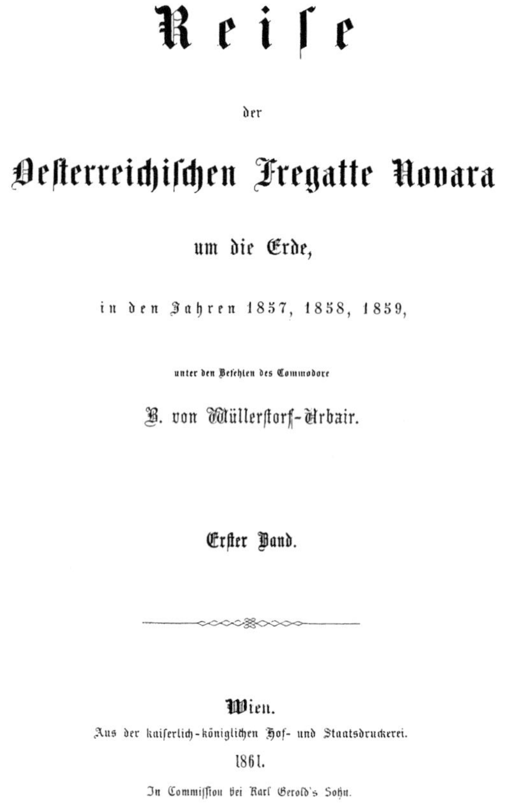
Reise
der
Oesterreichischen Fregatte Novara
um die Erde,
in den Jahren 1857, 1858, 1859,
unter den Befehlen des Commodore
B. von Wüllerstorf-Urbair.
Erster Band.
Wien.
Aus der kaiserlich-königlichen Hof- und Staatsdruckerei.
1861.
In Commission bei Karl Gerold's Sohn.

239. Scherzer

Captain Bernard von Wüllerstorf-Urbair. Archduke Maximilian was patron of the voyage.

There was also an abridged version published in the same year, and the numerous later editions included English and Italian versions.

Ferguson, 15471a; Hill, pp. 513 & 565. **$1250**

240. Reynolds

VOYAGE OF THE POTOMAC

240. REYNOLDS, J.N. Voyage of the United States Frigate Potomac.

Octavo, nine folding aquatint plates and a map; some scattered foxing; original blind-stamped cloth, spine repaired. New York, Harper & Brothers, 1835.

First edition: this American circumnavigation, under the command of John Downes, was performed in 1831 to 1834, mainly to check on American commercial interests in the east. The author sets out to "say something of our commercial interests in the east, of its history, present condition, and means of its further extension; of sailing directions and the monsoons... of the Sandwich Islands... of the great Pacific whaling fleet... of the people of South America... or of Rio and the empire of Brazil...".

Hill, p. 252; Judd, 149. **$850**

THE FIRST SWEDISH CIRCUMNAVIGATION

241. SKOGMAN, Carl. Fregatten Eugenies Resa Omkring Jorden aren 1851-1853, under befal af C. A. Virgin.

Two volumes in one, octavo, with three folding tinted maps, and 20 lithographs (18 coloured) and numerous woodcuts; contemporary quarter black calf, flat spine gilt, joints repaired. Stockholm, Adolf Bonnier, [1854-1855].

First edition: the official account of the voyage of the *Eugenie*, the first Swedish circumnavigation, with particularly good accounts of the expedition's stops in Honolulu, Tahiti, San Francisco, Sydney and Manila, and shorter notices on various South American ports, the Galapagos, Hong Kong, Singapore and the Cape of Good Hope. The plates include four of Tahiti, three of Hawaii, two of Manila and a view of Mission Dolores, San Francisco. These coloured lithographic illustrations are both naïve and attractive.

Borba de Moraes, p. 815; not in Ferguson; Hill, p. 573; Kroepelien, 1196. **$1100**

242. Jenkins

THE U.S. EXPLORING EXPEDITION

242. [WILKES] JENKINS, John S. Voyage of the U.S. Exploring Squadron, Commanded by Captain Charles Wilkes, of the United States Navy, in 1838, 1839, 1840, 1841, and 1842: Together with Explorations and Discoveries Made By Admiral D'Urville, Captain Ross, and other Navigators and Travellers...

Octavo, illustrated with engravings; some browning and spotting; original full calf, spine repaired. Auburn, J.M. Alden, 1850.

A condensation for the general reader of the Wilkes expedition and several earlier and important voyages. Part one contains expeditions in the Pacific and the South Seas, with much on Australia. Gold discoveries in California are noted. Part two covers Lynch's expedition to the Dead Sea which left the Brooklyn Navy Yard on the *Supply*, in 1847.

Ferguson 5404 (National Library copy only). **$985**

241. Skogman

CORAL REEFS

243. [WILKES] DANA, James D. Corals and Coral Islands.

Large octavo, colour frontispiece, numerous illustrations, folding maps; a fine copy in the original decorated green cloth. London, Sampson Low, 1875.

The second edition, with additions, of this comprehensive study of Pacific coral reefs, based on studies made during the Wilkes voyage. Dana explains in the Preface that "our cruise led us along the course followed by Mr. Charles Darwin... in the Beagle, under Captain Fitzroy... Soon after reaching Sydney, Australia, in 1839, a brief statement was found in the papers of Mr. Darwin's theory with respect to the origin of the atoll and barrier forms of reefs. The paragraph threw a flood of light over the subject, and called forth feelings of peculiar satisfaction, and of gratefulness to Mr. Darwin...".

This edition not in Ferguson (other editions are 8941a-8943). **$250**

VOLCANOES

244. DANA, James D. Characteristics of Volcanoes...

Octavo, illustrations; original black and red cloth, a very good copy. New York, Dodd, Mead, 1890.

"Including historical review of Hawaiian volcanic action for the past sixty-seven years, a discussion of the relations of volcanic islands to deep-sea topography, and a chapter on volcanic-island denudation".

Dana was the geologist on the Wilkes expedition. Chapter four contains portions of his work in New South Wales first described in the *Geological Report of the Wilkes United States Exploring Expedition*, which had been published in a very limited way fifty-one years previously.

Not recorded by Ferguson, despite Australian content. **$550**

10: ENGLISH VOYAGES

BOTANY & ZOOLOGY OF THE SULPHUR

245. BELCHER, Sir Edward. The Botany... [*and*] The Zoology of the Voyage of HMS Sulphur...

Two volumes, quarto, containing a total of 124 lithographic plates, many of the plates aged and stained; loose as issued in original portfolios of quarter cloth and printed boards. London, Smith, Elder, 1843-1846.

In 1835 Captain Belcher set out from Plymouth on the *Sulphur* to continue the exploration and hydrographic survey of the west coast of America that had been started by King and Fitzroy in the late 1820s. Belcher's voyage was to last an astonishing seven years. The surgeon on the expedition, Richard Brinsley Hinds, prepared the zoology findings for publication and also worked on the botanical atlas with the naturalist G. Bentham. These full-page lithographs were printed by Hullmandel; there are sixty lithographs in the Botanical atlas, and the Zoological atlas contains sixty-four lithographs of mammals, birds and ichthyology.

Sabin, 4390. **$4250**

245. Belcher

CHAPTER II.

WHALES OF THE SOUTHERN OCEANS.

The Cachalot, or Sperm Whale (*Catodon macrocephalus*, Lacep. *Physeter macrocephalus*, Shaw)—General and anatomical description of the species—Dissection of a *fœtus*—Diseases and deformities—Parasitic occupants.

THE Sperm Whale, or Cachalot, is the largest, as well as the most valuable of the southern whales; and has many points of interest in its structure and economy. It is known to the English as the Sperm- or Anvil-headed Whale; the French name it the *cachalot;* and the Germans, the *pottfisch.*

246. Bennett

THE LIVING SPERM WHALE

246. BENNETT, Frederick Debell. Narrative of a Whaling Voyage Round the Globe... Comprising sketches of Polynesia, California, the Indian Archipelago... with an account of Southern whales, the sperm whale fishery, and the natural history of the climates visited.

Two volumes, octavo, with frontispieces and a folding map, illustrations in the text; an attractive copy in contemporary polished crimson calf gilt with double labels, joints neatly restored; bookplates. London, Richard Bentley, 1840.

First edition of a significant book, with much on Hawaii and Tahiti. This is a very attractive copy, given as a leaving present at Eton in 1844.

Herman Melville believed that this and Beale's *Natural History of the Sperm Whale* were "the only two books which at all pretend to put the living sperm whale before you, and at the same time, in the remotest degree succeed in the attempt..." (*Moby-Dick*).

The *Tuscan* left London in 1833, bound for the Society and Marquesas Islands. The ship made stops at Pitcairn, Tahiti and the Marquesas, and Hawaii was visited three times (1834-35) in the course of the voyage. Bennett was specifically assigned to observe southern whales in order to determine improvements in the sperm whale fishery business, and collected important information, as well as natural history specimens. Most of the second volume contains notes on his scientific research.

Hill, p. 22; Judd, 19. **$4750**

COLOUR LITHOGRAPHS OF HMS MÆANDER

247. BRIERLY, Sir Oswald Walter. HMS "Mæander" 44 guns, in a heavy squall [*and*] Shortening sail for anchoring…

Pair of colour lithographs, 375 x 535 mm.; mounted. London, Ackermann, 1852.

Colour lithographs by T.G. Dutton after original watercolours by Oswald Brierly, printed by Day and Son for Ackermanns. The first print shows the *Mæander* in the Pacific, shortening sail in heavy weather; the caption dates the events to 9 July 1850; in the second the ship is coming slowly to anchor in Rio de Janeiro, and the scene is dated 9 June 1851. Both prints are dedicated to Henry Keppel, and the ship's officers.

Oswald Brierly (1817-1894) was a leading marine painter, and he had also studied naval architecture. He sailed on numerous expeditions including the *Rattlesnake* voyage surveying the Barrier Reef (see catalogue number 201), with Benjamin Boyd on the *Wanderer*, and on the *Galatea* with the Duke of Edinburgh in 1867 (see catalogue number 257). Brierley joined Henry Keppel on the *Mæander* after his voyage on the *Rattlesnake* and visited New Zealand, Tahiti and South America, returning to England in July 1851. A description of the voyage is given in Admiral Keppel's *A Sailor's Life Under Four Sovereigns* (London, 1899).

A fine pair of portraits of a spendid ship. **$12,500**

247. Brierly's lithographs of HMS Mæander

VOYAGE OF THE BLONDE

248. BYRON, Lord George Anson. Voyage of HMS Blonde to the Sandwich Islands.

Quarto, with 14 charts and aquatint plates, some folding; a few spots, one leaf stained and another with marginal defect; generally a good copy in modern calf with old label. London, John Murray, 1826.

First edition. Kamehameha II of Hawaii and his Queen Kamamalu were on a visit to London in 1824, when they both died of measles, against which they had no immunity. This voyage – "one of the most gracious acts that one country has ever extended to another" (Buck) – was commissioned by the British government specifically to return their bodies to the Hawaiian Islands, with the cousin of the poet Byron in command. The book contains a history of their visit to London, a résumé of the discovery, and visits to Hawaii by British explorers, as well as the main narrative, compiled by Maria Graham from journals kept by the naturalist Bloxam as well as from official papers.

The stay in Hawaii contains an interesting account of travels to Hilo, and the Volcano, Kealakekua, Lahaina and Honolulu. The fine aquatint views by Robert Dampier include a large view of Kilauea crater, two of Honolulu, four of Hilo, and one each of the City of Refuge near Kealakekua, and Lahaina on Maui. This is a fine and early illustrated book on Hawaii.

Not in Abbey; Hill, p. 309; Judd, 76. **$3600**

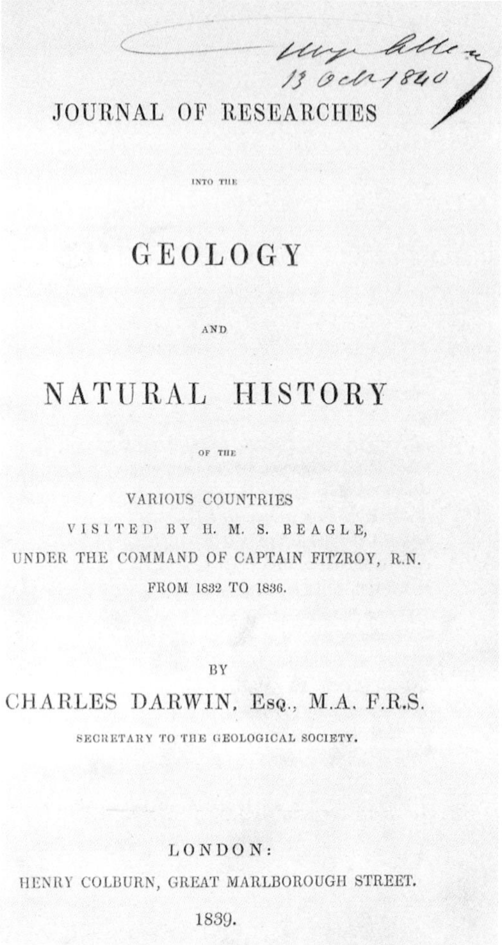
13 Octr 1840

JOURNAL OF RESEARCHES

INTO THE

GEOLOGY

AND

NATURAL HISTORY

OF THE

VARIOUS COUNTRIES
VISITED BY H. M. S. BEAGLE,
UNDER THE COMMAND OF CAPTAIN FITZROY, R.N.
FROM 1832 TO 1836.

BY

CHARLES DARWIN, Esq., M.A. F.R.S.

SECRETARY TO THE GEOLOGICAL SOCIETY.

LONDON:
HENRY COLBURN, GREAT MARLBOROUGH STREET.
1839.

249. Darwin

EARLY AUSTRALIAN LIBRARY

249. DARWIN, Charles. Journal of Researches into the Geology and Natural History of the various countries visited by HMS Beagle...

Octavo, with the two folding maps bound in as issued (both offset); a spot or two but a very good copy in the original blue-green cloth, the head and tail of the spine well restored. London, Henry Colburn, 1839.

First edition, first separate issue; this was also published in the same year as the third volume of the full official account of the *Beagle* voyage. Darwin's famous account is "one of the most interesting records of natural history exploration ever written and... one of the most important, for it was on this voyage that Darwin prepared for his lifework" (Hill).

This copy has a distinguished Australian provenance, from the library of George Allen of Toxteth Park, the Australian bibliophile and merchant. The title-page is neatly signed by him and dated 13 October 1840, and it has his bookplate and shelf marks. Allen's library was probably the most extensive in the colony but unfortunately there is no complete record of it. This copy has an eight-page catalogue bound in after the August 1839 catalogue noted by Freeman, advertising Colburn's Standard Novelists series, listing precisely the same novels as appear in the August catalogue.

Ferguson, 2739; Freeman, 11 (a binding variant). **$3650**

SECOND EDITION

250. DARWIN, Charles. Journal of Researches... Voyage of HMS Beagle.

Octavo; a good copy in contemporary half roan, a little rubbed. London, John Murray, 1845.

This second edition results from Darwin's substantial corrections and additions to the first edition "in order to render the volume more fitted for popular reading". In the Preface he cautions the naturalists among his readers to consult the fuller work and the scientific results of the expedition. These works are listed in the advertisement for works on the voyage of the *Beagle* on the last page.

Ferguson, 4025; Freeman, 14. **$485**

FIRST AMERICAN EDITION

251. DARWIN, Charles. Journal of Researches into the Natural History and Geology of the countries visited during the voyage of HMS Beagle round the World, under the Command of Capt. Fitz Roy, R.N....

Two volumes, small octavo; original brown cloth, uncut; an excellent copy. New York, Harper & Bros., 1846.

First American edition of Darwin's narrative, based on the second edition of the text as published by Murray's *Home and Colonial Library* in the previous year. This copy was once in the library of the New Hampshire Historical Society, with their neat blind-stamp on titles. Not in the Hill collection; Ferguson records only the Mitchell Library copy.

Ferguson, 4025; Freeman, 16. **$1550**

252. Dixon

NATURAL HISTORY PLATES COLOURED

252. DIXON, Captain George. A Voyage Round the World; but more particularly to the North-West Coast of America...

Quarto, with 21 maps and plates (a number folding), the seven natural history plates coloured; a little light off-

setting; old half calf and marbled boards, spine and corners renewed. London, George Goulding, 1789.

First edition: the better issue with the seven natural history plates by Lewin in original colouring. Dixon's account of his voyage in the *Queen Charlotte* is dedicated to Joseph Banks, and is a companion to Portlock's account of the same voyage; both men had voyaged with Cook, Dixon as armourer of the *Discovery*. They sailed together as far as Prince William Sound, Dixon then following the coast making a series of landfalls. He discovered and closely observed Queen Charlotte's Island, and entered Dixon's Straits, before ultimately arriving at Nootka where he joined both Portlock and Meares. The book is "an excellent authority for the early days of fur trading on the northwest coast" (Streeter).

The *Queen Charlotte* made visits to Hawaii in 1786 and 1787, trading at Oahu and Kauai. The book also includes a long account of commercial transactions at Canton. Though often catalogued as the work of William Beresford, whose letters to a friend signed W.B. form the basis of the work, Dixon added substantially to the text and edited the whole.

Hill, p. 23; Judd, 53; Lada-Mocarski, 43; Sabin, 20364 (confused); Streeter, 3484; Wagner, I, pp. 207 etc., II, 732-735. **$11,500**

CONVERSION OF THE HEATHENS

253. [DUFF VOYAGE] HAWEIS, Thomas, editor. A Missionary Voyage to the Southern Pacific Ocean, performed in the years 1796, 1797, 1798, in the Ship Duff, commanded by Captain James Wilson. Compiled from Journals of the Officers and the Missionaries...

Quarto, with six plates and seven maps (five folding; tear to one fold); paper repairs to margin of title-page, some foxing, but a good, large copy in contemporary dark calf, rebacked. London, S. Gosnell, for T. Chapman, 1799.

First edition of the full official account of the first missionary voyage to the South Pacific. The *Duff* set out for Tahiti in 1796, but visited many island groups, including particularly Tonga and the Marquesas. A new group of islands, the Duff Group, was discovered among the Santa Cruz islands. The maps here include a large chart of the Fiji islands as well as charts of Tongataboo, the Gambier Islands, the Marquesas, Tahiti and the Duff Group; the plates include an engraved view of Rio de Janeiro.

Ferguson, 301; Hill, p. 184; Kroepelien, 528. **$2100**

DEPARTURE OF THE DUFF

254. [DUFF VOYAGE] MEMORIAL MEDALLION. The Duff Capt. James Wilson. Sail'd 10th Aug. 1796.

Medal, white metal, 37 mm. London, 1796.

Rare medal, commemorating the departure of the famous missionary voyage; the inscription on the reverse reads "With missionaries to the South Seas for the Conversion of the Heathens". Captain Wilson (1760-1814) was master of the missionary ship *Duff*, despatched by the newly formed London Missionary Society in 1796.

From the William Yeates and Milford Haven collections. **$1325**

254. Duff voyage

FINE IN ORIGINAL WRAPPERS

255. FITZROY, Admiral Robert. Passage Table and General Sailing Directions.

Octavo; marginal defect to one leaf clear of text, otherwise a fine copy in the original printed dark blue wrappers, respined. London, Eyre and Spottiswoode, 1859.

Scarce sailing directions (in untypically fine condition) by the celebrated navigator and commander of the *Beagle*. Fitzroy, co-author with Darwin of the account of the *Beagle* voyage published in 1839, and governor of New Zealand 1843-45 amid great controversy regarding his support of the Maori population, was greatly respected as an author on maritime and scientific subjects, finishing his career as chief of the meteorological department of the Board of Trade. **$475**

256. Forrest

VOYAGE TO NEW GUINEA

256. FORREST, Thomas. A Voyage to New Guinea, and the Moluccas, from Balambangan: including an Account of Magindano, Sooloo, and other Islands... performed in the Tartar Galley...

Quarto, with 31 engraved maps and plates, many double-page or folding; a few plates slightly toned, frontispiece offsetting onto title; a good copy in modern quarter calf. London, G. Scott, 1780.

Second edition of an important and early work on New Guinea, with extensive mapping. Forrest was involved in the 1770 settlement of Balambangan, which was based on recommendations by Alexander Dalrymple; from 1774 he was in the service of the East India Company, who wanted to extend their sources of trade in the direction of New Guinea. Forrest – with two English officers and a crew of eighteen Malays – pushed the exploration as far as Geelvink Bay in western New Guinea, present-day Irian Jaya.

The numerous maps and coastal profiles published here show the detailed observations that Forrest was making on this voyage "of examination and enquiry rather than of discovery" (Hill).

Hill, p. 108. **$3850**

257. Brierly

GALATEA AMONG THE ICEBERGS

257. [GALATEA] BRIERLY, Sir Oswald Walter. HMS "Galatea"... Amongst the icebergs in the Southern Oceans 23rd April, 1868.

Chromolithograph, 380 x 680 mm., tipped onto original card mount; framed. London, John B. Day, n.d. but c. 1870.

Lithograph by John Day after an original watercolour by Oswald Brierly. Brierly was invited by the Duke of Edinburgh to join his expedition round the world on the *Galatea* in 1867; the ship had been commissioned specially for a voyage to the Australian colonies. The voyage included stops at Lisbon, Madeira, Tristan de Cunha, the Cape, Adelaide, Melbourne, Hobart and Sydney among many ports-of-call.

Oswald Brierly (1817-1894) was a leading marine painter, and had studied naval architecture. He sailed with several earlier expeditions including the *Rattlesnake* voyage on a survey of the Barrier Reef (catalogue number 201), and with Benjamin Boyd on the *Wanderer*. He also sailed with Keppel on the *Mæander* (catalogue number 247). He subsequently wrote, with Rev. John Milner, the narrative published as *Cruise of the Galatea* (London, 1869).

The event portrayed in this picture is recorded in Brierly and Milner's final chapter describing their passage home from Farm Cove, Sydney to England. On 23 April, off the southernmost tip of Western Australia, "...at 4 p.m. an iceberg was observed on the starboard bow... it appeared to be about 450 feet high, and about three quarters of a mile long... the ship was braced up on the port tack and stood to the northward...".

Not long after this publication Brierly was appointed marine painter to Queen Victoria and to the Royal Yacht Squadron. **$6000**

KEATE'S PELEW ISLANDS

258. KEATE, George. An Account of the Pelew Islands, situated in the western part of the Pacific Ocean...

Quarto, engraved portrait frontispiece, folding maps and views; a good copy in contemporary marbled calf, sides a little worn but attractive, rebacked. London, G. Nicol, 1788.

Keate's *Pelew Islands* was one of the most popular eighteenth-century books on the Pacific. In 1783 the *Antelope* was wrecked on an uncharted reef near Palau; the crew reached shore and were well treated by the natives. From the wreck they built a small boat which they managed to get to Macao, taking with them Prince Lee Boo, the son of King Abba Thule. Lee Boo travelled with them to England, but died soon after of smallpox (see catalogue numbers 261-3). This account did much to reinforce the idea of the noble savage, but was also the main source for early knowledge of the Palau Islands in Micronesia.

Hill, p. 160. **$2450**

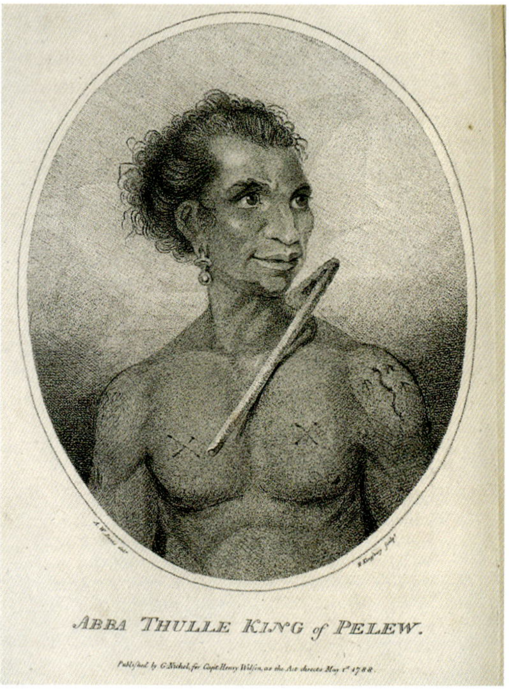

258. Keate

FIRST OCTAVO EDITION

259. KEATE, George. An Account of the Pelew Islands...

Octavo, with two portraits and a folding map; a good uncut copy in early twentieth-century olive half calf, gilt. London, printed for Captain Wilson, 1789.

Fourth edition, the first in octavo, published for Captain Wilson himself (i.e. for his benefit) by George Nicol. **$785**

RARE PHILADELPHIA IMPRINT

260. KEATE, George. An Account of the Pelew Islands...

Duodecimo; contemporary sheep with red label; front joint cracked but sound, an attractive copy. Philadelphia, Joseph Crukshank [sic], 1789.

An unusual Philadelphia voyage imprint, and – like all American voyage printings of the period – seldom seen on the market. The Hill catalogue describes this as "the rare American edition".

Hill, p. 161. **$875**

FIRST FULLY-ILLUSTRATED VERSION

261. [LEE BOO] NEWBERY, Publisher. The History of Prince Lee Boo, a Native of the Pelew Islands brought to England by Capt. Wilson. A New Edition.

Duodecimo, engraved portrait frontispiece and five engraved plates, with final leaf of advertisements; original vellum-backed boards with printed paper label on spine; contemporary engraved circular label of "Crockett's Lilliputian Library Norwich" on front pastedown, and ink inscription of "Geo. Harvey Feb. 21st, 1794". London, Newbery, c. 1793.

The immensely popular story of the the native prince who was brought back to England after the wreck of the *Antelope* in 1783, and who later died from smallpox. This is probably the second edition of this chapbook, and certainly the first fully-illustrated version. Newbery, the original publisher of the title, issued several editions: the first listed by Roscoe, the bibliographer of the press, is dated 1789 on the title-page and has a portrait-frontispiece only, while the present, undated, edition is listed next and is the first to include the five plates; copies can have up to six pages of advertisements at the end.

Hill, p. 146 (an 1844 edition only); Osborne, I, p. 166; Roscoe, J170/2. **$2100**

SPCK

262. LEE BOO. The History of Prince Lee Boo, a man of colour.

Duodecimo, with four engravings; a fine copy in the original marbled calf. London, W. Wetton, n.d., c. 1825.

A delightful copy of one of the numerous editions of the Lee Boo yarn, with a blind-stamp on the front cover of the Society for Promoting Christian Knowledge within a wreath, and a (?bookseller's) stamp inside the front cover "Russell Feb. 1825". **$525**

FRENCH LEE BOO

263. [LEE BOO] Histoire du Prince Lee Boo, ou Aventures d'un jeune Indien… ouvrage récréatif et moral…

Duodecimo, with four engravings; a few spots; pretty original binding of glazed boards, decorated in blind and gilt. Paris, F. Denn, n.d. but c. 1820.

Scarce French edition, described as translated by "F. D * * *" from the sixteenth London edition. **$420**

DISCOVERY OF FRANKLIN

264. McCLINTOCK, Captain Francis L. A. Narrative of the Discovery of the Fate of Sir John Franklin…

Octavo, plates and folding maps; contemporary ownership inscription, extremities a little worn and text a bit shaken, but a good copy in original blue cloth. London, John Murray, 1859.

The search to determine the fate of Franklin's party on board the *Erebus* and *Terror* was one of the most extensive in maritime history. The expedition had been last seen on 26 July 1845, leaving Lancaster Sound, but their three years of provisioning delayed fears for their safety until the winter of 1846-7. In all twenty-nine separate expeditions were mounted up to and including that of McClintock in the *Fox* of 1857-9.

McClintock, who had served with Ross in the Arctic, succeeded where the others had failed and discovered numerous skeletons and relics from the ships. He was able to complete the story from the eyewitness accounts of the native people, and established beyond doubt that Franklin had indeed been the discoverer of the north-west Passage.

Hill, p. 483. **$950**

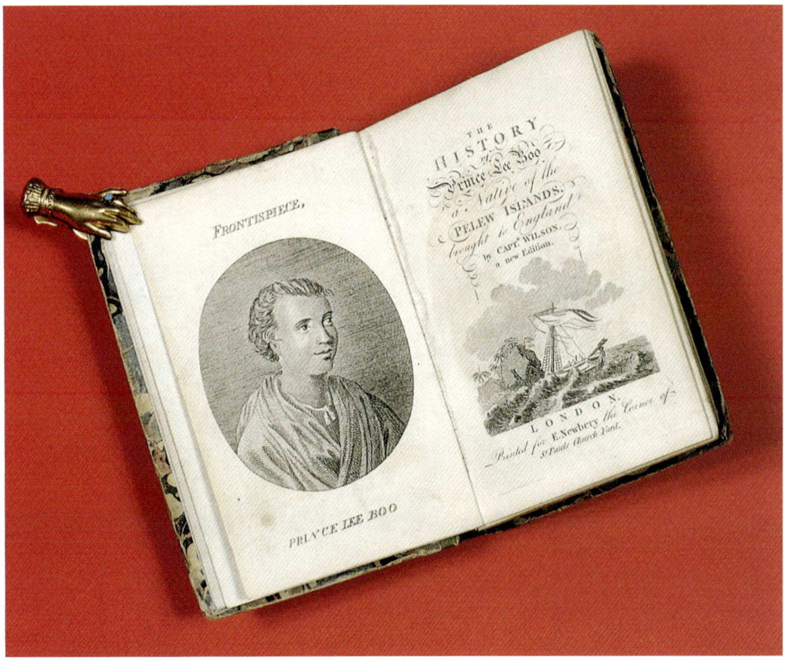

261. Lee Boo

FIRST SWEDISH EDITION

265. MEARES, John. Tvänne Resor från Ostindien till Americas nordvästra kust, åren 1786, 1788 och 1789…

Octavo; contemporary half calf, slight wear but a very good copy. Stockholm, J.S. Ekmanson, 1797.

Scarce: the first Swedish edition, translated by S. Odmann. "One of the early and fundamental books on the Northwest coast of America in general and on Alaska in particular" (Lada-Mocarski). Meares made two fur trading voyages, the first reaching Alaska and Kodiak, but frustrated by the presence of the Russians. On the Northwest coast he met Portlock and Dixon, later sailing to Hawaii and continuing to Canton.

His second voyage to the American coast was to alter the course of history, with his actions leading to the "Nootka Controversy" and nearly precipitating a war between England and Spain. The appendices contain letters and instructions, Duffin's journal while exploring the Straits of Juan de Fuca, and Meares's *Memorial* of 13 May 1790, claiming exclusive rights to Nootka and the prior raising of the British flag.

Lada-Mocarski, 46 (English edition). **$1250**

NORTH POLE ROSS

266. ROSS, Captain. An Account of Capt. Ross's Voyage to the North Pole.

Small octavo chapbook, with woodcut of a ship on the title-page; page repairs and generally aged; modern wrappers. Newcastle, W. Fordyce, n.d. [c. 1835].

Rare chapbook account of Ross's 1829 "expedition on his own resources, with the view of affecting a passage into the Polar Sea, and to reach Bhering's Straits along the northern coast of the American continent" (Preface). **$450**

Rossbank Observatory — Hobart Town. Page 126.

267. Ross

SOUTH POLE ROSS

267. ROSS, Sir James Clark. A Voyage of Discovery and Research in the Southern and Antarctic Regions, during the years 1839-43.

Two volumes, octavo, with 26 plates (including 17 engraved chapter-titles) and eight maps (two folding), one with folds reinforced with linen, some foxing to maps, bound without advertisements; a good copy, in contemporary half blue calf. London, John Murray, 1847.

First edition of "one of the most important works in the history of Antarctic exploration..." (Hill), the official narrative of Ross's famous expedition to the Antarctic in the *Erebus* and *Terror*.

In the course of the expedition Ross reached the highest latitude yet attained, determined the position of the Magnetic Pole and reached within 160 miles of it. It was also during this expedition that he named Mounts Erebus and Terror.

Abbey, 610; Ferguson, 4636; Hill, p. 260; Spence, 993. **$4500**

268. Thomson

VOYAGE OF THE CHALLENGER

268. THOMSON, Sir C. Wyville. The Voyage of the Challenger. The Atlantic. A preliminary account of the general results of the voyage... during the year 1873 and the early part of the year 1876.

Two volumes, octavo, folding maps, numerous illustrations; a fine copy in the original green cloth. New York, Harper, 1878.

In 1872 the first modern scientific voyage around the world was made in the *Challenger*, a steamship that replaced the earlier sailing vessels. This is the first American edition of this account of the voyage: "chiefly an abstract of the less technical portions of my own journal for the first year of the voyage of the *Challenger* (1873), and the early part of the fourth year (1876), when she again traversed the Atlantic on her way home..." (Preface).

Borba de Moraes, p. 861. **$850**

BEST EDITION OF TURNBULL

269. TURNBULL, John. A Voyage Round the World... In which the author visited... the English Settlements of Botany Bay and Norfolk Island, and the principal Islands in the Pacific Ocean.

Quarto; a good copy in old calf, joints neatly repaired. London, A. Maxwell, 1813.

Second and best edition of an important voyage: the earlier edition was much slighter and this edition is always preferred. It contains substantial additions to all chapters, revising and expanding each by eight years. An additional chapter on New South Wales, an account of New Zealand, a narrative of Baudin's voyage on the *Géographe* and *Naturaliste*, the destruction of the ship *Boyd* and a description of the Fiji Islands all appear here for the first time.

Turnbull and his partner John Buyers, attracted by the lucrative profits made by American traders, purchased and fitted out the *Margaret* for a speculative trading venture. They arrived in Sydney via the Cape of Good Hope in February 1801, and the author describes the town, trading activities and agricultural prospects on more than 98 pages.

Borba de Moraes, p. 871; Ferguson, 570; Hill, p. 295; Judd, 176; Kroepelien, 1306; O'Reilly-Reitman, 718. **$3600**

THE FULL FRENCH VANCOUVER

270. VANCOUVER, George. Voyage de Découvertes, à l'océan Pacifique du Nord, et autour du Monde, dans lequel la côte Nord-Ouest de l'Amerique a été soigneusement reconnue et exactement relevée... et exécuté en 1790, 1791, 1792, 1793, 1794 et 1795.

Three volumes, quarto, with engraved plates and maps, and large folio atlas volume with 16 maps, charts and elevations; some spotting to fore-margins of atlas; text in vellum-backed marbled boards, atlas in contemporary red glazed boards, morocco spine. Paris, de l'Imprimerie de la République, An VIII [1800].

A particularly good set of the first French edition of Vancouver's great narrative description of "one of the most important voyages ever made" (Hill). Vancouver, who had served on Cook's second and third voyages, sailed to the Pacific via Australia where he discovered and charted King George's Sound and Cape Hood, passed Van Diemen's Land, then visited New Zealand and Hawaii, and the Northwest Coast. During the course of three seasons, he surveyed Alaska and the Northwest Coast, investigated the Straits of Juan de Fuca, discovered the Strait of Georgia, and circumnavigated Vancouver Island. He visited San Francisco, Monterey, and other Spanish settlements in Alta California, and made three visits to the Hawaiian Islands where he introduced cattle from Monterey. A substantial portion of the text relates to these visits.

Ferguson noted this edition, but could not cite any copies (the 1986 *Addenda* does however now locate Mitchell and State Library of NSW copies).

Cowan, pp. 654-5; Ferguson, 320b; Hocken, p. 30; this edition not in Hill; Judd, 179; O'Reilly-Reitman, 637; Sabin, 98441. **$6750**

SPARRMAN ON VANCOUVER

271. [VANCOUVER, George] SPARRMAN, Anders. En Upptäckts-Resa till Norra Stilla Hafvet, och kring Jordklotet... förrättad åhren 1790, 1791, 1792, 1793, 1794, 1795, under commando af Capitain George Vancouver...

Two volumes in one, octavo, with a folding map and eight engraved plates; an excellent copy in original light blue boards and paper spine. Stockholm, Anders Zetterberg, 1800-1801.

Very rare. The first separate issue of the first edition of an especially interesting version of Vancouver's voyage account. It is not strictly speaking a translation, but rather an abridged account of the voyage written by Anders Sparrman, closely based on the published Vancouver narrative. Sparrman knew Vancouver well: they had spent a couple of years together on the *Resolution* on Cook's second voyage, and he gives his personal impressions of his former shipmate on pp. 2-3 in the first volume.

The folding map is of some interest: drawn and engraved by Sparrman himself, it is a revised impression of the map of the North Pacific and the Northwest coast of America that first appeared in his extremely rare account of Cook's third voyage, *Strödda Underrättelser...* (Stockholm, 1781). In this more detailed, revised version Sparrman has added the discoveries of both La Pérouse and Vancouver to those of Cook. The eight engraved plates depict native inhabitants of Tasmania, Tonga, Alaska, and Kamchatka. Furthermore, Sparrman has added an appendix on the Patagonians.

271. Sparrman

This work was also issued as volumes II and III of Sparrman's three-volume *Utvalda Allmänt nyttiga...* (Stockholm, 1797-1801), the first volume of which is devoted to articles on medicine, pharmacy, chemistry and other scientific topics.

As a version of Vancouver – especially under the editorship of Anders Sparrman – this is of obvious Australian, Hawaiian and Northwest coast interest, and it is very rare. Ferguson knew only the Kroepelien copy, and none is added in the *Addenda*.

Ferguson, 320a (citing the Kroepelien copy only); not in Hill nor Judd; Kroepelien, 1314; O'Reilly-Reitman, 642. **$3850**

11: FRENCH VOYAGES

OFFICIAL ACCOUNT OF THE BAUDIN VOYAGE

272. Baudin

272. [BAUDIN] PERON, François & Louis FREYCINET. Voyage de Découvertes aux Terres Australes… Sur les corvettes le Géographe, le Naturaliste, et la goélette le Casuarina, pendant les années 1800, 1801, 1802, 1803 et 1804.

Two volumes, quarto, and large quarto atlas in two parts containing 41 full-page engravings, many in original handcolouring, and 14 fine charts; small tear to the Australian map; the text volumes uncut; recent quarter morocco and marbled boards, a fine set. Paris, Imprimerie Royale, 1807-1816.

The official account of the important Baudin voyage to Australia and the Pacific.

As Baudin died in the course of the voyage, the narrative was begun by Péron, the expedition's naturalist, and completed by Freycinet after Péron's death. This celebrated voyage took back to France the most important collection of natural history specimens in the history of the French Museum, as well as a mass of geographical information. A navigational text and atlas, issued separately by a different publisher a few years later, is only rarely found together with this narrative account.

The exceptional illustrations in the atlas volume are the work of the remarkable artist Charles-Alexandre Lesueur: "Lesueur's scientific work runs parallel to artistic work of great importance. He was at one and the same time draughtsman and painter, naturalist and landscape artist. His talent was recognised in France in his own lifetime, by the award… of the silver medal of the Société des Beaux-Arts…" (*Baudin in Australian Waters*, p. 26).

272. Baudin

The French and English completed their circumnavigations of the Australian continent at approximately the same time, but because of Flinders's imprisonment by the French on Mauritius it was this French account which contained the first complete and detailed chart of the Australian continent to appear in print. This map, often said to have been at least partly based on Flinders's charts and papers seized by the French, is one of the most beautiful, as well as one of the most famous of all maps of Australia.

The well-equipped Baudin expedition was one of the great French voyages of discovery, and made significant visits to Western Australia, Tasmania and Sydney. The coastal explorations are commemorated by numerous place-names along the Australian coast, especially in Tasmania and Western Australia. French interest in the continent was motivated by military reconnaissance as much as scientific discovery.

Ferguson, 449, 536; Davidson, pp. 108-10; Dunmore, 'French Explorers in the Pacific' II, pp. 9–40; Hill, pp. 229-230; Sharp, 'Discovery of Australia', pp. 232-9; Wantrup, 78a & 79a. **$24,000**

ONLY EDITION IN ENGLISH

273. [BAUDIN VOYAGE] PERON, François Auguste. Voyage of Discovery to the Southern Hemisphere…

Octavo, with two folding engraved plates; a good copy apart from some scattered browning in contemporary polished calf, rebacked. London, Richard Phillips, 1809.

The only narrative of the important Baudin voyage to appear in English, produced by the prolific publisher Phillips who specialised in abridgements for a mass market. This Baudin account is longer than many of his publications, and of more interest than usual as the only version in English.

Péron, the naturalist on the Baudin expedition, was commissioned to prepare his narrative after Baudin's death late in the voyage; a good part of it covers the expedition's findings on the Australian mainland. There are chapters covering the time spent in Tasmania, with a full account of their contact with the Aborigines. The description, here published for the first time in English, of the stay in Sydney, includes a wonderful description of the buildings, a visit to Government House and an excellent account of the natural history and agriculture of the colony. The two plates show a group of Aborigines, and a fine view of Sydney Harbour.

Ferguson, 485; Hill, p. 230. **$750**

274. Baudin

DEPARTURE OF THE BAUDIN VOYAGE

274. [BAUDIN] BONAPARTE, Napoleon. Medal commemorating the Baudin expedition. *Obverse*: Portrait of Napoleon. *Reverse*: "Les Corvettes Le Géographe et Le Naturaliste, Commandées par le Captaine Baudin".

Medal struck in silver, 38 mm. diameter; in fine condition. France, Pierre Antoine Montagny, An. 9 [1801].

Rare medal issued to commemorate the departure of the Baudin voyage to Australia and the Pacific in the *Géographe* and the *Naturaliste.*

Milford Haven, 'British and Foreign Naval Medals. **$2000**

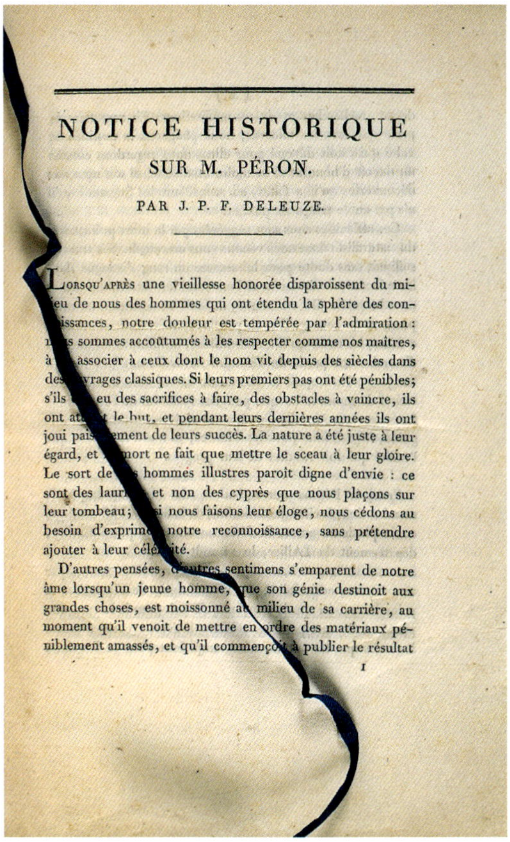

NOTICE HISTORIQUE

SUR M. PÉRON.

PAR J. P. F. DELEUZE.

Lorsqu'après une vieillesse honorée disparoissent du mi-[illegible]eu de nous des hommes qui ont étendu la sphère des con-[illegible]issances, notre douleur est tempérée par l'admiration: [illegible]s sommes accoutumés à les respecter comme nos maîtres, à [illegible] associer à ceux dont le nom vit depuis des siècles dans de[illegible]vrages classiques. Si leurs premiers pas ont été pénibles; s'ils [illegible] eu des sacrifices à faire, des obstacles à vaincre, ils ont at[illegible] le but, et pendant leurs dernières années ils ont joui pai[illegible]ment de leurs succès. La nature a été juste à leur égard, et [illegible]ort ne fait que mettre le sceau à leur gloire. Le sort de[illegible] hommes illustres paroît digne d'envie : ce sont des laur[illegible] et non des cyprès que nous plaçons sur leur tombeau; [illegible]i nous faisons leur éloge, nous cédons au besoin d'exprim[illegible] notre reconnoissance, sans prétendre ajouter à leur célé[illegible]ité.

D'autres pensées, [illegible] sentimens s'emparent de notre âme lorsqu'un jeune homme, [illegible]e son génie destinoit aux grandes choses, est moissonné a[illegible] milieu de sa carrière, au moment qu'il venoit de mettre en [illegible]rdre des matériaux pé-niblement amassés, et qu'il commenço[illegible] à publier le résultat

I

275. Deleuze

OBITUARY OF PERON

275. [BAUDIN VOYAGE] [PERON] DELEUZE, J.P.F. Notice historique sur M. Péron.

Quarto, [ii], 28 pp. and a final blank leaf; uncut; a fine copy in modern boards, leather label. N.p., n.d. [Paris, c. 1810].

Rare separately-published obituary of François Péron. Editor of the first portion of the official account of Baudin's voyage (which was completed after his death by the explorer Freycinet), Péron had travelled as zoologist-anthropologist on the voyage, and was effectively the leader of those who took issue with the commander's anti-scientific strictures. Péron certainly took his revenge after Baudin's death: nowhere in the text of the account of the voyage is Baudin's name mentioned; and while no Australian placename commemorates the commander's name, Peron Peninsula is a sizeable memorial to the scientist.

Not in Ferguson or Hill. **$2400**

276. Borget

"VERY RARE AND LITTLE KNOWN"

276. BORGET, Auguste. Fragments d'un voyage autour du Monde.

Oblong quarto, comprising a lithograph title-page and 12 tinted lithograph plates on china paper mounted, all but one with a leaf of descriptive text; a few occasional spots but overall an excellent copy in the original glazed boards, slightly rubbed, worn at edges, but preserving the original decorative design and lettering on the front cover. Moulins, P.A. Desrosiers, [1850].

An exceptionally rare album of voyage views. We have sold another copy in recent years, but otherwise have not seen the book on the market. This copy is complete except for a leaf of descriptive text for plate 10, evidently never present in this copy. There is a copy of the book in the Peabody Museum, and another in the Hawaiian State Archives, but it is not recorded by Cordier, nor does it appear in the Hill catalogue, Sabin, Judd, or Kroepelien.

The book, produced in the provincial town of

Moulins (between Lyons and Paris), describes and illustrates scenes evidently observed by Auguste Borget during an actual voyage made in the 1840s. A painter who specialised in oriental subjects, Borget was born at Issoudun in 1809 and died in Châteauroux in 1877. According to Bénézit, he was known as a book-illustrator. Some of his illustrations were signed with the pseudonym "Old Nick".

The twelve fine plates are printed in two tints, lithographed by Borget himself. The subjects are:- 1. Moulin à vent, sur les bords de l'Hudson, en face New York; 2. Notre Dame de gloire (à Rio de Janeiro); 3. Une rue de Buenos Ayres; 4. Un Ravin dans la Sierra de Cordova (Amérique du Sud); 5. Une rue à Lima (Pérou); 6. Halte de Chiliens dans la Plaine de Santiago (Chili); 7. Un Abreuvoir à Aréquipa (Pérou); 8. Rue et marché à Canton (Chine); 9. La Plage d'Honoloulou (Iles Sandwich); 10. Balzas (bateaux de pêche) sur la côte de Bolivie; 11. Pont et village de Passig, à 6 lieues de Manille (Iles Philippines); 12. Rue de Clives à Calcutta.

In his text facing the view of Honolulu, Borget praises the simplicity of the way of life of the islanders, and notices their lack of nostalgia for a quickly disappearing past. He voices fears for the future: "Alas! I fear that the simple and primitive beings who live on the shores of this beautiful sea, in modest huts at the foot of these beautiful palm trees, may pay dearly for their rushed initiation into our European ways, and that our civilisation may slaughter them just as it did the Indians of South America...".

Borba de Moraes, 112 ("very rare and little known"). **$9500**

277. Bougainville

THE FIRST FRENCH CIRCUMNAVIGATION

277. BOUGAINVILLE, Louis Antoine de. Voyage autour du Monde, par la frégate du Roi, La Boudeuse, et la Flûte l'Etoile; en 1766, 1767, 1768 & 1769.

Quarto, with 20 engraved charts (18 of them folding) and three plates of boats; contemporary ink annotations on half-title and title, one tear to a text-leaf repaired; French red morocco, gilt arms at centre of each side, gilt edges. Paris, Saillant & Nyon, 1771.

First edition: the first French circumnavigation. Bougainville's account of his travels in the Pacific created enormous interest in France, and was largely responsible for building up the romantic vision of a South Sea paradise where Rousseau's noble savage lived in a state of blissful innocence.

Bougainville sailed from Nantes in November 1766 to the Falkland Islands. They picked up a supply ship, the *Etoile*, and both ships passed through the Straits of Magellan in January 1768, spent time looking for the mythical "Davis Land", said to be off the Chilean coast, and then started on a direct route across the Pacific. Bougainville discovered the Tuamotus, and in April sighted and thought he had discovered Tahiti, not knowing that Wallis had been there the previous year. He visited Samoa, sailed through Melanesia, sighted the Great Barrier Reef, and passed through the Solomons, and New Britain, to Batavia.

It was in fact only the Great Barrier Reef that prevented Bougainville landing on Australian soil. He had specifically set out to reach "New Holland" by running west from Quiros's "Espiritu Santo", but turned away from the obvious barrier presented by the Reef. Had he made it, "he would have come to the Australian coast near Cooktown, and would, likely enough, have been wrecked where Cook was wrecked two years later..." (Wood).

From the library of the Australian collector F.G. Coles. This attractive French morocco binding is not contemporary, but has on the covers the gilt arms of the eighteenth-century collector Charles Gravier de Vergennes.

Hill, p. 31; Kroepelien, 109; O'Reilly-Reitman, 283. **$7250**

TRANSLATED BY THE FORSTERS

278. BOUGAINVILLE, Louis Antoine de. A Voyage round the World...

Quarto, with five folding maps and a folding plate of canoes; contemporary calf, slightly rubbed, small defect at base of spine, joints just splitting but the sides firmly held. London, Printed for J. Nourse and T. Davies, 1772.

First English edition of the first French circumnavigation, translated by Johann Reinhold Forster and varying from the French original by the addition of an introduction by him (according to the preface, although the translation at least is now thought to have been the work of Forster's son Georg).

Forster senior calls the voyage "a work written by a learned, intelligent, and judicious traveller, which abounds with remarkable events and curious observations...", and says that he has edited "and partially vindicated the British nation where we thought the author had been unjustly partial".

Hill, p. 32; Kroepelien, 113; O'Reilly-Reitman, 285. **$5500**

VOYAGE OF THE THETIS

279. BOUGAINVILLE, Baron Hyacinthe de. Journal de la navigation autour du globe de la frégate La Thétis et de la corvette l'Espérance, pendant les années 1824, 1825, et 1826.

Two volumes, quarto, text, with a folio atlas containing 56 engraved or lithographed plates and maps, 12 of them coloured; the paper stock of the coloured plates slightly browned as always, small erasures from title-pages; an excellent set with the atlas in contemporary French marbled calf gilt (neat repairs to spine and front cover) and the text in modern French quarter calf. Paris, Arthus Bertrand, 1837.

A rare and most attractive French voyage account. Hyacinthe de Bougainville sailed as an eighteen-year-old ensign on the Baudin voyage. After distinguished service in the Napoleonic Wars, he was given command of the *Thétis*, only the second French frigate to be commissioned for a circumnavigation, the first having been his father's vessel the *Boudeuse*.

The voyage took twenty-eight months, visiting amongst other places Pondicherry, Manila, Macao, Surabaya, Sydney (a stay of almost three months), Valparaiso and Rio. Bougainville returned to France with a fine collection of natural history specimens, and the official account of the voyage was handsomely published in this form after a delay of some eleven years.

The fine plates in the atlas include a number of Australian views which are not generally as well known as the images from other French voyages of the nineteenth century. The topographical views include several of Sydney (Fort Macquarie, Government Stables, La Pérouse monument) and others of the countryside around (two of the Nepean, and a Blue Mountains view, as well as a fine group portrait of Aborigines of Camden Shire). There is also a fine plan of Port Jackson, as well as a number of natural history plates, with those of the male and female gang-gang particularly striking.

Ferguson, 2236; Hill, p. 31. **$22,500**

SIBERIA

280. CHAPPE D'AUTEROCHE, Abbé Jean. A Journey into Siberia, made by order of the King of France... containing an Account of the Manners and Customs of the Russians...

Quarto, folding map, chart, and eight engraved plates; wormhole in outer margins of most leaves, only touching text on about 20 leaves; otherwise a good copy in contemporary calf, rebacked. London, Printed for T. Jefferys, 1770.

First English edition of Chappe d'Auteroche's classic work on Siberia, published by Jefferys, the Royal Geographer. Priest and astronomer, Chappe d'Auteroche was closely involved with the Pacific from both sides, and observed the transits of Venus from Siberia in 1761 and from California in 1769.

Not in the Hill catalogue. **$2400**

279. Bougainville

THE FRENCH IN NEW ZEALAND

281. CROZET, Julien Marie. ROCHON, Alexis Marie, editor. Nouveau Voyage à la Mer du Sud, commencé sous les ordres de M. Marion... On a joint a cet voyage un Extrait de celui de M. de Surville dans les mêmes Parages.

Octavo, with seven plates (including two maps, one folding); a few stains but a good copy in old French marbled boards, later calf spine. Paris, Barrois l'ainé, 1783.

First edition of an important voyage to the Pacific, the first French visit to New Zealand and the second French visit to Australia, preceded only by Allouarn's brief stop in the West. This is "an exceedingly rare item... seldom available" (Davidson).

The expedition, initially under the command of Marion du Fresne, was a sequel to Bougainville's in that it set out to return the Tahitian Mayoa, who had been brought to France by Bougainville, to his homeland. Mayoa died on the voyage. Their other instructions were to search for "Terra Australis": they actually discovered the Crozet Islands, visited Tasmania – exploring and mapping some of its southern coast – and made a lengthy stay in New Zealand, where the captain and twenty-six crew were massacred at the Bay of Islands. Crozet took over the command: they returned to France via Guam, the Philippines, and Mauritius. "Crozet's narrative, apart from the drama of its story, has much careful observation on Maori life and custom and, with the reports of Cook and his officers, was virtually the only source material available for 40 years..." (NZNB).

An abridged version of Surville's voyage, in which he reached the New Zealand coast and just missed meeting Cook, appears at the end of the volume.

Davidson, pp. 98-99; Hill, p. 72; Hocken, pp. 21-22; Kroepelien, 1104; NZNB, 1502; Sabin, 72371. **$13,500**

282. Astrolabe medal

DEPARTURE OF DUMONT D'URVILLE

282. [DUMONT D'URVILLE, J.S.] Medal for the departure of the first voyage of the Astrolabe. *Obverse*: Portrait of Charles X, "Roi de France et de Nav."; *Reverse*: "Voyage de Decouvertes de la Corvette L'Astrolabe...".

Silver medal, 50 mm.; in fine condition. Paris, Depaulis, 1826.

Rare original medallion commemorating the departure on 25 April 1826 of the first voyage of the *Astrolabe* to the South Seas, under the command of Dumont d'Urville.

The *Astrolabe* (Duperrey's old ship the *Coquille* renamed in honour of La Pérouse) was instructed to explore the principal island groups in the South Pacific, completing the work of the Duperrey voyage, on which the commander had been a naturalist. The expedition sailed via the Cape of Good Hope, through Bass Strait, stopped at Port Phillip, and arrived at Sydney on 1 December 1828. They later sailed via the northwest coast of Australia to Tasmania, from where they proceeded to Vanikoro in search of traces of La Pérouse.

Marquess of Milford Haven, 'British and Foreign Naval Medals', London, 1919-1928. **$2400**

SOME PLATES IN TWO STATES

283. DUMONT D'URVILLE. Voyage de la corvette l'Astrolabe... An album of plates from the Atlas Historique.

Oblong folio, with 68 images, 13 of them in two states (coloured and uncoloured) and one in three states, comprising altogether 82 lithographs, 16 of them in colour, the others on china paper mounted; the plates in beautiful condition, bound as an oblong folio album in plain limp boards. Paris, J. Tastu, 1833.

A splendid, specially-assembled series of eighty-two lithographs from the *Astrolabe* voyage, obviously gathered together by a nineteenth-century collector. Those chosen include all of the Australian subjects and most of the New Zealand ones.

The full account of Dumont d'Urville's famous Pacific voyage, rarely found complete, consists of sixteen volumes and four atlases. Our album has a healthy assortment of beautiful coloured and uncoloured plates from the *Atlas Historique*, including all of the plates of Australian subjects and most of the New Zealand ones. The anthropological depictions are remarkable in that they are present mainly in two states, uncoloured (on china paper, mounted) and coloured. The colour plates are especially striking. A full listing of the plates is available on request.

The existence of this "tailor-made" album illustrates the immediate and deep interest in the results of Pacific exploration, and in particular the major ethnographic discoveries and observations being made.

Ferguson, 1341 (not noting any copies with plates in both states); Hill, p. 88. **$21,000**

283. Dumont d'Urville

RARE ABRIDGED ATLAS

284. DUMONT D'URVILLE, Jules Sébastien. Voyage de la corvette l'Astrolabe exécuté pendant les années 1826, 1827, 1828, 1829…

Folio, engraved title with vignette, lithograph portrait, eight maps (one folding, five double-page) and 12 lithograph plates, six of them coloured by hand; two plates spotted; generally in excellent condition in the original quarter calf and marbled boards, several pieces chipped from spine. Paris, Tastu, 1833.

A rare abridged form of the historical atlas to the *Astrolabe* voyage. The full publication of the voyage ran to more than twenty volumes, with 237 plates in the historical atlas alone. Here, the publisher has selected plates of Australia, New Zealand, Tonga, New Guinea and Vanikoro (including the inauguration of the monument to La Pérouse).

Apparently unrecorded in this form. **$5850**

HYDROGRAPHICAL ATLAS

285. DUMONT D'URVILLE, Jules Sebastien. Voyage de la corvette l'Astrolabe… Atlas.

Large folio, vignette on title (La Pérouse monument at Vanikoro), 42 maps, many double-page, three plates of coastal profiles partly coloured, and a double-page table; the table slightly stained, otherwise all in fine condition; later half calf over original marbled boards, rubbed. Paris, Tastu, 1833.

The superb hydrographical atlas of Dumont d'Urville's *Astrolabe* voyage, which formed part of the immense official account of the voyage, but was available quite separately as a series of charts. These magnificent and very large maps of the Pacific are among the finest ever produced, and include an important series of charts of Australia (especially of King George's Sound, Bass Strait, Western Port and Jervis Bay) and New Zealand, as well as a close examination of Vanikoro, of particular interest to the French since the identification of it as La Pérouse's final port-of-call. There are also very good maps of New Guinea and of Tonga and other Pacific islands.

The final three plates comprise an unusual series of coloured coastal-profile views: one is a sheet of New Zealand scenes and the others show a variety of coasts including Tasmania and Vanikoro.

Ferguson, 1341; Hill, p. 88. **$7500**

BOTANICAL ATLAS

286. DUMONT D'URVILLE, Jules Sébastien. Voyage de la corvette l'Astrolabe… Atlas Botanique.

Folio atlas, with 80 engraved plates (including two 'bis' numbers); some light foxing mainly confined to tissue guards and margins; an excellent copy in the original quarter blue calf. Paris, J. Tastu, 1833.

The finely illustrated botanical atlas from the voyage of the *Astrolabe*, containing forty-one beautifully printed full-page engravings of the flora of New Zealand, some of them printed in colour, and thirty-nine plates of the "Sertum Astrolabianum", the second part of the botanical publication, some of these printed in bistre.

Ferguson, 1341; Hill, p. 88. **$3000**

RAPPORT

SUR LES OPÉRATIONS DE LA CAMPAGNE

DE LA

CORVETTE L'ASTROLABE,

DEPUIS LE DÉPART DE RIO-JANEIRO

JUSQU'A L'ARRIVÉE A VALPARAISO.

287. Dumont d'Urville

FIRST DESCRIPTION

287. DUMONT D'URVILLE, Jules Sébastien. Rapport sur les opérations de la campagne de la corvette l'Astrolabe, depuis le départ de Rio-Janeiro jusqu'à l'arrivée à Valparaiso…

Octavo, 28 pp., with two large folding charts (small tear to one); title-page slightly stained; uncut, old quarter cloth with leather label. Paris, Imprimérie Royale, 1838.

Very rare: the first description of any part of Dumont d'Urville's second, Antarctic expedition to appear in print. It takes the form of a long formal communication to the minister for the marine written by the explorer aboard the *Astrolabe* in May 1838. This is the separate issue of this report, produced as a separately-paginated offprint (pp. 1-28) from the *Annales Maritimes et Coloniales* for 1838 (series II, vol. II, pp. 1146-1170). This version would have been issued well before distribution of the periodical, to a relatively limited audience.

The report covers the period from November 1837, during which time the expedition left Rio and headed south, spending time in Patagonia and investigating the Straits of Magellan and Lemaire Strait before heading further south towards the polar ice-pack, which they reached during January. Dumont d'Urville gives a long description of their experiences in the ice, some of it hair-raising with both ships at one time blocked in the ice for five days. When they finally escaped, "Les marins de *l'Astrolabe* s'écrièrent: *Nous sommes sauvés! Nous voilà revenus sur le liquide!*".

They could get no further south than the 64th parallel, and turning to the east they headed for the South Orkney and South Shetland Islands, and exploring around them made the important discoveries of Joinville Island and Louis Philippe Land. In his letter written on the way back to Valparaiso, the commander specifically requests the minister to send the accompanying maps to the *Annales Maritimes* for publication: there is thus a general map showing the expedition's progress from January to March 1838 which includes the South Orkney and South Shetland visit and a separate, more detailed map of Terre-Louis Philippe itself. Both maps are based on work by Vincendon-Dumoulin.

Apparently unrecorded in the standard bibliographies. **$3400**

288. *Dumont d'Urville*

NATURAL HISTORY

288. DUMONT D'URVILLE, Jules Sébastien. Voyage au Pole Sud et dans l'Océanie sur les corvettes l'Astrolabe et la Zelée... Natural History Atlas.

Folio, series of 61 plates, mostly coloured; seven plates uniformly browned, otherwise in excellent condition, in contemporary quarter blue morocco, spine worn at top. Paris, Gide, 1842.

The splendid series of thirty-six ornithological plates and twenty-five plates of insects, published as one of the three natural history atlases in the official account of Dumont d'Urville's second expedition (1837-40). The birds illustrated include specimens from all over the Pacific and a number of Australian species, including the famous plate of the cockatoos.

Ferguson, 3184 (part); not in the Hill catalogue. **$9500**

IMAGES OF THE DUPERREY VOYAGE

289. DUPERREY, Louis Isidore. Voyage autour du monde... [A series of lithograph plates illustrating the voyage].

Eight original parts, folio, containing a total of 72 plates, 50 of which are in original colouring; completely unbound as issued, and in fine condition in the original printed wrappers, some fraying to spines and edges. Paris, Arthus Bertrand, c. 1826.

A substantial series of plates from the finely illustrated Duperrey voyage, published in Paris between 1826 and 1830; publication was never in fact completed. The plates here are from three of the four atlases of the full account (the fourth was the hydrographical atlas); they are in the original 'livraison' loose printed wrappers although, typically, they do not necessarily correspond with the lists of plates on the front wrappers.

A full list of the plates may be had on request: broadly, this set comprises twelve coloured plates from the *Atlas Historique* (which was finally issued with a total of sixty plates); twenty-eight plates, of which six are coloured, from the *Atlas Botanique* (issued with one hundred and six plates); and thirty-two coloured plates from the *Atlas Zoologique* (issued with one hundred and fifty-seven plates).

Complete sets of the full account are rare, and often incomplete. This assembly represents a chance to acquire a significant portion – over a fifth – of the graphic record of the great voyage.

See Borba de Moraes, p. 276; Ferguson, 941; Hill, p. 90. **$8500**

VINCENDON-DUMOULIN ON TAHITI

290. [DUPERREY] VINCENDON-DUMOULIN, C.A. and C. DESGRAZ. Iles Taïti. Esquisse Historique et Géographique précédée de considérations générales sur la colonisation Française dans l'Océanie.

Two volumes, octavo, with two folding maps (one of Tahiti from information on the Artemise voyage); recent French quarter morocco, marbled boards. Paris, Arthus Bertrand, 1844.

A very scarce title: one of the most extensive early works on Tahiti. Vincendon-Dumoulin was the hydrographer on the voyage of the *Coquille*, which left France in 1822 with a crew of sixty-five men under Duperrey, with instructions to complete the work started by the *Uranie* under Freycinet. This work is an in-depth study of the expedition's discoveries in Tahiti, particularly the history of the islands based on information acquired during their visit, the first French visit to the islands since that of Bougainville.

The interesting comments in ink that appear throughout this copy were made by a descendant of Pomare, the Tahitian King.

Kroepelien, 1328; O'Reilly-Reitman, 754, 4243, 4751, 6269, 6590. **$1850**

DEPARTURE OF THE COQUILLE

291. [DUPERREY] Medal for the voyage of the Coquille. *Obverse*: Portrait of "Ludovicus XVIII. Franc et Nav. Rex". *Reverse*: Voyage Autour du Monde de la Corvette La Coquille.

Medal struck in bronze, 50 mm. diameter; in good condition. Andrieu, 1822.

Medallion struck to commemorate the departure of the Duperrey voyage on the *Coquille*.

Marquess of Milford Haven, 'British and Foreign Naval Medals', London, 1919-1928. **$2200**

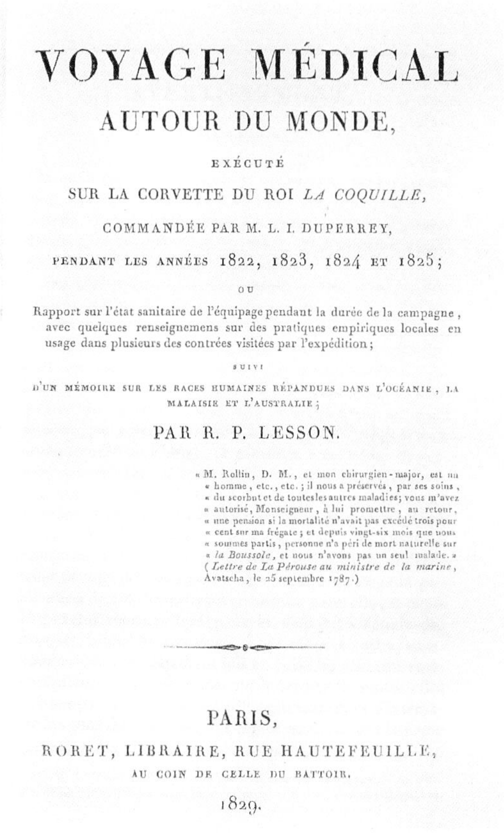
VOYAGE MÉDICAL
AUTOUR DU MONDE,
EXÉCUTÉ
SUR LA CORVETTE DU ROI *LA COQUILLE*,
COMMANDÉE PAR M. L. I. DUPERREY,
PENDANT LES ANNÉES 1822, 1823, 1824 ET 1825;
OU
Rapport sur l'état sanitaire de l'équipage pendant la durée de la campagne, avec quelques renseignemens sur des pratiques empiriques locales en usage dans plusieurs des contrées visitées par l'expédition;
SUIVI
D'UN MÉMOIRE SUR LES RACES HUMAINES RÉPANDUES DANS L'OCÉANIE, LA MALAISIE ET L'AUSTRALIE;
PAR R. P. LESSON.

« M. Rollin, D. M., et mon chirurgien-major, est un « homme, etc., etc.; il nous a préservés, par ses soins, « du scorbut et de toutes les autres maladies; vous m'avez « autorisé, Monseigneur, à lui promettre, au retour, « une pension si la mortalité n'avait pas excédé trois pour « cent sur ma frégate; et depuis vingt-six mois que nous « sommes partis, personne n'a péri de mort naturelle sur « *la Boussole*, et nous n'avons pas un seul malade. »
(*Lettre de La Pérouse au ministre de la marine*, Avatscha, le 25 septembre 1787.)

PARIS,
RORET, LIBRAIRE, RUE HAUTEFEUILLE,
AU COIN DE CELLE DU BATTOIR.
1829.

292. Lesson

BY THE SHIP'S SURGEON

292. [DUPERREY VOYAGE] LESSON, René Primavère. Voyage Médical autour du monde, exécuté sur la corvette du roi La Coquille…

Octavo; a fine copy, with the half-title, in later French quarter calf. Paris, Roret, 1829.

A surprisingly rare voyage account, by the ship's surgeon on the Duperrey expedition. Lesson succeeded to the post after Garnot, *chirurgien-major* at the start of the voyage, had to be put ashore in Sydney for ill health. He prepared a report, as was normal practice on French exploring voyages, to hand to the health department at the ministry of marine as soon as the expedition reached France. Lesson explains that he later showed the report to a number of his medical colleagues and was persuaded to publish it.

This is not a dry, medical report so much as a narrative of the voyage, interspersed with the scientific observations of a medically-educated man. Lesson includes as a supplement his lengthy anthropological memoir on the races of Oceania, Malaysia and Australia. The book is particularly interesting as Lesson – in this and his later *Voyage autour du monde* (1838; see below) – provided the only separately-published narrative of the voyage: Duperrey's own account was never completed.

Ferguson, 1279; not in Hill; not in O'Reilly-Reitman. **$7850**

LESSON'S SECOND
BOOK ON THE VOYAGE

293. [DUPERREY VOYAGE] LESSON, René Primavère. Voyage autour du monde entrepris par ordre du Gouvernement sur la corvette La Coquille…

Two volumes, octavo, with a portrait and 42 plates, 19 of which are coloured; some light spotting but overall a handsome copy in contemporary purple polished and grained roan, the ornately gilt spines sunned, gilt wreaths and borders to sides. Paris, P. Pourrat frères, 1838-9.

The original Paris edition (there were later Paris and Brussels versions). Lesson's account is of special interest as this and his *Voyage médical* (see above) were the only separate narrative accounts to be published of Duperrey's voyage in the *Coquille* as the official history was never completed. Otherwise only the "Observations personelles" of Dumont d'Urville, an officer on the voyage, appeared (in volume III of the *Astrolabe* Voyage).

Lesson gives detailed descriptions of their calls at Hawaii, the Falklands, New Guinea, New South Wales and New Zealand, as well as Tahiti and other stops in the South Seas. "The expedition achieved notable scientific results and corrections in maps, much meteorological data, and many rock samples and botanical specimens were brought back. Lesson was the naturalist of this expedition, and his account of the voyage supplies the details which Duperrey failed to publish" (Hill). The coloured plates are particularly attractive, and include a number of Australian species.

Ferguson, 2533; Hill, p. 179; O'Reilly-Reitman, 828. **$4250**

293. Lesson

294. Freycinet

VOYAGE OF THE URANIE

294. FREYCINET, Louis Claude Desaulces de. Voyage autour du monde, enterpris par ordre du Roi… exécuté sur les corvettes de S. M. l'Uranie et la Physicienne, pendant les années 1817, 1818, 1819 et 1820.

Together nine volumes of text bound as twelve volumes, quarto and four folio atlases; occasional spotting in the text (as expected), a couple of old ownership stamps; the atlases with some faint browning or spotting (mainly marginal); the whole set overall in good condition, in a uniform binding of quarter French dark blue straight-grained morocco and marbled boards. Paris, 1824-1844.

One of the most important of all voyages to the Pacific: this magnificent publication, which appeared over twenty years under the direct supervision of Freycinet himself, is among the most handsome of all voyage accounts.

Twelve volumes of text are accompanied by four large atlases, containing a total of about 320 engraved plates and 35 maps. The plates are partly derived from original paintings and drawings done on the expedition by the two official artists, Jacques Arago (who also wrote his own account of the expedition – see catalogue numbers 297-9) and Alphonse Pellion, and partly on natural history specimens taken back to Paris. The *Atlas Historique* contains sixteen views and plates of Australia, thirty-two of Guam and the Marianas, nine of Hawaii and sixteen of Timor.

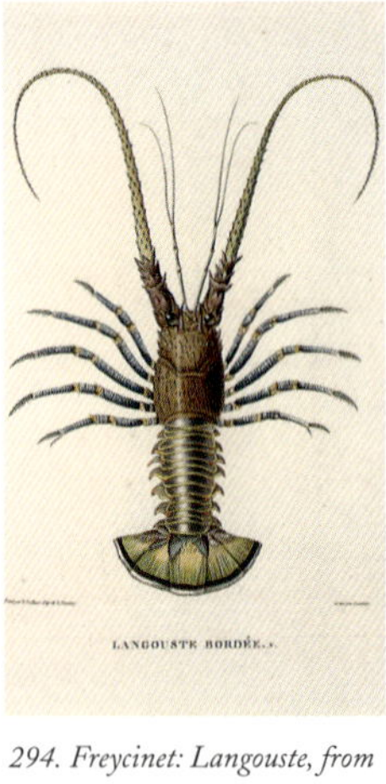

294. Freycinet: Langouste, from the Atlas zoologique

Freycinet's major Pacific expedition was commissioned by the French government partly to make observations on geography, magnetism and meteorology, though its natural history discoveries were to prove of equal value, and partly to examine distant territory, one of the most important areas to be observed being the Australian continent, both its settled areas and some of the virgin territory.

Freycinet had previously sailed to the Pacific with Baudin – in fact he co-authored the official account of the voyage (see catalogue number 272) – so had already made close examinations of large parts of the Australian coastline during the survey on which they met Flinders at Encounter Bay, and had carried out detailed charting in Tasmania. On this expedition, they concentrated on Western Australia, of obvious interest for the possibilities that it presented. Their visit to Shark Bay is beautifully recorded in the *Atlas Historique* here with a famous series of views and maps, including a wonderful colourplate of the expedition's encampment on the Peron peninsula, and the meeting with the West Australian Aborigines.

A long examination of parts of the East Indies – particularly Timor, the Mariana and Caroline Islands and Guam – produced important scientific results and again superb topographical art. This was followed by the important visit to Hawaii, at a crucial time in the history of the islands. The expedition arrived in August 1819, and visited Maui, Hawaii and Oahu. Theirs was the last careful examination of the native culture before it was dismantled by western influences. The *Atlas Historique* has a series of plates of exceptional interest and beauty: the nine named portraits of inhabitants include one with a tattoo commemorating the death of Kamehameha I, while there is a magnificent study of an officer in a feathered cloak as well as other important ethnographic plates, views of the *morai* which would not be so accessible to later visitors, and the superb coloured plate of the baptism of the king's prime minister aboard the French ship.

The New South Wales section has several fine maps, an important series of portrait-studies of named Aborigines, a well-known and beautiful series of views in the Blue Mountains, as well as images of Government House, Parramatta and the Barracks in Sydney.

The rarely seen *Atlas Hydrographique* contains seven maps of Guam, four of Hawaii (including Kealakekua Bay, Lahaina, Maui and Honolulu), a fine map of Shark's Bay in Western Australia and four of New Guinea.

The scientific atlases are splendid too. Despite the expedition's losses after the wreck of the ship at the Falklands, they still took back literally masses of natural history specimens, of which forty-five of the birds (among which were three new genera), more than thirty reptiles and perhaps one hundred and twenty fish were entirely new to science. And of the approximately three thousand plant specimens (largely from Australia, the Marianas and Hawaii) some one thousand two hundred were hitherto unknown. This remarkable catch is reflected in the botanical atlas and, even more spectacularly, in the superb series of zoological plates and especially the magnificent coloured bird plates. The fish plates too are particularly beautiful.

Freycinet, in advance of modern scientific method, passed out standardised questionnaires among his staff, requesting them to examine everything which "embraced the physical, moral, and political condition of man". The voyage has also become famous by the fact that Freycinet, contravening orders, was clandestinely accompanied by his wife, she thereby becoming the first white woman to land in Western Australia and to see many Pacific Islands, including Hawaii.

The publication is composed as follows:

Zoologie. Text by Quoy & Gaimard, one volume quarto. With Atlas, folio, with engraved title and 96 engraved plates (all but 19 printed in colour and finished by hand). Paris, Langlois, 1824.

Historique. Text by Freycinet, two volumes in five, quarto, with a colour-printed engraved plate of Chinese tomb stones at Timor. With Atlas Historique, folio, with engraved title and 112 plates comprising 13 maps and plans (two folding) and 99 engraved plates (41 printed in colour and finished by hand). Paris, Pillet aîné, 1825-39.

Botanique. Text by Charles Gaudichaud, one volume quarto. With Atlas Botanique, folio, with engraved title and 120 engraved plates. Paris, Pillet aîné, 1826.

Navigation et Hydrographie. Text by Freycinet, two volumes, quarto, with three engraved plates. With Atlas, folio, with 22 engraved sheets of maps and charts (10 of which are folding). Paris, Pillet aîné, 1826.

Observations du Pendule. Text by Freycinet, one volume quarto. Paris, Pillet aîné, 1826.

Magnetisme Terrestre. Text by Freycinet, one volume quarto, with folding engraved chart. Paris, Imprimerie Royale, 1842.

294. Freycinet: meeting the Portuguese at Diely

294. Freycinet: the encampment at Shark Bay

Météorologie. Text by Freycinet, one volume quarto. Paris, Imprimerie Royale, 1844.

This is one of the most important of all major voyage accounts. Ferguson records only two sets – the Mitchell and National Library copies, the latter incomplete.

Ferguson, 941; Hill, 425 (Historique only). **$128,000**

297. Arago

BOTANICAL ATLAS

295. [FREYCINET] POIRET, A., engraver, after various artists. Voyage autour du monde. Histoire Naturelle: Botanique. Planches.

Folio atlas, with 120 engraved botanical plates; some sporadic foxing as almost always with this book, but a decent copy in contemporary brown morocco, ornately decorated in blind and gilt, gilt edges. Paris, chez Pillet aîné, 1826.

The very scarce botanical atlas – forming part of the full official account of the Freycinet expedition. "Gaudichaud was the first botanist of the French expeditions to botanise in the Blue Mountains and Bathurst, which he did on a brief journey, with the assistance of Cunningham and Fraser... Gaudichaud's fine work (as principal contributor) and a splendid folio atlas of 120 plates... are valuable to the botanist for the specific localities of plants collected. Port Jackson, Botany Bay, the Blue Mountains and Shark's Bay in Western Australia were visited..." (Maiden, "Records of the Earlier French Botanists", *Journal and Proceedings of the Royal Society of NSW*, vol. XLIV, pp. 137-8). The expedition made a special study of the botany of all the countries they visited. The atlas includes thirty-two plates of Hawaiian plants (including the Sandalwood – plate 45), and there are twenty plates of newly described plants from New Holland ("a charming country", thought Gaudichaud, and the only place visited on the voyage that reminded him of "notre belle France").

Ferguson, 941; Hill, p. 7. **$4750**

DEPARTURE OF THE EXPEDITION

296. [FREYCINET] LOUIS XVIII. Medal for the voyage of the Uranie.
***Obverse*: profile portrait of Louis XVIII. *Reverse*: "Hemisphere Austral. Physique Astronomie. La Corvette l'Uranie Mr. Ls. de Freycinet Commandt...".**

Copper medal, 41 mm. diameter; in very good condition. France, Gayrard, 1817.

Medal struck to commemorate the sailing of the *Uranie* expedition from Toulon in 1817 under Freycinet.

Marquess of Milford Haven, 'British and Foreign Naval Medals', London, 1919-1928. **$1650**

FIRST EDITION IN ENGLISH

297. [FREYCINET] ARAGO, Jacques. Narrative of a Voyage Round the World, in the "Uranie" and "Physicienne" corvettes, commanded by Captain Freycinet during the years 1817, 1818, 1819 and 1820... In a series of letters to a friend...

Quarto, with a folding map as frontispiece and 25 engraved plates from original drawings by Arago; a few spots, but a very good copy in half calf, extremities rubbed. London, Treuttel and Wurtz, 1823.

The first edition in English of this private narrative of the 1817-1820 Freycinet expedition to Australia and the Pacific – in fact the first appearance in English of any account of the voyage. Originally published in French, Arago's book in its many subsequent editions became one of the voyage best-sellers of the nineteenth century.

Arago was the official artist on the voyage, and the lithograph plates here are all after his own drawings.

He writes in an entertaining style with his text reflecting his keen powers of observation. His narrative is readable, not least because he entirely avoids the conventional forms of the voyage narrative, ignoring the "eternal repetition of winds, currents, longitude and lattitude". Full of wry humour, it takes the form of a series of letters to a friend. Long portions relate to Australia, firstly the visit to Western Australia and later descriptions of Sydney, the Blue Mountains, and meetings with Governor Macquarie and John Oxley. Arago's work provides an independent account of the time spent in Australian waters, with an excellent summary of life in Sydney and colonial society at the close of the Macquarie era. The French were entertained handsomely by the Governor and other members of the local aristocracy, with expeditions to the country estates of the Macarthurs, the Oxleys and the Kings being the order of the day.

There is also a long account (almost 100 pages) of their stay in Hawaii at a crucial period in the history of the islands.

Ferguson, 885 (not noting the map in error); Hill, p. 295; Judd, 4. $6850

BEST EDITION OF ARAGO, WITH COLOURED PLATES

298. [FREYCINET] ARAGO, Jacques. Souvenirs d'un aveugle, Voyage autour du monde...

Four volumes, octavo, with two portraits and 60 coloured lithographs; a fine set in contemporary quarter calf, flat spines panelled in gilt. Paris, Gayet et Lebrun, 1840.

The best edition of the artist Jacques Arago's narrative of the Freycinet voyage, with a superb suite of sixty coloured lithographs, especially interesting for Western Australia, New South Wales, Hawaii and the islands of Micronesia. It is only this third edition – of one of the most frequently published of all nineteenth-century voyage accounts – in which the plates appear in this coloured form. As the Hill catalogue notes, "this richly illustrated edition is much prized"; it is seldom seen on the market, and should be ranked as one of the rarest Pacific or Australian colourplate books.

By the time this version of his book reached the public, the remarkable Arago had lost his sight: hence the new title used for the work.

Hill, p. 10. $8850

WITH WOOD-ENGRAVINGS

299. [FREYCINET] ARAGO, Jacques. Souvenirs d'un aveugle. Voyage autour du monde...

Large octavo, with frontispiece portraits on china paper mounted, and many wood-engraved illustrations; an excellent copy in contemporary quarter morocco. Paris, H. Lebrun, 1868.

Fifth edition under this title, and one of the more handsome of the many editions published in the mid-nineteenth century. $675

298. Arago

TOWARDS THE ANTARCTIC

300. [KERGUELEN] PAGES, Pierre-Marie François de. Voyages autour du monde, et vers les deux Pôles par terre et par mer, pendant les années 1767... 1776...

Two volumes, octavo, with ten folding maps and plates; a fine copy in contemporary French marbled calf gilt. Paris, chez Moutard, 1782.

First edition: the rare original account of Pagès's considerable travelling, of most interest for the second volume which recounts his voyage towards the South Pole in search of a Southern Continent under Kerguelen in 1773-74. This was Kerguelen's second voyage, and a disastrous one: Pagès was to be an important witness in the subsequent court-martial of the commander for the events surrounding the dismasting of the *Rolland* during a storm. Kerguelen was dismissed from the service, and his printed account of the voyage, which was officially suppressed, is today very rare. This second volume also describes Pagès's experiences on a whaling voyage north of Spitzbergen in 1776.

Kerguelen's voyages finally proved to the French what Cook had already generally demonstrated – that the southern continent was a fiction – but more importantly, their disastrous result ensured that promotion in the French navy was henceforth based on merit rather than breeding. "La Pérouse may not have been a well-connected courtier, but he was honest and capable: that he was given the opportunity to lead the most comprehensive French expedition of the eighteenth century was a direct result of the Kerguelen episode..." (J. Dunmore, *French Explorers in the Pacific*, p. 249; pp. 196-249 give a full history of Kerguelen and his significance in the history of French exploration).

Hill, pp. 526-7 (English edition only); Sabin, 58168; Spence, 887. $6750

VOYAGE OF THE FAVORITE

301. LAPLACE, Cyrille Pierre Theodore. Reis rondom de wereld door de zeeën van Indië en China; uitgevoerd met Z.M. Korvet van Oorlog La Favorite…

Seven volumes, octavo, lithograph title-page to each volume, each with a vignette view, 14 lithograph maps and plates; a very attractive set, complete with half-titles and tissue-guards to the plates, in contemporary quarter black roan, flat spines gilt, pale green endpapers. Zaltbommel, Johannes Noman en Zoon, 1834-1836.

First Dutch edition of the *Favorite* voyage. The whole of the sixth volume is devoted to New South Wales and Tasmania, and there are two very good Sydney views, as well as a vignette of the Derwent River.

The purpose of this voyage was to show the French flag in eastern and other waters, in order to re-establish French influence over Indo-China and the Pacific. The voyage was very successful scientifically. The hydrographic work was thorough and reliable, the work done in the Anambana and Natuna groups of Malaysia was valuable, and a good collection of natural history specimens was brought back. Laplace visited Port Jackson and Hobart, as well as the Bay of Islands, and Singapore, Manila, Canton, Batavia and Chile and other ports. The voyage's activities in the East Indies area would have made it particularly interesting to the Dutch reader – hence this rather elegant publication of the Dutch translation.

Borba de Moraes, I, 458 (French edition); Ferguson, 1810 (Mitchell and State Library of Victoria copies only); Hill, p. 175 (French edition). **$3850**

303. Sainson

VOOLOOMOLOO

302. [LAPLACE] SAINSON, M. de, after Barthelemy LAUVERGNE. Vooloo-moloo au Port Jackson.

Aquatint view, 220 x 315 mm.; in fine condition, in bird's eye maple frame. Paris, Imprimerie Royale, 1833.

This fine view of Woolloomooloo was made during Laplace's voyage. In the foreground can be seen Woolloomooloo House, owned by the first fleeter, John Palmer. Present day Potts Point and Garden Island as well as a sweeping view of Sydney Harbour are also recorded.

This and the following views were made on the voyage of the *Favorite* under Laplace's command. After a long and difficult voyage through Asia, they moored in Port Jackson during August and September 1831 for a recreational visit. The relaxed stay, in which "banquets and balls took up our time", gave the voyage's artist, Barthelemy Lauvergne, the opportunity to draw these beautiful, detailed views of Sydney. **$4750**

SYDNEY COVE

303. [LAPLACE] SAINSON, M. de, after Barthelemy LAUVERGNE. Sydney. Etablissement principale de la Nouvelle Galles du Sud.

Aquatint view, 220 x 315 mm.; in fine condition, in bird's eye maple frame. Paris, Imprimerie Royale, 1833.

A splendid view of Sydney Cove, made during the Laplace voyage. It shows Circular Quay brimming with ships, the well developed Rocks area and, in detail, Fort Macquarie, the present site of the Sydney Opera House. **$4750**

304. Sainson

SYDNEY HEADS

304. [LAPLACE] SAINSON, M. de, after Barthelemy LAUVERGNE. Vue de L'Entrée de Port Jackson.

Aquatint view, 330 x 520 mm.; in fine condition, mounted. Paris, Imprimerie Royale, 1833.

This fine view shows French naval officers, presumably the personnel of the *Favorite*, elegantly positioned at the Heads. **$2850**

DERWENT RIVER

305. [LAPLACE] SAINSON, M. de, after Barthelemy LAUVERGNE. Rivière de Derwent (Van Diemen).

Aquatint view, 220 x 315 mm.; in fine condition, mounted. Paris, Imprimerie Royale, 1833.

A fine view of the Derwent River, one of only two Tasmanian illustrations published in the Laplace voyage. The *Favorite* arrived in Hobart on 11 July 1831, both ship and crew in a weakened condition. Their stay was almost a month; several of the crew died whilst others spent the time in hospital. "Those who were fit were able to relax and enjoy the hospitality of the local citizens, whom they found courteous but somewhat solemn and dull" (Dunmore, II, p. 251). **$1800**

PACIFIC WHALING

306. LECOMTE, Jules. Pratique de la pèche de la Baleine dans les mers du sud.

Octavo, with the half-title; some spotting, upper fore-corners of last four leaves torn (well clear of text); contemporary quarter calf, spine ornately gilt. Le Havre, J. Morlent, 1833.

First edition of a very scarce description of Pacific whaling. Lecomte, born in 1812, was the son of a naval officer and travelled widely throughout the world. Around 1832, he decided to abandon his sea-going career and turned to writing. Melville was familiar with his work and Lecomte's description of the great white whale was one of the principal sources for *Moby Dick.*

This work gives essential details for the preparation of a whaling expedition and examines the different species of whales, the best fishing regions and ways to approach and capture the whales. It ends with a list of "Lois et Ordonnances sur la pèche de la Baleine". This is one of two issues of the first edition; we have also seen the book with the imprint of Paris publishers Lecointe and Pougin.

Allen, 792; Forster, 212. **$2400**

VOYAGE TO NEW GUINEA

307. SONNERAT, Pierre. Voyage à la Nouvelle Guinée, dans lequel on trouve la description des Lieux, des Observations physiques & morales, & des détails relatifs à l'Histoire Naturelle...

Quarto, with 119 plates (numbered to 120), some folding; an excellent copy, bound without the half-title in contemporary polished calf gilt with anchor devices at corners, neatly rebacked. Paris, chez Ruault, 1776.

First edition of Sonnerat's important book on New Guinea, the Philippines and the Dutch East Indies, superbly illustrated with natural history engravings after his own original drawings.

Sonnerat, whose uncle Pierre Poivre was the Intendant of Mauritius, was sent on a spice- and plant-hunting voyage, and his first stop was the Seychelles. The famous frontispiece to the book shows him sitting under a coco-de-mer palm, and the first four plates in the book show details of the palm and its botany, while his text includes one of the earliest descriptions of the palm, and of its habitat in the early days of the colony. The first folding plate is a coastal panorama of the Seychelles and Coétivy Island, the first to appear in any publication.

Another engraving, remarkably, is one of the very earliest printed depictions of an Australian bird – the laughing kookaburra, wrongly entitled by Sonnerat the "Grand Martin-Pêcheur de la nouvelle Guinée" (plate 106).

Nissen, IVB, 885; Stafleu, TL2, 12.451a; Whittell pt. 2, p. 676; Zimmer, pp. 596-7. **$6500**

307. Sonnerat

VOYAGE FOR CHILDREN

308. ST. AULAIRE, Alphonse. Récréations Instructives – Instructive Recreations. Voyage pittoresque à travers le Monde...

Quarto, title-leaf and 24 lithographic plates; a fine copy in the original decorated green boards, rubbed but very good. Paris, chez Aubert & Cie., c. 1840.

Rare French illustrated account for children of a supposed voyage around the world, beginning with France, and finishing with New Zealand. Each lithographed leaf deals with one geographical area, and each has text in French and English as well as four or five vignettes, most of which derive from various of the "grands voyages". The Australian lithograph has images of Aborigines of King George's Sound and of Jervis Bay.

Forster identifies St. Aulaire as a lieutenant on the *Adventure,* a French whaleship out of Le Havre, but Pasquier does not note the ship in his census of French sailing vessels. St. Aulaire also published an expanded version of this as *Voyage autour du monde* (see below); his works are all rare, and found in few of the standard references.

This is a very attractive copy of the book, in its original binding of green glazed boards decorated "à la cathédrale" by St. Aulaire.

Forster, 402; not in Ferguson; Gumuchian, 5037. **$3000**

309. St. Aulaire: Récréations

310. St. Aulaire: Voyage

RARE COLOURED VERSION

309. ST. AULAIRE, Alphonse. Récréations Instructives – Instructive Recreations…

Quarto, title-leaf and 24 lithograph plates, highly coloured and finished with gum arabic; a few spots but a very good copy in the original ornately decorated cloth. Paris, Aubert, c. 1845.

A rare coloured version of St. Aulaire's *Récréations*.

Not in Ferguson; Forster, 402 (not noting coloured issue); Gumuchian, 5037 (not noting coloured issue). **$3500**

EXPANDED EDITION

310. ST. AULAIRE, Achille. Voyage autour du Monde.

Quarto, with 24 lithograph plates; a fine copy in original quarter red morocco, all edges gilt. Paris, Arnauld de Vresse, [c. 1845].

This is the expanded version of the book with 54 pages of descriptive text: the Aubert issue was composed simply of the lithographed plates. Although the publisher and the format are different, the plot remains unchanged: the round-the-world voyage begins in France, and finishes in New Zealand. There are two pages of text and a plate covering Australia.

This version not in Ferguson, Forster or Gumuchian. **$2850**

DOCTOR ON A WHALER

311. THIERCELIN, Dr. Louis. Journal d'un Baleinier. Voyages en Océanie…

Two volumes, octavo; a nice set, with the half-titles, in a later neat binding of sprinkled boards with leather labels. Paris, Hachette, 1866.

A very scarce narrative of whaling in the Pacific based on direct observation: Thiercelin served as doctor aboard a number of whaling vessels. His book "offers an interesting account of whaling vessels and their crews, a detailed description of the different types of whales, details on the actual whale hunt, and explanations of some of the newer whaling techniques such as poisoned harpoons, and harpoons with explosive charges propelled by gun-like mechanisms…" (Hill). There are chapters on Tasmania, New Zealand, Tahiti and Hawaii.

Ferguson, 16990-1 (he records the two volumes as separate works; Mitchell and State Library of Victoria copies only recorded of each); Hill, pp. 290-291; O'Reilly-Reitman, 1246; not in Forster. **$6600**

IV: *The East*

12: FAR EAST & INDIA

CHINA VIEWS

312. ALLOM, Thomas & Rev. G.N. WRIGHT. China, in a Series of Views, displaying the Scenery, Architecture, and Social Habits...

Four volumes in two, quarto, with 124 steel-engraved plates; some spotting, mainly marginal; a good set well bound in modern quarter blue morocco. London, Fisher, Son, & Co., [1843].

Thomas Allom (1804-1872) trained and worked as an architect but was best known as an artist, exhibiting regularly at the Royal Academy "where his charming pencil usually gained a place of honour" (DNB).

Allom produced a number of illustrated books of picturesque views of the British countryside and Egypt, and for this magnificent work on China drew heavily on the original Chinese drawings from Sir George Staunton's collection by Captain Stoddart and others.

The fine views are mainly of Peking, Nanking, the south-east and the south. The production of the book was a major undertaking: a team of thirty engravers worked on Allom's drawings. The book was a great commercial success but it has become scarce in good condition as so many copies have been broken for the decorative value of the plates. It is a fine example of steel-engraved book illustration, from the period of its greatest popularity.

Cordier, 80-1; Lust, 363. **$1950**

312. Allom

THE MACARTNEY EMBASSY

313. BARROW, Sir John. Travels in China, containing descriptions, observations, and comparisons, made and collected in the course of a short residence at the Imperial Palace of Yuen-Min-Yuen, and on a subsequent journey through the country from Pekin to Canton...

Quarto, with eight plates (five handcoloured, one folding); text lightly browned, occasional foxing; contemporary half calf, rubbed, front joint split but sound. London, A. Strahan, 1804.

First edition of one of the best early illustrated English accounts of China.

Barrow was a member of the Macartney embassy, which reached the court of China in 1792. The fine plates are after drawings by William Alexander who travelled with the embassy (and later published his own work).

Abbey, 531; Cordier, 2388; Hill, p. 15; Lust, 365. **$3500**

313. Barrow

SECOND EDITION

314. BARROW, John. Travels in China...

Quarto, with eight plates (five handcoloured, one folding); some light foxing and offsetting; inner hinge strengthened, a very good copy in recent half calf. London, T. Cadell and W. Davies, 1806.

Second edition, published two years after the first and virtually unchanged.

Abbey, 531; Lust, 366. **$2000**

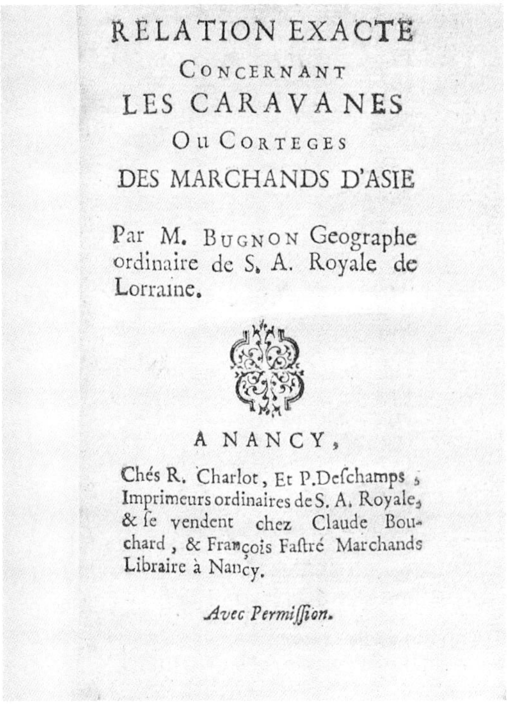
RELATION EXACTE
CONCERNANT
LES CARAVANES
OU CORTEGES
DES MARCHANDS D'ASIE

Par M. BUGNON Geographe ordinaire de S. A. Royale de Lorraine.

A NANCY,

Chés R. Charlot, Et P.Defchamps, Imprimeurs ordinaires de S. A. Royale, & se vendent chez Claude Bouchard, & François Faftré Marchands Libraire à Nancy.

Avec Permiſſion.

315. Bugnon

SILK ROAD

315. BUGNON, Didier. Relation exacte concernant les Caravanes ou Cortèges des Marchands d'Asie...

Duodecimo, with three engraved plans; a fine, crisp copy in contemporary French calf gilt. Nancy, R. Charlot, P. Deschamps [et al.], 1707.

Rare and early study of the caravans of Asia which brought trade via camel-trains to the market places of the Middle East. This must be one of the earliest studies of an important aspect of the Silk Road. The three plans show the organisation of caravan encampments used along the trade routes, quite complex defensive structures.

Inscribed on the title-page "Pour Monsieur le comte de Brionne", and with a few manuscript corrections in the text. $7500

PANORAMA OF NANKING

316. BURFORD, Robert. Description of a View of the City of Nanking.

Octavo, folding frontispiece and one plate; a very good copy in modern brown wrappers. London, T. Bretell, 1845.

The scarce guidebook which accompanied the public display of Burford's panorama of Nanking. Burford's "Panorama" was one of the leading attractions of mid-nineteenth-century London with its ever-changing array of visions from around the world, from "The Battle of Waterloo" to the "Ruins of Pompeii". Ruskin visited the impressive exhibit as a boy, and sang its praises in his *Præterita* (1885).

This panorama, painted by Burford from "drawings taken on the spot", commemorates a meeting between the British plenipotentiary, the commanders-in-chief of the army and navy, Chinese imperial commissioners, the Tartar general, and other high ranking officials, set against the great beauty of Nanking.

Not in Cordier or Lust (who record Burford's Canton, Macao and Hong Kong panoramas). $550

COAST-TO-COAST DERRING-DO

317. BURNES, Lieutenant Alexander. Travels into Bokhara being the Account of a Journey from India to Cabool, Tartary, and Persia; also, Narrative of a Voyage on the Indus from the Sea to Lahore...

Three volumes, octavo, three lithographs (one folding) and five engravings; a few spots; old half red morocco. London, John Murray, 1834.

First edition: an attractive copy of this scarce travel book. Burnes writes in his preface that "I was deputed in a political capacity to the Court of Lahore, charged with a letter from the King of England, and a present of some horses, to the ruler of that country. The principal object of my journey was to trace the course of the Indus... which had never been surveyed but between Tatta and Hydrabad". His successful mission which made him the first European in modern times to have navigated the Indus from its mouth in the Arabian Sea, spurred him on to extend his journey – this time in the guise of a private individual – from the Punjab across Afghanistan to the Caspian Sea. Thus his incredible trek extended from coast to coast.

Yakushi, B302. $3600

317. Burnes

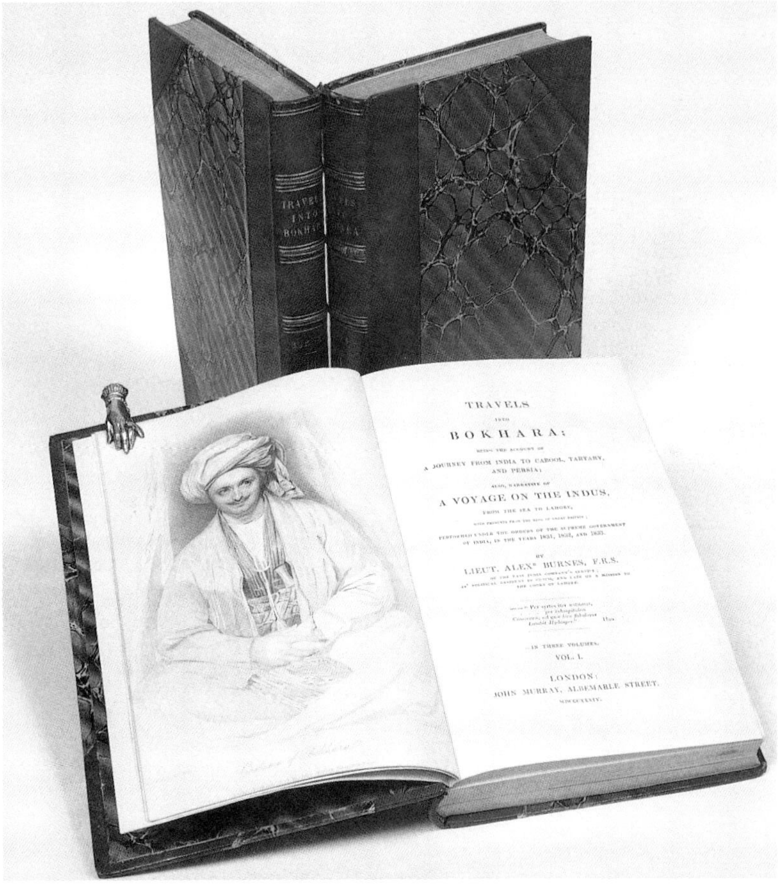

OBSERVATIONS FROM COOK'S SHIP

318. BURNEY, James. Memoir, Explanatory of a Chart, of the Coast of China, and the Sea Eastward, from the River of Canton, to the Southern Islands of Japan.

Quarto, pp. [iv], 23; an excellent copy in modern boards, morocco label. London, Nicol, Payne, et al., 1811.

First separate edition, and rare: Burney's memoir, like others of its kind (Flinders's *Observations on the Coasts of Van Diemen's Land* is an example; see also catalogue numbers 353 and 358), was intended to accompany a map, but the book is complete as it stands, although it could be purchased for seven shillings with the chart. These memoirs were frequently bound separately while the charts joined a composite atlas. The text of this reappears in the third volume of Burney's great *History*; it demonstrates the breadth of his knowledge, drawing from sources as varied as Hessel Gerritz, Dalrymple, a Japanese chart acquired by Broughton, and La Pérouse as well as other published voyagers.

The book itself also has a place in a Cook collection: in a paragraph here Burney describes the Pratas shoal in the China Sea from direct observations made during Cook's third voyage, on which he served as a lieutenant at first on the *Discovery* and then on the *Resolution*.

Not in Beddie, Hill, Holmes, Lust or Cordier. **$3850**

319. Ellis

THE AMHERST EMBASSY

319. ELLIS, Sir Henry. Journal of the Proceedings of the late Embassy to China... the voyage to and from China, and of the journey from the mouth of the Pei-Ho to the return to Canton...

Quarto, with a portrait frontispiece, seven coloured plates, and three maps (one folding); a few spots and two maps foxed, but the coloured plates fine and fresh; contemporary polished calf, front hinge cracked but firm, spine rubbed. London, John Murray, 1817.

First edition: the author was "a noted diplomat and historian. On Earl Amherst's embassy to China in 1816, sent out by George III to protest at the ill-treatment of British subjects, he served as third commissioner. Unfortunately this honour was short-lived. Amherst and his retinue were sent home in disgrace after he refused to 'kow-tow' (nine strikings of the fore-head on the ground) at his presentation to the Emperor Khien Lung in Peking. As if this humiliation was not enough, their ship, the *Alceste*, was wrecked off the coast of Sumatra on the return home. Happily, all hands survived, and another ship was found to carry them home again. On the return voyage the ship stopped at St. Helena; included in the text is Sir Henry's interview with Napoleon Bonaparte. On the journey out, the *Alceste* had visited Madeira, Rio de Janeiro, Cape Town, Java, and Macao. This edition, with hand-colored aquatints, is much prized" (Hill).

Abbey, 536; Cordier, 2393; Hill, p. 413; Lust, 509. **$3600**

CHINA TEA

320. FORTUNE, Robert. A Journey to the Tea Countries of China; including Sung-Lo and the Bohea Hills; with a Short Notice of the East India Company's Tea Plantations in the Himalaya Mountains.

Octavo, map, tinted lithograph frontispiece, and one double-page tinted lithograph, 13 engravings; corners bumped and some wear to extremities, but a very good copy in the original decorative cloth, neatly rebacked, retaining original spine. London, John Murray, 1852.

Robert Fortune (1813-1880) was sent to China in 1842 to collect botanical specimens for the Royal Horticultural Society's garden at Chiswick. He spent four years abroad and experienced many adventures such as pirate attack, shipwreck and entering the city of Loo-chow, then closed to Europeans, in Chinese disguise. He was responsible for the introduction to England of the kumquat, the double yellow rose, many varieties of tree peonies, azaleas and chrysanthemums. He visited China again in 1847 to collect tea plants on behalf of the East-India Company, successfully introducing tea to the Himalayas. This account of his second journey to China "is to give a peep into the Celestial Empire, to show its strange hills and romantic valleys... and its strange and interesting people".

Cordier, 2115. **$750**

THIRD JOURNEY TO CHINA

321. FORTUNE, Robert. A Residence among the Chinese: Inland, on the Coast, and at Sea. Being a Narrative of Scenes and Adventures during a Third Visit to China, from 1853 to 1856. Including notices of many natural productions and works of art, the culture of silk, &c...

Octavo, frontispiece and illustrations; a very good copy well bound in later half calf. London, 1857.

Fortune's third journey to China, important for its description of the culture of the silkworm.

Cordier, 2115. **$950**

PLANT-COLLECTING IN THE EAST

322. FORTUNE, Robert. Yedo and Peking. A Narrative of a Journey to the Capitals of Japan and China. With Notices of the Natural Productions, Agriculture, Horticulture, and Trade of those Countries, and other things met with by the way.

Octavo, folding map, folding frontispiece and illustrations; a very good copy bound in contemporary half blue calf. London, John Murray, 1863.

This, Fortune's last work, was completed after his retirement and describes his commission to travel to China and Japan on behalf of the United States Government to collect tea-shrubs and other plants.

Cordier, 2115. **$1450**

THIRD & BEST EDITION

323. GROSIER, Abbé J.B.G.A. De la Chine, ou Description Générale de cet empire, rédigée d'après les mémoires de la mission de Pé-kin...

Seven volumes, octavo, with two folding maps, one edge of the second map slightly torn; an excellent set in contemporary marbled sheep, crimson labels, red edges. Paris, Pillet, 1818-1820.

Third and best edition, revised and substantially enlarged, of this standard work on China, particularly rich in detailed information collected by the Jesuits. This was a successful book, with five European editions before 1820. It gives a "picture of an attractive country before semi-colonization set in" (Lust). Grosier (1743-1823) was the Royal Librarian; his preface dwells at length on a dispute with de Guignes, and finishes with a long letter from Klaproth. This was originally intended to be published as part of De Mailla's history of China, which is the source for the maps.

See Lust, 30; See Cordier, 61. **$1550**

EIGHTEENTH-CENTURY CHINA

324. GUIGNES, Christian Louis Joseph de. Voyages à Peking, Manille, et l'Ile de France faits dans l'intervalle des années 1784 à 1801...

Three volumes, octavo and folio atlas; 97 engravings in the atlas, some folding, some two to a page; a little very pale spotting to some plates, mainly marginal; text volumes in contemporary calf, the atlas in original papered boards. Paris, Imprimerie Impériale, 1808.

A very good set of this very scarce account of late eighteenth-century China. De Guignes was sent to China under the ancien régime in 1784 as an official representative, although it seems most unlikely that the Chinese saw him in that guise. Forgotten by successive governments in Paris after the fall of the monarchy, he remained in China until 1800. As he became a more familiar figure, he was increasingly able to observe the details of ordinary Chinese daily life, as well as the more exotic rituals of ceremonial life that were more usually the subject of travel literature.

Cordier, 2351-2; Lust, 336. **$7500**

CANTON

325. [HERALD] ANDERSON, J.B. and Berthold SEEMANN. Original sketches of Canton.

26 small drawings and a large sketch on five sheets (with another small sketch of St. Helena from the sea); pencil or ink on paper. HMS Herald at Canton, 1852.

A slight but charming series of sketches of Canton mostly by Anderson, a crew member on the *Herald*, Captain Henry Kellett, during her voyage originally in search of Franklin's *Erebus* and *Terror* and subsequently on an extensive circumnavigation.

The smaller drawings are by Anderson, signed by him on the sheets, but the large sketch of "European residences, Canton" is signed by Berthold Seeman, naturalist on the expedition. **$2250**

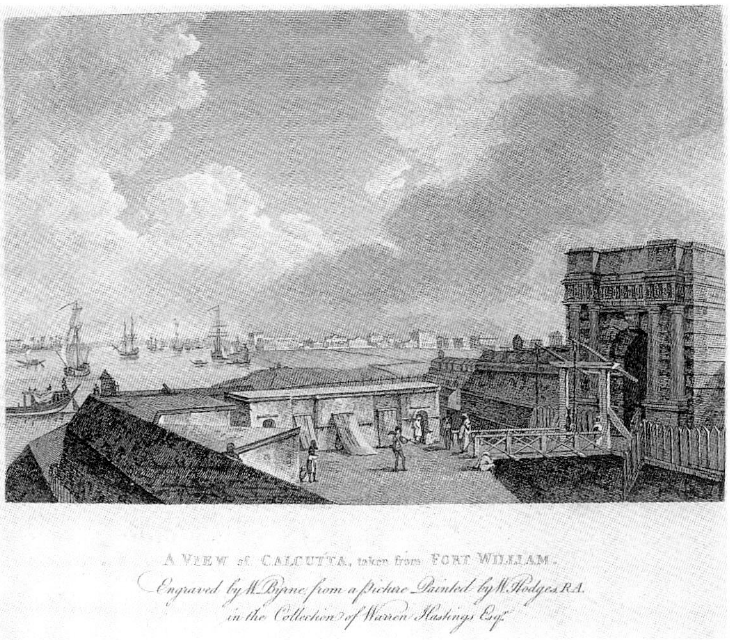

326. Hodges

COOK'S ARTIST IN INDIA

326. HODGES, William. Travels in India during the years 1780, 1781, 1782 and 1783.

Quarto, folding map, engraved plates after the author; contemporary red morocco by H. Walther (with his ticket), some rubbing but a fine copy with wide margins. London, J. Edwards, 1793.

First edition of this fine view-book by a famous travel-artist. William Hodges, best known as the artist on Cook's second voyage, travelled to India in 1778 under the patronage of Warren Hastings. He stayed six years, and painted a number of views.

"Hodges' book, one of the very few written by an English 18th-century artist... has little of the form of an explanation of plates skilfully interwoven into a personal journal. The *Travels* are the first account of India that flash verbal as well as visual pictures on the eye of the reader" (Archer/Lightbown).

Hodges covers towns, people, temples, forts, art and architecture, palaces, the Taj Mahal, Sikandra, and Fatehpur among his many views and descriptions.

Archer/Lightbown, 'India Observed', p. 74. **$1850**

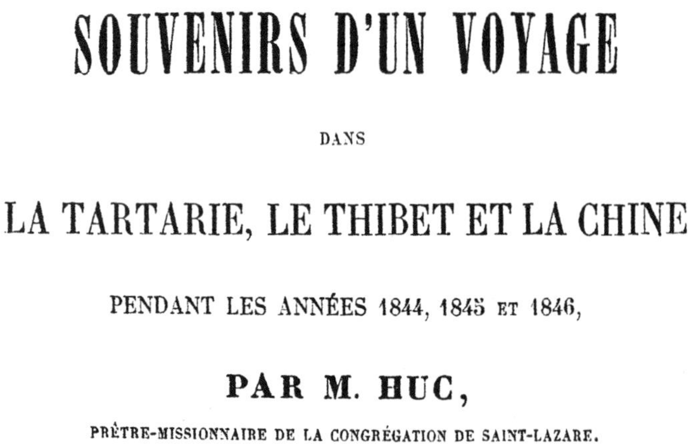

SOUVENIRS D'UN VOYAGE

DANS

LA TARTARIE, LE THIBET ET LA CHINE

PENDANT LES ANNÉES 1844, 1845 ET 1846,

PAR M. HUC,

PRÊTRE-MISSIONNAIRE DE LA CONGRÉGATION DE SAINT-LAZARE.

327. Huc

CENTRAL ASIA

327. HUC, M. Souvenirs d'un voyage dans la Tartarie, le Thibet et la Chine.

Two volumes, octavo, with a folding map; a very nice copy in contemporary half morocco. Paris, Adrien Le Clere, 1850.

Scare first edition of this important account of travels in Central Asia by a French missionary based at Si-Wang. In 1844 Huc had been sent by Mgr. Mouly, the Vicar Apostolic of Mongolia, on a long journey to Lhasa. According to Mouly, Huc and his companion Gabet were "both fairly well acquainted with the Manchu and Mongol languages, and knowing enough Tibetan to enable them to carry on their ministry usefully among the nomadic Mongols, and to attempt to found a mission in their midst".

The journey – lasting some eighteen months – took them from Dolon Nor, across the Yellow River, to Sining, Kumbun Monastery near Lake Koko Nor to Lhasa, where they became the first missionaries to enter the city since the Capuchins had been expelled some one hundred years earlier. Huc's fascinating account of his travels was an immediate success.

Yakushi, H249. **$1400**

THE CHINA SEA

328. KING, Commander J.W. and J.W. REED. The China Sea Directory. Vol. I [and II]. Containing directions for the approaches to the China Sea and to Singapore, by the Straits of Sunda, Banka, Gaspa, Carimata, Rhio, Varella, Durian, and Singapore [Directions for the navigation of the China Sea, between Singapore and Hong Kong].

Octavo, two volumes; slight wear to joints and a neat repair; an excellent set in the original bright yellow cloth. London, Admiralty, 1867-8.

Scarce set of sailing directions for the China Sea, in unusually good condition. Such collections of navigational information are based on experiences reported by mariners in the area, often dating back to quite early voyages. **$2100**

WALL MAP

329. LAURIE, Robert and James WHITTLE. Asia and its Islands according to d'Anville; divided into Empires, Kingdoms, States, Regions, &c., with the European Possessions and Settlements in the East Indies and an exact delineation of all the Discoveries made in the Eastern Parts by the English under Captn. Cook, Vancouver & Peyrouse.

Engraved map, 1450 x 1200 mm., assembled from three sheets; in good handcolouring; well framed. London, Laurie and Whittle, 1799.

A large and imposing wall map – with a most attractive orientalist cartouche – by the famous partnership of Laurie and Whittle. As well as Asia, the map shows the East Indies, Australia and the north-west Pacific in excellent detail, based on the observations of the navigators at the end of the eighteenth century.

Tooley, 854 (a 1794 issue). **$6500**

A DANCE FOR LIFE.

330. Lindley

SHANGHAI TO MANILA

330. LINDLEY, Captain Augustus F. The Log of Fortuna. A Cruise in Chinese Waters being Tales of Adventure by Sea and By Shore.

Octavo, plates and illustrations; fine in original pictorial cloth. London, Cassell Petter & Galpin, [1880].

An entertaining tale of a journey through China from Shanghai to Manila aboard a clipper, incorporating anecdotes of a number of other journeys including one down the Yangtze-kiang and one in Australia ("Stuck Up: An adventure with Australian bushrangers", which describes a hold up in West Maitland). With four plates of Australian scenes.

Ferguson, 11649 (variant binding). **$875**

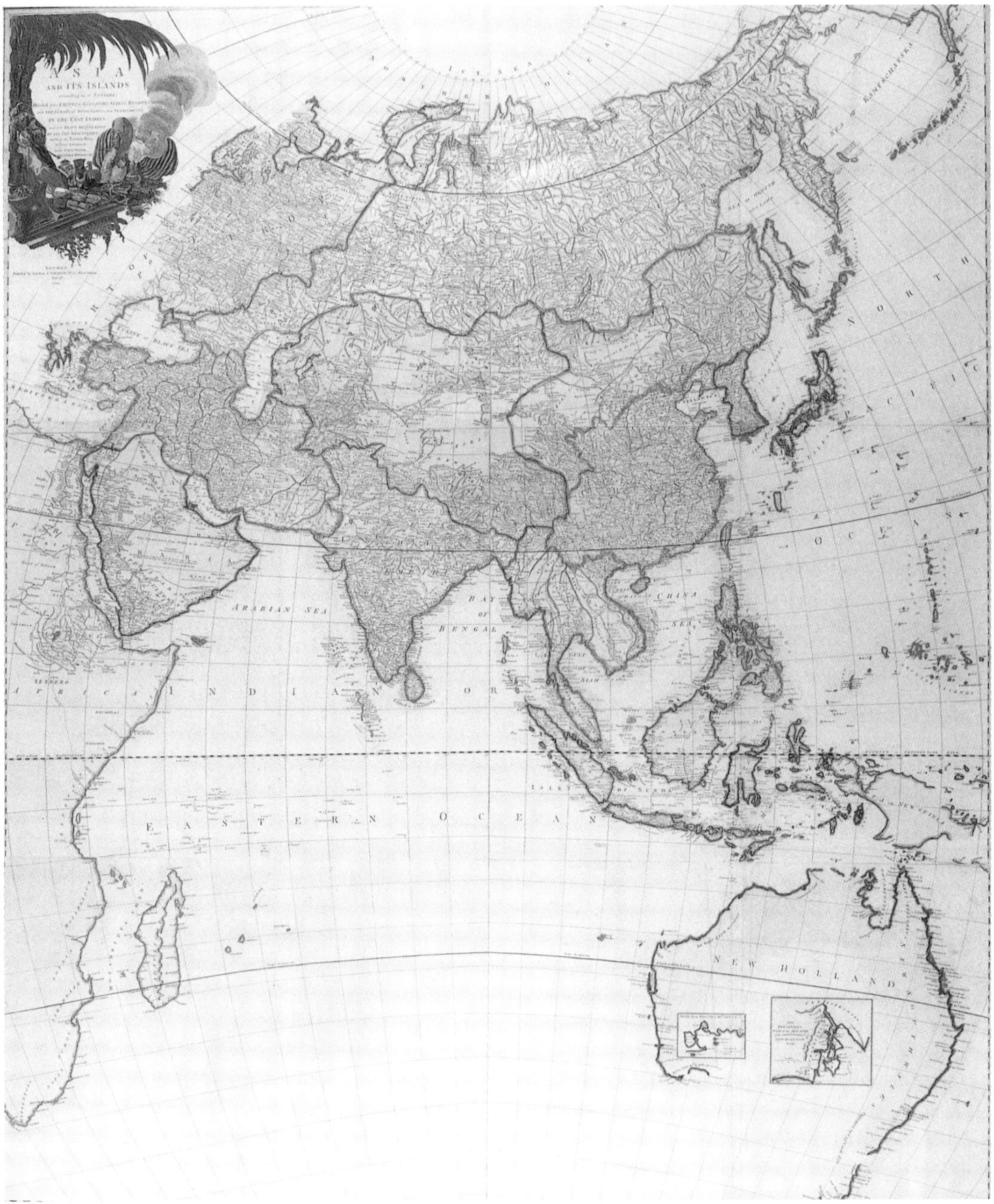

329. Laurie & Whittle

OPIUM TRADE

331. MACAULEY, W. Hastings. Kathay: A Cruise in the China Seas.

Octavo, with advertisement leaf; original cloth, gilt vignette on upper board, slight defect to head of spine. New York, Putnam & Co., 1852.

First edition. The author left Charlestown, Massachusetts, in January 1850 and sailed the world for two years visiting Rio de Janeiro, the Cape of Good Hope (twice), Christmas Island, Macao, Hong Kong, Canton, Manila, Shanghai, Amoy, Formosa, Batavia, Java and the Mascarene Islands. "Much on Chinese cultural life and the opium trade" (Hill).

Hill, pp. 482-3; not in Smith, 'American Travellers Abroad'.

$850

SUPERCARGO

332. [R., J.] Diary of a Journey Overland, through the Maritime Provinces of China, from Manchao, on the South Coast of Hainan, to Canton, in the Years 1819 and 1820.

Octavo; a very good copy bound in later cloth. London, Printed for Sir Richard Phillips and Co., 1822.

An interesting anonymous account (the Preface is signed "J.R., Supercargo") of an early journey through China from Macao to Canton on board the English vessel *Friendship*, captain Ross. The book also includes several separate articles of interest including an official edict on the banning of opium and Milne's description of Chinese printing.

This was published as part of Sir Richard Phillips's *New Voyages and Travels* (London, 1820-23), and like the other titles in the series was also available separately.

See Cordier, 303; Lust, 227. **$1500**

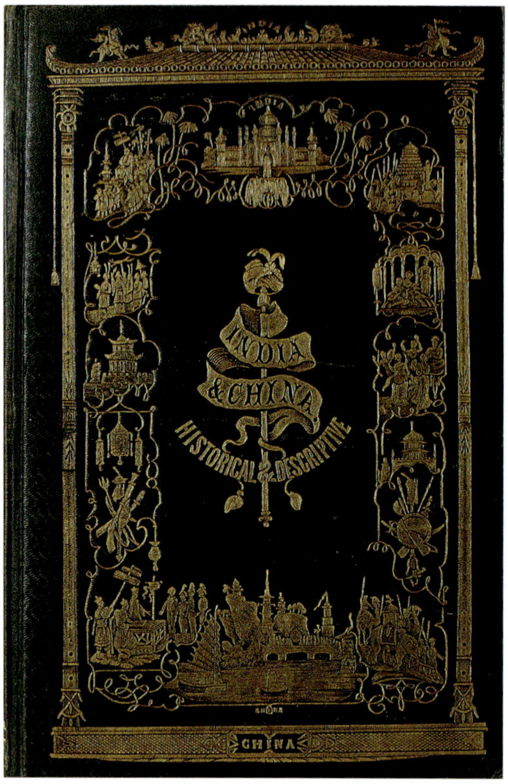

333. Sales

PRINTED FOR PRIVATE DISTRIBUTION

333. SALES, William de Fane de. Reminiscences of Travel in China and India in 1848.

Quarto, with 32 full-page lithographs; in good condition, in original half roan, cloth sides a little faded and worn. London, Waterlow and Sons Limited, 1892.

Printed only for private distribution, this is a rare travel memoir with an excellent series of lithograph views by a gifted amateur and sometime resident of Sydney. The book is certainly scarce on the market: we have not seen a copy before. The illustrations include views of Hong Kong, Singapore, Cochin, Ceylon, the Middle East, Egypt, the Mediterranean, and the English Channel.

De Sales (1812-1896) was an influential Sydney businessman. Born in England, educated at Eton and Oxford and later called to the Bar, he became involved with the trading empire of Jardine Matheson through family connections, and made a number of journeys to Australia and the Far East. He became a partner in Jardine Matheson's Sydney office, and lived in Fort Street, Sydney for some years. He was also a director of P. & O. for four decades, from 1851 to 1895. **$2400**

CHINA & INDIA

334. SEARS, Robert (editor). Pictorial History of China and India; Comprising a Description of those Countries and their Inhabitants...

Octavo, two folding maps, profusely illustrated, with advertisements; original pictorial cloth, gilt, some browning and spotting; contemporary ownership inscription. New York, Robert Sears, 1853.

First published in 1851, this is a most attractive and comprehensive study of China and India covering general history from ancient times to the mid-nineteenth century, the culture, philosophy, and language of both countries; it also includes chapters on the opium trade and even one on Chinese jugglers.

See Cordier, 3027. **$750**

CASED MAP

335. STANFORD, Edward, publisher. Stanford's Map of Eastern China, Japan and Korea...

Colour lithographed map, 670 x 560 mm.; backed on linen, folding into original small octavo cloth case, fine. London, Edward Stanford, 1904.

Scarce map of the far east, in attractive condition. The map extends north to Mongolia, east to Japan and south to the Philippines. There is a detailed inset map of Korea. **$275**

THE MACARTNEY EMBASSY

336. STAUNTON, Sir George. An Authentic Account of an Embassy from the King of Great Britain to the Emperor of China...

Three volumes, octavo, with three folding maps; front hinges reinforced, a good set in contemporary sprinkled calf, neatly rebacked. London, G. Nicol, 1797.

Second (first octavo) edition of the official narrative of the important Macartney embassy to China of 1792, by which the British sought to establish formal diplomatic and commercial relations. Staunton, a medical doctor and friend of Dr. Johnson, had already served in many diplomatic posts, some as aide-de-camp to Lord Macartney, when after a period of retirement he was again called to serve Macartney, as Secretary to the embassy to China. He compiled this book chiefly from the papers of the Ambassador and his fellow envoys. His work was remarkably successful: fifteen editions were issued in seven countries in thirty years.

Although largely concerned with events in China, a description of the Mission's stay at Tourane in Vietnam appears in chapter VIII of the first volume.

Cordier IV, 2383; Lust; 546; Morrison I, 695-696. **$2450**

BIBLIOTHEQUE PORTATIVE

337. STAUNTON, Sir George. Voyage en Chine et en Tartarie... traduit de l'anglais...

Six volumes, small octavo, and an atlas of uniform size, with twenty double-page plates; a few marginal wormholes in the last text volume; original blind-stamped calf. Paris, la veuve Lepetit, 1804.

A fine set of the delightful 'Bibliothèque portative' edition, a diminutive French version of the official narrative of the important Macartney embassy to China.

Lust, 545 (note). **$1150**

13: SOUTH-EAST ASIA & EAST INDIES

EAST INDIES HERBAL

338. ACOSTA, Christoval. Tractado de las drogas, y medicanas de las Indias Orientales, con sus Plantas debuxadas al bivo.

Small quarto, title in architectural woodcut border, woodcut portrait of the author, and 44 full-page woodcuts of plants and two of elephants; discreet stamp in lower right margin of title; faint waterstain in upper margin of first few leaves; some leaves lightly browned, occasional marginal foxing; overall a fresh and attractive unsophisticated copy in contemporary limp vellum. Burgos, Martin de Victoria, 1578.

Drugs, herbs and spices of the East Indies: this is a lovely copy of the first edition of this influential herbal, important for disseminating botanical and pharmaceutical knowledge of recently discovered species from newly discovered lands, particularly in the East Indies.

The handsome woodcuts, largely taken from drawings by the author, illustrate many species from the East relatively unknown in Europe at the time, including ginger, nutmeg, cinnamon, mace, aloes, tamarind, cloves and china root – all of which became increasingly important commodities on the trade routes. The depictions of the elephant are the earliest representations of that animal in a scientific (rather than an antiquarian or festival) publication (Lach).

A physician from Portuguese Mozambique, Acosta (c. 1515-1580) accompanied the Portuguese viceroy to Goa, and also travelled to Persia and China. His book was reprinted in Spanish in 1582 and 1596, and incorporated into the many botanical compilations by L'Ecluse and others which followed.

He makes some notable attempts at comparative botany, in which the characteristics of a species known in both Asia and America are contrasted. As Colmeiro and other authorities point out, some of Acosta's material derives directly from Orta's *Colloquios dos Simples e Drogas he Cousas Medicinais da India* (Goa, 1563), but Acosta has substantially clarified and augmented the earlier work, which anyway had no illustrations and was badly organised and poorly printed.

Lach, I, pp. 194-5; Palau, 1962; Pritzel, 13; Sabin, 113; Stafleu & Cowan, 23. **$27,500**

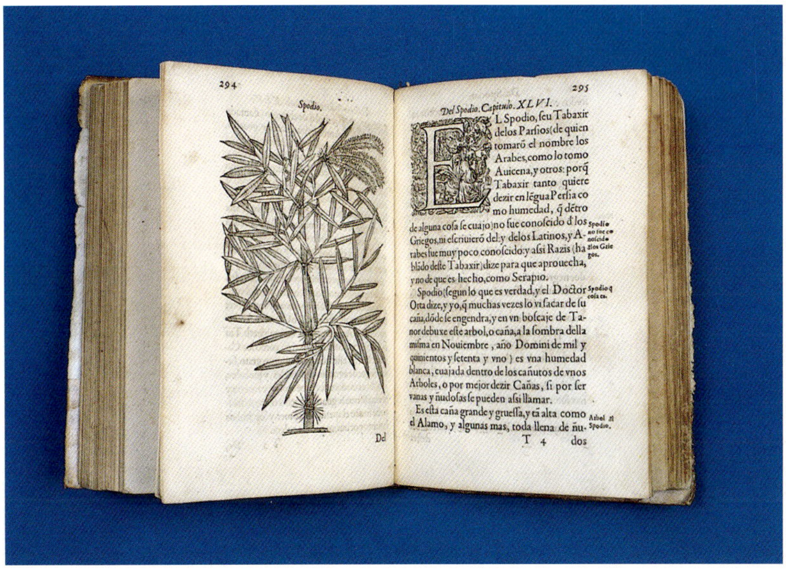

338. Acosta

SOUTH VIETNAM

339. BARROW, Sir John. A Voyage to Cochinchina, in the years 1792 and 1793…

Quarto, two folding maps and 19 coloured aquatint plates; text lightly browned; contemporary half calf, slightly rubbed, front joint cracked but strong. London, T. Cadell, 1806.

First edition: the first illustrated English work on South Vietnam, particularly the region of Da Nang and Hue: Barrow gives the first accurate description of the region and its inhabitants by an Englishman. Dampier had touched in Vietnam on his circumnavigation more than one hundred years prior to this, but his visit was largely confined to the areas around the Gulf of Tonkin.

"The voyage visited Madeira, the Canary Islands, and Rio de Janeiro; a description of that city and of Brazil in general is given. Touching at Tristan da Cunha, the ship rounded the Cape and eventually reached Cochin China via the city of Batavia on Java. The volume is also of Cook interest as it describes finding Captain Cook's *Resolution* transformed into a smuggling whaler under the French flag" (Abbey).

"This book is the account of Barrow's voyage on the way to China and is dedicated to his travelling companion, Sir George Staunton. The detailed information on Cochin China is taken from a manuscript memoir drawn up by Captain Barissy, a French naval officer, who had the means and opportunity of collecting accurate information…" (Hill).

"The aquatinting is of excellent quality, [plate] number 10 in particular being technically interesting, since it appears to be printed in as many as three colours, with one colour added by hand, while no. 11 is printed in green, with other colours added by hand" (Abbey).

Abbey, 514; Borba de Moraes, p. 88; Cordier, 'Indosinica', 2424 (& 'Sinica', 2390); Hill, p. 16. **$3250**

MALAYSIA

340. CAMERON, John. Our Tropical Possessions in Malayan India: Being a Descriptive Account of Singapore, Penang, Province Wellesley, and Malacca; their Peoples, Products, Commerce, and Government.

Octavo, with seven colour lithographed plates (one double-page); one plate with sealed tear, general signs of use but a good ex-library copy, rebound in modern half calf. London, Smith, Elder, 1865.

Quite scarce: a history of Malaysia from 1160, through the colonial period and British possession.

$1100

LONG VOYAGES TO HOT COUNTRIES

341. CLARK, John, M.D. Observations on the Diseases in Long Voyages to Hot Countries, and particularly on those which prevail in the East Indies... Published by order of the Court of the Directors of the Honourable East India Company...

Octavo; a fine copy in contemporary calf, flat spine gilt in compartments, red label. London, D. Wilson & G. Nicol, 1775.

First edition, second issue with cancel title (the book had first appeared in 1773, and was re-issued like this in 1775 quite unchanged, with the exception of the re-edited title-page itself. This second issue is not recorded in any of the standard medical bibliographies).

Clarke, who had been a surgeon aboard the *Talbot*, an Indiaman, discusses health measures aboard East Indiamen, from actual experiences on the ships themselves; he also describes experiments with anti-scorbutic measures at approximately the time of Cook's medal-winning work on scurvy.

Keevil, 'Medicine and the Navy', vol. III, p. 382; Carpenter, 'History of Scurvy', p. 258. **$1850**

341. Clark

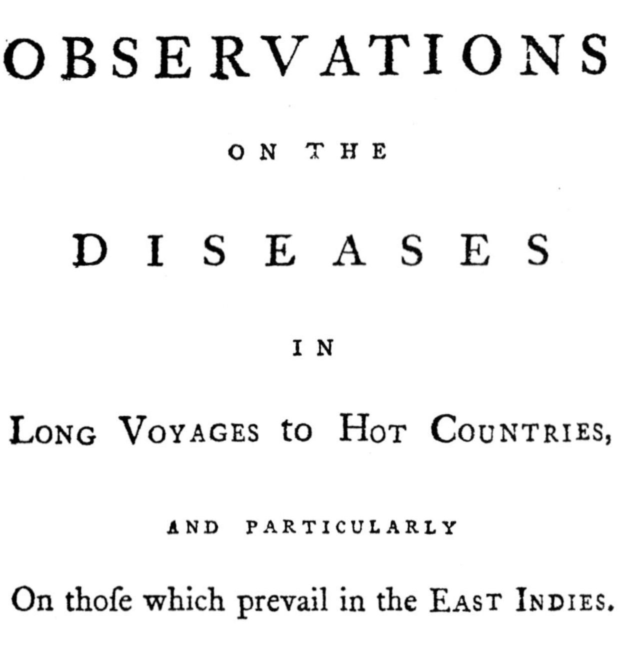

OBSERVATIONS

ON THE

DISEASES

IN

LONG VOYAGES to HOT COUNTRIES,

AND PARTICULARLY

On thoſe which prevail in the EAST INDIES.

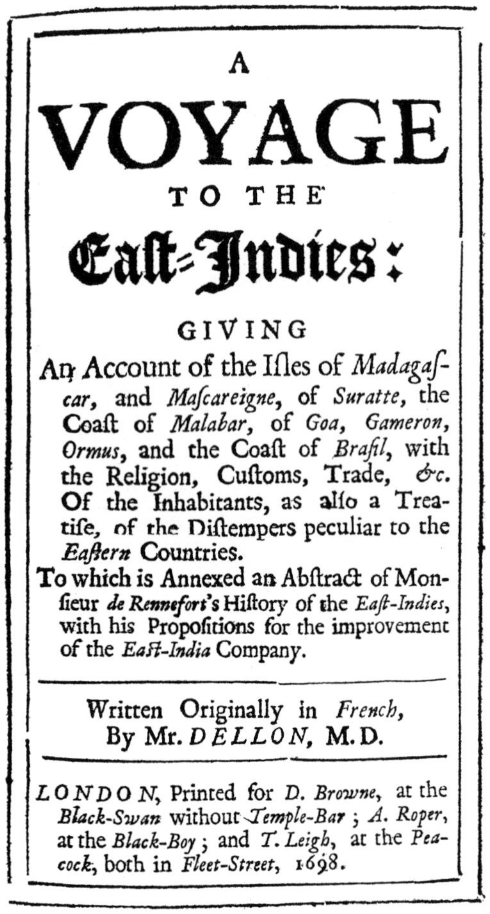

A

VOYAGE

TO THE

East-Indies:

GIVING

An Account of the Iſles of *Madagaſcar*, and *Maſcareigne*, of *Suratte*, the Coaſt of *Malabar*, of *Goa*, *Gameron*, *Ormus*, and the Coaſt of *Braſil*, with the Religion, Cuſtoms, Trade, &c. Of the Inhabitants, as alſo a Treatiſe, of the Diſtempers peculiar to the *Eaſtern* Countries.

To which is Annexed an Abſtract of Monſieur *de Rennefort's* Hiſtory of the *Eaſt-Indies*, with his Propoſitions for the improvement of the *Eaſt-India* Company.

Written Originally in *French*,
By Mr. DELLON, M.D.

LONDON, Printed for *D. Browne*, at the *Black-Swan* without *Temple-Bar*; *A. Roper*, at the *Black-Boy*; and *T. Leigh*, at the *Peacock*, both in *Fleet-Street*, 1698.

342. Dellon

MEDICINE FOR TRAVELLERS

342. DELLON, Charles. A Voyage to the East-Indies: giving an Account of the Isles of Madagascar, and Mascareigne, of Suratte, the Coast of Malabar, of Goa, Gameron, Ormus, and the Coast of Brasil... To which is annexed an Abstract of Monsieur de Rennefort's History of the East-Indies, with his Propositions for the Improvement of the East-India Company.

Octavo; some staining and embrowning, but overall a good copy in contemporary calf, rebacked. London, D. Browne, A. Roper and T. Leigh, 1698.

First English edition, originally published in French in 1685. This is the variant issue with three publishers listed in the imprint (another variant identifies Browne alone as publisher). This is an unusually good, early narrative of experiences in India; on the way back, Dellon visited Brazil: "the description of Bahia is to be found between pages 197 and 206. In the *Supplement* there is a chapter on Pernambuco" (Borba de Moraes).

An Appendix contains Rennefort's proposals for a French settlement in Madagascar, with some discussion of sailing via Brazil and the voyage around the Cape of Good Hope, with suggestions for the improvement of the French East-India Company, including the sending of an envoy to Japan. There is also, in a separate section with its own title-page, a very interesting "Treatise of the Distempers relating in particular to the Eastern Countries, and such other Diseases as commonly happen in the Voyages into those Parts, with their proper Remedies" translated from the French. It starts simply – "Vomiting is the first inconveniency Travellers are afflicted with at Sea" – but gets more technical.

Borba de Moraes I, p. 216; Wing, D943. **$2300**

DELIVERED FROM THE INQUISITION

343. DELLON, Charles. The History of the Inquisition, as it is exercised at Goa, in the East-Indies, under the Dominion of the King of Portugal... by Monsieur Dellon, who laboured five years under the Severities of that Inquisition. With an Account of His Deliverance.

Octavo, [iv], 76 pp.; modern half morocco. London, R. Baldwin, n.d., c. 1750.

Translated by Henry Wharton and first published in 1688 but quickly suppressed "on Pretence of it's being against the King's Religion...". The publisher went to jail and the edition was impounded, but there were various printings done both then and later, even an edition in Hull as late as 1812. **$850**

THE
HISTORY
OF THE
INQUISITION,
AS IT IS
Exercifed at GOA,
In the EAST-INDIES,
Under the Dominion of the KING of PORTUGAL.

Written in FRENCH by
Monfieur DELLON,
Who laboured Five Years under the Severities of that INQUISITION.

WITH
An Account of his Deliverance.

Tranflated into ENGLISH.

LONDON:
Printed for R. BALDWIN in *Pater-Nofter-Row.*
[Price One Shilling.]

343. Dellon

THE
EAST-INDIA REGISTER
AND
DIRECTORY,
FOR
1806;
CORRECTED TO THE 14TH NOVEMBER, 1805.
CONTAINING
Complete Lists of the Company's Servants;
CIVIL, MILITARY, AND MARINE,
At the different Presidencies in the East-Indies;
With Indexes to the same, and Lists of Casualties during the last Year.
TOGETHER WITH
LISTS of the EUROPEANS, MARINERS, &c.
Not in the Service of the East-India Company;
AND
MERCHANT VESSELS
EMPLOYED IN THE COUNTRY TRADE.
Compiled, by Permission of the Honorable the East-India Company,
FROM THE
OFFICIAL RETURNS RECEIVED AT THE EAST-INDIA HOUSE:
BY
JOHN MATHISON & ALEXANDER WAY MASON,
Of the Secretary's Office, East-India House.

LONDON:
Printed by COX, SON, and BAYLIS, Great Queen-Street.
Sold by BLACKS and PARRY, Leadenhall Street, and JOHN HATCHARD No. 190, Piccadilly.

344. Register

EAST-INDIA REGISTER

344. EAST-INDIA REGISTER. The East-India Register and Directory for 1806...

Small octavo, engraved folding map and three folding tables; a very attractive copy in contemporary red morocco gilt. London, Cox, Son, and Baylis, 1805.

The "annual report" of the East-India Company, containing all manner of information including lists of merchant vessels and civil and military personnel based in England and throughout India, China, Sumatra and on St. Helena. **$1250**

PRINCE GIOLO

345. [GIOLO] SAVAGE, F., Engraver. Prince Giolo, Son to the King of Moangis or Gilolo: lying under the Equator in the Long: of 152 Deg. 30 Min. a fruitful Island abounding with rich Spices and other valuable Commodities...

Engraving, 380 x 275 mm., with a fifteen-line caption; neat paper repairs to upper corners; two small stains (where once fixed in an album), otherwise in very good condition; mounted. London, Sold by Joseph Medley, n.d., c. 1692.

Of exceptional rarity and interest, this separately-published engraving depicts the tragic, tattooed East-Indian native brought back to England by William Dampier in late 1691 (the voyage on which an Englishman first set foot on the Australian continent). Dampier called him Jeoly, while this and some other references have him as Giolo.

Jeoly must have been among the most exotic visitors to London in the seventeenth century. Appallingly cruel as it may seem today, a number of early voyagers had brought back human beings from distant countries, as "curiosities" alongside animals large and small. None can have been as startling as Dampier's "Painted Prince" or "Amiable Savage" as he called him.

Dampier met the "Painted Prince" during his trading voyage of 1690 (described in *New Voyage round the World*; see catalogue numbers 18-19). He purchased a half-interest in the Prince and his mother from Moody, an English trader, who had acquired them at Mindanao where they had been sold as slaves after being captured at Jailolo (Gilolo in the title) on Halmahera, the largest island of the Moluccas (Gill identifies Moangis, mentioned in the title, as Meangis or Kepulauan Talaud).

Jeoly's mother died during the subsequent voyage; Dampier kept a good eye on Jeoly, and evidently became fond of him, although his overriding concern was salvaging the one remaining chance to make money from his complex and unlucky ventures. But as Dampier tells it, "I was no sooner arrived in the Thames, but he was sent ashore to be seen by some eminent Persons; and I being in want of Money, was prevailed upon to sell first, part of my share in him, and by degrees all of it..." (*New Voyage*, p. 549).

When Jeoly's new owners acquired him from Dampier, they were quick to capitalise on their investment, displaying him to paying customers. A broadside of the time noted that "He is exposed to view every day... at his lodgings at the Blue Boar's Head in Fleet Street... But if any persons of quality, gentlemen or ladies, do desire to see this noble person, at their own houses, or any other convenient places, in or about this City of London, they are desired to send timely notice...". Tragically, as Dampier notes briefly, "After this I heard he was carried about to be shown as a Sight, and that he died of the Small-pox at Oxford...".

Jeoly must have indeed been the "Wonder of the Age", as the caption to this portrait claimed. Apart from anything else, tattoos were hardly known in Europe (the naval tradition of tattooing dates to much later, actually to the Cook period). Dampier and some other adventurers had seen tattooed people in their travels, and had described them in print, but for Londoners to see a tattooed man in the flesh, or even in an accurate engraving, must have been exotic. "His whole Body (except Face Hands and Feet) is curiously and most exquisitely Painted or stained full of Variety and Invention with prodigious Art and Skill..." (caption). The diarist John Evelyn had dinner with the diarist Samuel Pepys that year and recorded that there "was Captain Dampier, who had been a famous buccaneer, had brought hither the painted prince Job [sic], and printed a relation of his very strange adventure..." (Gill, p. 238).

Giolo is depicted in an exotic landscape, surrounded by snakes and scorpions, and the caption teases us that "The Pictures & those other engraven Figures copied from him & now dispersed abroad serve only to describe as much as they can the Fore-parts of this inimitable Piece of Workmanship: The more admirable Back-parts afford us a Represent-ation of one quarter part of the Sphere upon & betwixt his Shoulders where the Arctick & Tropic Circles center in the North Pole of his Neck...".

Of all the Europeans who came into contact with Jeoly, Dampier seems to have had better if not his best interests at heart; but he had brought him halfway across the world to make money, and even a wish to send him home was completely commercial: "I proposed no small advantage to my self from my Painted Prince... For beside what might be gained by shewing him in England, I was in hopes that when I had got some money, I might there obtain what I had in vain sought for in the Indies, viz. A Ship from the Merchants wherewith to carry him back to Meangis, and reinstate him there in his own Country, and by his favour and negotiation to establish a traffick for the Spices and other products of those Islands...".

In the later period, there are many more images of natives of far-away places brought back to Europe: the Reynolds portrait of Omai, the poignant image of Sophie Elau, Darwin's Patagonians – even the King and Queen of Hawaii, are all examples. The 'noble savage' period made the whole genre – whether graphic or textual – more understandable; the individual images are mostly of enormous interest today, and as it happens most are very rare.

This image of Jeoly is a noteworthy precursor of the 'noble savage' tradition, before it had been fully articulated by the philosophers of the eighteenth century. Jeoly is depicted in a classical pose, with elegance of stance and hand gestures, certainly not as the "travelling freak" which a number of later unscrupulous entrepreneurs in the nineteenth century were promoting (the "Wild Man of Botany Bay" is an example), nor as a scientific exhibit (like Sophie Elau).

This seventeenth-century image of the mysterious inhabitant of the fabulous East Indies must be regarded as one of the rarest of all voyage prints.

Dampier, 'New Voyage' (1697), pp. 492, 511-520, 549; Gill, 'The Devil's Mariner: William Dampier', pp. 214-22; Nan Kivell & Spence, pp. 126, 177 (illust.). **$24,850**

345. Giolo, or Jeoly:
Dampier's "Painted Prince"

LIMITED EDITION

346. HAMILTON, Alexander. A New Account of the East Indies. Now edited with Introduction and Notes by Sir William Foster.

Two volumes, quarto, with plates and maps; uncut and partly unopened in original quarter vellum, one corner bumped, a little wear to cloth, but a good set. London, Argonaut Press, 1930.

One of only 975 copies. Hamilton describes the natives, trade goods, weather, settlements, and many other features of his route from the Cape of Good Hope to Japan via India and Burma. Based on the first edition of 1727, this work "remains one of high value as a source of information; while its raciness, and the numerous anecdotes interspersed in the text, render it exceedingly attractive reading" (Preface). The editor, Sir William Foster, was president of the Hakluyt Society. **$450**

WITH SUPPLEMENTARY INFORMATION IN MANUSCRIPT

347. HORSBURGH, James. Directions for sailing to & from the East Indies, China, New Holland, Cape of Good Hope, and the interjacent ports...

Two volumes in one, thick quarto, old inscription on the title-page; some light browning or staining, nowhere offensive; attractive contemporary binding of dark calf gilt, spine gilt in compartments with ships and anchors, joints neatly strengthened; old canvas chemise inscribed "USN Bencoolen 1832"; boxed. London, Printed for the author, 1809-1811.

An excellent copy of the first edition of this navigational classic, which contained the essential information for navigation in the East Indies, Australian waters and the South Seas based on navigators' reports. The work was in continual revision, and the numerous editions and separately printed additions reflect current information. Perhaps because of their shipboard use, most early editions are scarce.

With a ten-page manuscript at the end of supplementary information, apparently compiled in the 1820s from actual observations mainly on the *Sprightly*, Captain George Norris, sometimes accompanied by the *Lively*, evidently South Sea whalers belonging to the Enderby empire. The writer – perhaps the James Goldie who has signed the title-page – includes mentions of Enderby's Islands ("... were approached close on February 27th 1826 and the Inhabitants came off in canoes with cocoanuts and some flying fishes..."), and "Two Islands... thought by Capt. Kennick to be a new discovery but they are placed in Admiral Krusenstern's Atlas of the South Pacific Ocean under the name of Kama...".

Ferguson, 481. **$3200**

BATAVIA

348. [JAVA]. Carte de l'Ile de Java...

Handcoloured engraved map, 900 x 400 mm.; original folds, some wear and discolouration, but generally in very good condition. Paris, 18th century.

A detailed and informative eighteenth-century French map of the island of Java, with extensive information about the Dutch presence. Of particular interest is the inset detailed plan of the Port of Batavia, the Dutch colonial city on the northern coast of the island. Batavia, modern-day Jakarta, was strategically placed to monopolise the sea traffic with the spice islands of east Indonesia. Whilst it was well located for economic reasons, its low-lying situation and lack of sanitation made it somewhat of a death-trap for visiting mariners. **$1500**

JAVA EARTHQUAKE IN PHOTOGRAPHS

349. [JAVA] LAAR, L. van. Beknopte Beschrijving, van de Aardbeving, die het eiland Java in den ochtend van den 10den Junij 1867 heeft geleisterd, benevens eenige historische herinneringen aan den oorsprong van Passar-Gedée...

Quarto, the half-title and title printed in gold, silver and colour; with four original photographs approx. 160 x 210 mm. and one photographic vignette; original glazed boards, lettered in gilt, later cloth spine; binding slightly stained; preserved in a cloth box; in good condition. Samarang, C.J. Morel, 1867.

Notably early photographs of Indonesia: this is a very rare book with a fine series of actual photographs, describing the earthquake that struck Java in 1867, killing thousands of people and causing enormous damage to buildings and plantations. The proceeds of this beneficial publication were to go to the victims. Published at Samarang only three months after the calamity, it was based on public reports and personal interviews with eye-witnesses, and illustrated with five original photographs. At the end is a short historical description of Passar-Gedée. **$3200**

RAFFLES & CRAWFORD

350. [JAVA] [RAFFLES & CRAWFORD] MARCHAL, M. Description Géographique, Historique et Commerciale de Java et des autres îles de l'Archipel Indien, par Min. Raffle, ancien Gouverneur-General Anglais, à Batavia et John Crawfurd, ancien Résident à la Cour du Sultan de Java...

Quarto, with two folding maps and 48 plates, many handcoloured; some plates embrowned and scattered foxing throughout but a very good copy in modern quarter calf. Brussels, Tarlier and Jobard, 1824.

An interesting combination in French translation of the two major writers on Java during the British occupation. Java fell to British East India Company forces under Lord Minto, the governor-general of India, who installed Raffles as lieutenant-governor. Raffles was keen to liberate the Javanese from the tyranny of the methods of agriculture imposed by the Dutch, and aimed at introducing liberal economic principles. He was a keen student of the Javanese culture, and was the discoverer of the great Buddhist temple of Borobudur. His *History of Java* was published in 1817, after his return to England.

Crawford was a long-time resident of Java, and his

350. Raffles & Crawford

History of the Indian Archipelago was the other source book for this present composite work, edited by J.F. Marshal. The fine lithographed plates are the work of J.B.A.M. Jobard, a pioneer of the art in Brussels.

Cordier, 1109. **$2400**

RAJAH BROOKE

351. KEPPEL, Captain The Hon. Henry. The Expedition to Borneo of HMS Dido for the Suppression of Piracy: with Extracts from the Journal of James Brooke, Esq. of Sarawak.

Two volumes, octavo, with 11 fine tinted lithographs (printed by Day & Haghe), a folding table and six folding maps; a good set in the original olive-green cloth, a bit stained, one endpaper renewed. London, Chapman and Hall, 1846.

First edition: one of the best accounts of Borneo, partly in the form of a memoir of James Brooke who had been appointed rajah of Sarawak by the Sultan of Brunei. Keppel was in the Malacca Straits in the early 1840s, acting against the Borneo pirates who dominated the area; he gives a geographical overview of Borneo, and a good description of the Dyaks, the aboriginal Bornean tribes.

Not in Abbey; Hill, p. 162. **$800**

FRENCH INDOCHINA

352. LEMIRE, Charles. Cochinchine Française. Royaume de Cambodge, Royaume d'Annam, et Tonkin. L'Indo-Chine.

Octavo, eight plates, two folding maps; a very good copy in contemporary quarter calf gilt, all edges gilt. Paris, Challamel ainé, 1884.

For many years Lemire's study remained one of the fundamental references dealing with the French colonies in Indochina. This third edition is much expanded and updated to include an account of the French advance into northern Vietnam.

Not in Cordier. **$750**

STRAITS OF MALACCA

353. LINDSAY, J. Directions to Accompany J. Lindsay's Charts of the Straits of Malacca; with two Journals from the Isle of Mauritius to India.

Quarto, pp. [ii], 30; an excellent copy in modern boards. Printed by R. Morrison, [London, c. 1796].

Rare set of sailing directions based on actual observations by Lindsay himself, and on the journals of other ships. Two brief journals are printed at the end of the book, from the *America* in 1795 and the *Fort Lewis* in 1792.

As with other similar works (e.g. catalogue numbers 318 and 358), this was published to accompany the charts of the area, but exists quite legitimately as a printed book, separately from the charts which would normally have been bound in an atlas or stored with the ship's charts. **$2750**

SARAWAK

354. LOW, Hugh. Sarawak; Its Inhabitants and Productions: being Notes during a Residence in that Country with His Excellency Mr. Brooke.

Octavo, seven engraved plates; good in contemporary half calf, respined. London, Richard Bentley, 1848.

The first issue (the title-page was subsequently amended to read "H.H. The Rajah Brooke") of Low's account of Sarawak at the beginning of a century of rule by the British Raj.

James Brooke, a former military officer of the East India Company, had visited the territory (now part of Malaysia) in 1839 and assisted the sultan in suppressing a revolt. He was rewarded in 1841 by being granted the position of rajah of Sarawak. Despite his endeavours to suppress head-hunting, Low's account provides detailed observations of this practice, as well as information about a number of tribes, to many of whom he represented the first point of white contact. **$1500**

VIETNAM TO MADAGASCAR

355. LYAUTEY, Louis-Hubert-Gonzalve. Lettres du Tonkin et de Madagascar...

Two volumes, 28 drawings by the author (some folding), 14 maps in colour; a very good uncut copy in contemporary blue half calf, original wrappers bound in. Paris, Armand Colin, 1920.

Author's signed presentation copy to Gerald Selous. Lyautey was a renowned military leader, whose career spanned both Vietnam and Madagascar, and later Morocco. This collection of letters covers the period of his posting in Indochina, where he was sent in 1894, and his triumphant conquest of Madagascar. **$950**

SIAM & CAMBODIA

356. MACGREGOR, Surgeon-Major John. Through the Buffer State. A Record of Recent Travels through Borneo, Siam and Cambodia.

Octavo, 11 plates and a plan; a few odd spots but a very good uncut copy in original red cloth. London, F.V. White, 1896.

The author, an Indian Army Medical Officer, visited Siam and Cambodia while on leave in 1894. He appears to have had influential contacts in Bangkok and comments on the manner of the French acquisition of Western Cambodia from the Thais. **$400**

THE ALCESTE

357. McLEOD, John, surgeon. Voyage of His Majesty's Ship Alceste, along the Coast of Corea to the island of Lewchew; with an account of her subsequent shipwreck...

Octavo, frontispiece and coloured aquatint plates; a good copy in contemporary speckled calf, expertly rebacked. London, John Murray, 1818.

Second and best edition, with numerous additions and alterations, of this famous account of the voyage of the *Alceste* which brought Lord Amherst to China to serve as ambassador, and specifically to negotiate the China trade, at the time a matter of dispute between China and Great Britain. McLeod, who was surgeon on the ship, describes the voyage out, the various calls in the East Indies, surveying off Korea, the visit to Okinawa, another to Manila, and the shipwreck off the coast of Sumatra and subsequent rescue from Malay pirates. This was one of the most popular travel books of its time, portraying the Far East to a much wider audience than had been reached by earlier accounts.

Abbey, 559; Borba de Moraes, p. 507; Hill, pp. 188-9. **$800**

357. McLeod

SAILING DIRECTIONS

358. MOOR, Henry. Sailing Directions to Accompany a new Chart of the Moluccas and Eastern Islands... Surveyed, in the course of two years and six months station in those seas by Henry Moor, Acting First Lieutenant of His Majesty's Sloop, Hobart.

Quarto, 19 pp.; an excellent copy in modern boards. London, Laurie and Whittle, 1801.

First edition of this rare set of sailing directions based on actual observations by Moor around the Moluccas, surveying on HMS *Hobart*, captain V.V. Ballard.

Like other similar works (e.g. catalogue numbers 318 and 353), this was published to accompany a chart of the area, but exists quite legitimately as a separate printed book. **$2250**

Ueber
Alterthümer
des
Ostindischen Archipels
insbesondere
die Hindu-Alterthümer u. Tempelruinen auf
JAVA, MADURA u. BALI.
„Lasst Eurem Herzen die Bitte um Vernichtung Eurer thierischen Natur entkeimen, damit diese Pforte sich nur für stille Zufriedenheit öffne"
Im Jahre 1029 nach Saliwana, 1107 nach Christi Geburt
von
Dr. Johannes Müller
Medicinalrath in Berlin
Mit 21 Kunst-Beilagen.

359. Muller

HINDU TEMPLES IN JAVA, MADURA & BALI

359. MULLER, Dr. Johannes. Ueber Alterthuemer des Ostindischen Archipels insbesondere die

Hindu-Alterthuemer und Tempelruinen auf Java, Madura und Bali.

Octavo, with 21 plates (ten of which are in colour and one folding); light foxing, library stamp; original decorative wrappers, worn but very attractive. Berlin, published by the author, 1863.

Rare study of Hindu temples in Java, Madura and Bali, illustrated by beautiful atmospheric colour lithographs of the ruined temples, as well as detailed line-drawings of the decorations. **$1500**

EXPLORATION OF NEW GUINEA

360. [NEW GUINEA] AA, P.J.B.C. Robidé van der. Reizen naar Nederlandsch Nieuw-Guinea... 1871, 1872, 1875-1876...

Octavo, with two large folding maps; old quarter calf, worn. The Hague, Nijhoff, 1879.

This scarce work is an in-depth account of explorations throughout New Guinea during the 1870s, including those of Van der Crab, J.E. Teysmann and J. G. Coorengel. **$825**

ETHNOGRAPHISCHE BESCHRIJVING

VAN DE

WEST- EN NOORDKUST

VAN

NEDERLANDSCH NIEUW-GUINEA

DOOR

F. S. A. DE CLERCQ,

MET MEDEWERKING VAN

J. D. E. SCHMELTZ,

Conservator aan 's Rijks Ethnogr. Museum. — Redacteur van het „Internationales Archiv für Ethnographie". — Correspondeerend lid van het Kkl. Inst. voor de Taal- Land- en Volkenkunde van Nederl. Indië te 's Gravenhage, enz.

MET BIJVOEGING EENER SCHETS DER ETHNOGRAPHIE VAN DUITSCH EN BRITSCH NIEUW-GUINEA.

361. Clercq

IRIAN ETHNOGRAPHY

361. [NEW GUINEA] CLERCQ, Frederick de and Johannes SCHMELTZ. Ethnographische beschrijving van de West- en Noordkust van Nederlandsch Nieuw-Guinea.

Folio, with colour plates, a folding map and charts; a very good copy in original decorated cloth. Leiden, P.W.M. Trap, 1893.

First edition of this definitive work on Dutch New Guinea: one of the earliest publications on the material culture of Irian, illustrated with beautiful full-page coloured lithograph plates from the Trap printing press. The illustrations depict weapons, everyday utensils, jewellery and artefacts. De Clercq was a government official in Indonesia and Schmeltz was later the director of the Ethnographical Museum in Leiden. A fine copy of a rare book. **$4750**

362. Earl

NEW GUINEA ANTHROPOLOGY

362. [NEW GUINEA] EARL, George Windsor. The Native Races of the Indian Archipelago. Papuans.

Octavo, with five plates (four coloured) and two folding maps; a very good copy in half morocco. London, Hippolyte Bailliere, 1853.

First edition, separate issue (another version had the series-title "Ethnographical Library" which here only appears as an imprint on the plates). Important anthropological observations on the natives of New Guinea, but also on the North Australian Aborigines: chapter XII is devoted to Melville Island, Port Essington and North Australia in general, and one of the folding plates gives anthropometric comparisons between North Australian Aborigines and New Guinea natives. There are numerous discussions of Australian Aborigines elsewhere in the text, particularly those of Melville Island and Port Essington.

The fine full-page handcoloured lithographs, produced in London, were based on drawings done on the spot by the Dutch artists Van Oort and Van Raalten.

Ferguson, 9339. **$1375**

THE MARCHESA IN NEW GUINEA

363. [NEW GUINEA] GUILLEMARD, F.H.H. The Cruise of the Marchesa to Kamschatka and New Guinea, with notices of Formosa, Liu-kiu, and various islands of the Malay Archipelago.

Two volumes, octavo, coloured bird-plate as frontispiece to each volume, with maps (some folding and coloured) and illustrations; a very good set in the original blue-grey cloth, ornate designs in gilt on covers. London, John Murray, 1886.

First edition: a private, semi-scientific voyage in the 420-ton auxiliary screw schooner yacht *Marchesa*, of considerable importance for the ornithological collecting undertaken in New Guinea. **$950**

PORT MORESBY

364. [NEW GUINEA] MORESBY, Capt. John. Discoveries & Surveys in New Guinea and the Dentrecasteaux Islands. A Cruise in Polynesia and visits to the pearl-shelling stations in Torres Straits of HMS Basilisk...

Octavo, with a folding map (small tears to folds) and six plates and charts; a good copy in the original blue cloth gilt, spine slightly worn. London, John Murray, 1876.

The scarce first edition, in nice condition.

The expeditions of HMS *Basilisk* placed more than one hundred and forty islands and many excellent harbours on the chart. Moresby's book, which has always been popular, includes material on the Torres Strait, Black-Birding, a Corroboree, Norfolk Island, Rotumah, Santa Cruz, Ambrym, Pentecost, the discovery of Port Moresby and Fairfax Harbour, Yule Island, the D'Entrecasteaux Group, the discovery of China Straits, taking conditional possession of newly discovered islands, discovery of Milne Bay, Killerton Group, finding the true East Cape of New Guinea, and much more.

Ferguson, 12801. **$850**

366. Ogilby

365. Rosenberg

FINE LITHOGRAPH ILLUSTRATIONS

365. [NEW GUINEA] ROSENBERG, Carl Benjamin Hermann von. Reistochten naar de Geelvinkbaai op Nieuw-Guinea...

Quarto, 20 lithographic plates (16 in sepia and four in colour) and a folding map; a fine copy in original boards. The Hague, Martinus Nijhoff, 1875.

Original edition: very scarce. When Rosenberg (1817-1888) arrived in Indonesia he was appointed assistant to Franz Junghuhn at Padang. He travelled widely through several parts of the East-Indies between 1840-1870, exploring areas in Sumatra, the Moluccas, and certain regions of Teluk Cendrawasih (Geelvink Bay) of West Irian. In this finely-illustrated book he extensively records his exploration in New Guinea, discussing the natural environment of each region he visits, and including many references to the people and their customs. This is considered one of Rosenberg's best publications, and is illustrated with fine lithographs including four handcoloured prints of newly discovered birds. **$2850**

OGILBY'S ASIA MAP

366. OGILBY, John. A New Map of Asia.

Engraved map, 420 x 540 mm.; early folds, a couple of very slight marginal tears expertly sealed, but in very good condition. London, John Ogilby, [1673].

A fine map of Asia by John Ogilby (1600-76), cosmographer and geographic printer to the King, who had an excellent reputation for the fine quality of his maps, particularly his well-known road strip-maps. Dedicated to the East India Company, the map shows the extensive knowledge of this area at the time, extending from Africa to Japan, with a particularly detailed study of south-east Asia at an early date. **$2500**

VERY RARE, WITH THE APPENDIX

367. ORIENTAL NAVIGATOR. The Oriental Navigator; or, New Directions for sailing to and from the East Indies. Also for the use of the country ships, trading in the Indian and China Seas, to New Holland, &c. &c... [with Appendix].

Large quarto; some sporadic damp-stains, last leaf a little frayed; later plain wrappers. London, Laurie and Whittle, 1794[-97].

Rare sailing-guide, prepared "as a Companion to the East India Pilot... originally published by the late Mr. Robert Sayer". This substantial ship's reference work is based on voyagers' reports, and like other works of this nature successive editions get ever larger. This runs to 590 densely-printed quarto pages of specific and detailed observations, and is full of extraordinary detail.

The book is rare (Ferguson knew only the Mitchell Library copy, and none is added in the *Addenda*, although it is of great specific Australian interest). This copy is complete with the additional 18-page Appendix, printed in 1797, including "Observations by Captain Thomas Butler of the Walpole relative to his passage from the Cape of Good Hope to China, round New Holland, in 1794" (4 pp.). An extra slip is pasted to the Contents leaf dealing with this Appendix, and there is a pasted slip of errata.

Ferguson, 191a. **$2100**

THE FRENCH REACH SAIGON

368. PALLU DE LA BARRIERE, Leopold. Histoire de l'Expédition de Cochinchine en 1861.

Octavo, with two plans and a folding colour map; a very good copy in contemporary quarter calf. Paris, Berger-Levrault, 1888.

Author's presentation copy of the second, revised edition, inscribed to Camille Rousset. The French expedition to Cochinchina (southern Vietnam) of 1861, under the command of Vice-Admiral Léonard Charner, was organised to defeat the Vietnamese resistance which had prevented the French from advancing beyond Saigon since 1859. Charner successfully defeated the Vietnamese general Nguyen Tri Phuong at Chi Hoa, near Saigon (now Ho Chi Minh City). Under the terms of the peace treaty signed in 1862, the French received Saigon and three of the southern provinces of Cochinchina, the opening of three ports to trade, freedom of missionary activity, a vague protectorate over Vietnam's foreign relations, and a large cash indemnity. Five years later France annexed the rest of Cochinchina. **$450**

"TELLS ONE EVERYTHING IN THE WORLD"

369. RAYNAL, Abbé. JUSTAMOND, J. (translator). A Philosophical and Political History of the Settlements and Trade of the Europeans in the East and West Indies.

Five volumes, octavo, with a frontispiece and four folding maps; a few defects and repairs but an attractive set in contemporary calf, gilt. London, T. Cadell, 1777.

Third edition, with revisions and corrections: this was an enormously reprinted book; Lust notes that NUC records 33 editions before 1821. The collection includes detailed reports of Dutch, English and French trading in the East Indies, and a comprehensive study of the implications of European discovery and settlements in the Americas.

First published in Amsterdam in 1770, "its essential theme is the doctrine of our moral obligations to the lower races. It contains, besides much entertaining information on a great variety of subjects, many disquisitions against religion, despotism, slavery, etc., probably supplied in part by Diderot. It encouraged the revolutionary ardour of Toussaint Louverture among others... Horace Walpole, after having commended his work as "the most amusing book that tells one everything in the world", found the author himself a tiresome bore, and pretended to be deaf to escape his importunities" (Harvey & Heseltine).

Lust, 653 (note); Sabin, 68088. **$1600**

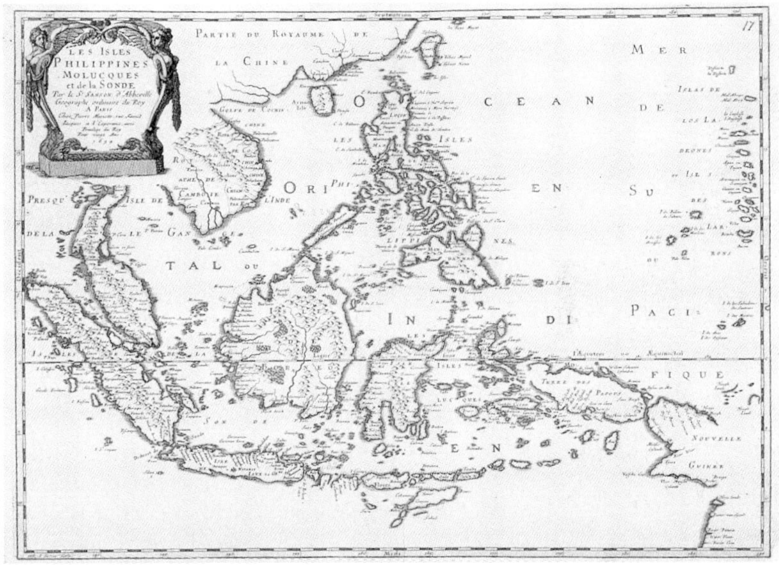

370. Sanson

SANSON'S SOUTH-EAST ASIA

370. SANSON, Nicolas. Les Isles Philippines Molucques et de la Sonde.

Engraved map, in original handcolouring, 398 x 560 mm.; in fine condition, mounted. Paris, Pierre Mariette, 1654.

Scarce chart of South East Asia showing Sumatra in the east, China and the Philippines in great detail in the north, and extending south to New Guinea and Cape York Peninsula. This is one of only few seventeenth-century maps to show Jansz's discoveries in New Guinea and the west coast of Cape York Peninsula. His voyage in the *Duyfken* in 1605 "to discover the great land Nova Guinea and other unknown east and south lands" took him from Java to the south coast of New Guinea through Torres Strait to land on the eastern coast of the Gulf of Carpentaria. However it was not until Hessel Gerritsz's 1622 chart that any record of these discoveries appeared cartographically. In Sanson's map New Guinea is shown as two separate islands.

This is one of the earliest maps produced by the Sanson family business in Paris. Nicolas Sanson was France's foremost cartographer; his distinctively elegant and restrained style helped his country dominate the field in the seventeenth century, effectively ending the Dutch domination of mapmaking.

Tooley, 1111. **$3600**

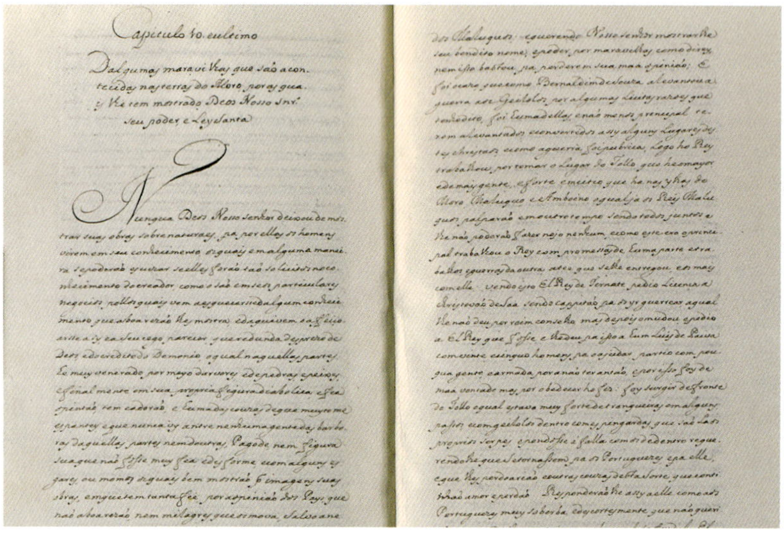

Capitulo 10. e ultimo

Dalgumas maravilhas que saõ acontecidas nas terras do Moro, por as quaes Vhe tem mostrado Deos Nosso Snr. seu poder e Ley Santa

371. Spice Islands

PORTUGUESE MANUSCRIPT DESCRIPTION OF THE MOLUCCAS

371. [SPICE ISLANDS] ANONYMOUS. An untitled eighteenth-century manuscript copy of a Portuguese account of the Moluccas, or Spice Islands, in the mid-sixteenth century.

Folio, 68 leaves of manuscript text and 8 leaves with pen, ink and wash drawings; contemporary mottled calf, spine gilt. N.p., n.d. [?Portugal, perhaps c. 1720-1750].

A handsome and intriguing manuscript report on the Moluccas, or Spice Islands, evidently deriving from a Portuguese original of the mid-sixteenth century.

Discoveries in the East Indies, and the development of colonial power and trading links there, were the stuff of intrigue and clandestine activity from the earliest days of discovery. The relationship between knowledge and power is evident throughout the history of the period; many materials were simply unavailable, and the Portuguese certainly had for some time a deliberate policy of not allowing sensitive material to be published; as early as 1504 King Manuel had "decreed that complete secrecy, under pain of death, should be kept with regard to southeastern and north-eastern navigation" (Lach, I, p. 152).

This secrecy may go some way towards explaining the existence of a manuscript like this, and the general background of the transmission in manuscript of many of the texts on the East Indies and the Far East,

371. Spice Islands

and their ultimate publication sometimes centuries later from manuscript originals.

The text is handsomely written in ink on a fine laid paper (the watermarks have not been identified) and is followed by the leaves of drawings which are on a thicker paper, better suited to the purpose, watermarked "GB" alternating with a large coat-of-arms.

The text is divided into two parts, of twenty and ten chapters respectively. The first part deals with the complicated encounters in the Moluccas between several parties: the Spanish interlopers led by Ruy Lopes de Villalobos, sent from Mexico in 1542 to assert Spanish imperial claims in the region; the Portuguese; the king of Gilolo, with whom the Portuguese were at war; and the king of Tidore.

The second part gives general information on the islands and their inhabitants, customs, natural history and so forth. It is followed by the series of attractive and well-executed illustrations, including several depictions of Moluccan ships (proa-type vessels with dragon's-head bows, some with cannon mounted) and an armed warrior, and drawings of a fish and a pair of small evidently marsupial animals.

The initial leaf, headed "Señor", is a dedication dated October 1561, but gives no clue to identities of either the dedicatee or the author. **$11,000**

371. *Spice Islands*

MALAY PENINSULA

372. STANFORD, Edward. Map of the Malay Peninsula.

Folding linen-backed lithographed map, 590 x 660 mm.; excellent in original red cloth lettered in gilt, folding to small octavo. London, Edward Stanford for The Singapore & Straits Printing Office, 1910.

Scarce map of the Malay Peninsula from Siam in the north to Singapore in the south, bounded by the Straits of Malacca in the west and the China Sea in the east. **$385**

BURMA'S POWER & GREATNESS

373. SYMES, Michael, Lieutenant-Colonel. An Account of an Embassy to the Kingdom of Ava, in the Year 1795... To which is now added, a Narrative of the Late Military and Political Operations in the Birmese Empire. With Some Account of the Present Condition of the Country, its Manners, Customs, and Inhabitants.

Two volumes, duodecimo, with finely engraved titles and a folding map; a very good set bound in contemporary half maroon calf. Edinburgh, Constable, 1827.

Originally published in 1800, this work on Burma was updated and reissued in 1827 as part of *Constable's Miscellany*; according to the Preface, "the able and interesting work of Colonel Symes on the Birman Empire, or Kingdom of Ava, has now become exceedingly scarce...the Birman Empire ...of all the Eastern nations, is second in power and greatness to China alone".

See Cordier, 445-446. **$650**

UNABRIDGED EDITION

374. TERRY, Edward. A voyage to East-India; wherein some things are taken notice of, in our passage thither, but many more in our abode there, within that rich and most spacious empire of the Great Mogul.

Octavo, engraved frontispiece and three plates, folding map (expertly repaired); some light browning but a good copy in old quarter calf. London, J. Wilkie, 1777.

Second and best edition, here published unabridged over a century after the first edition of 1655. Edward Terry (1590-1660) went out to India as chaplain with a fleet sent by the London East India Company. Shortly after his arrival he became chaplain to the ambassador Sir Thomas Roe, and his journal is a useful complement to Roe's own account of the embassy to the court of the Great Mogul.

Cox I, p. 271 ("Entertaining reading"). **$1250**

TIDEMARKS

375. TOMLINSON, H.M. Tidemarks: Some Records of a Journey to the Beaches of the Moluccas and the Forest of Malaya in 1923.

Octavo, with a photogravure frontispiece; two tiny rust-spots at inner hinge otherwise a very fine copy in the original brown cloth, boxed. London, Cassell, 1924.

First edition of this famous reminiscence of the East Indies: first issue (with the palm tree in gilt on the spine).

Muir, 'Points 1874-1930', p. 155. **$450**

376. Yule

COLOUR PLATES OF BURMA

376. YULE, Captain Henry. A Narrative of the Mission sent by the Governor-General of India to the Court of Ava in 1855, with notices of the Country, Government, and People.

Quarto, with 26 lithographed plates, some colour, four maps (one folding in front pocket); some spotting and general wear, but a good copy in the original gilt-decorated cloth, spine repaired. London, 1858.

Yule was secretary to Arthur Phayre's 1855 mission which was sent to improve Anglo-Burmese relations after the second Anglo-Burmese War. He was a careful observer and tireless gatherer of information, and was able to draw extensively on unpublished papers by other experts. The fine plates which depict exquisite temples and picturesque scenes of daily life, are after drawings by a number of artists including Captain Yule himself, and others are based on photographs.

Abbey, 408; Cordier, 457-458; see Herbert, 215. **$1850**

14: THE MEKONG RIVER EXPEDITION

KHMER ARCHITECTURE

377. DELAPORTE, L. Voyage au Cambodge. L'Architecture Khmer.

Large octavo, with 175 illustrations, including 14 double-page plates and a map; some scattered foxing, generally in the preliminaries, but a very good copy in the original pictorial gilt cloth, leather spine, elaborately gilt, a little rubbed, all edges gilt. Paris, Librairie Ch. Delagrave, 1880.

The first major study of Khmer architecture, and a very handsome book. Delaporte was a member of the 1866-68 Mekong Expedition and was responsible for the magnificent illustrations in the official account of that epic journey. He subsequently returned to Cambodia and prepared this important study of the country's architecture. He later became the first director of the museum that was to become the Musée Guimet. **$1800**

DELAPORTE'S STORY

378. [DELAPORTE] BEAUVAIS, René de. Louis Delaporte. Explorateur. Ses missions aux ruines khmères.

Quarto, with illustrations and folding maps; a fine copy, uncut with the original wrappers, modern buckram. Paris, Imprimerie des Orphelins d'Auteuil, 1929.

Intended for a young audience, this attractive book commemorates Delaporte's role in the Mekong expedition. He was, of course, the artist appointed to record the expedition, and his breathtaking images of Khmer architecture and the Mekong landscapes are the highlight of the lavish official account. **$385**

LAGREE VERSUS GARNIER

379. DE VILLEMEREUIL, M. A.-B. Doudart de la Grée, Capitaine de Frégate, Chef de l'Exploration du Mé-Kong et de l'Indo-Chine Exécutée en 1866-67-68 par ordre et aux frais du Gouvernement Français, et la Question du Tong-King.

Octavo, folding map; some foxing but a very good copy in original printed wrappers. Paris, aux Bureaux de L'Explorateur, 1875.

Second edition of this rare pamphlet: one of the resounding shots fired in the literary battle fought by supporters of both Lagrée and Garnier after the conclusion of the Mekong River Expedition and the death of both the principals. Lagrée died in 1868, Garnier five years later, but the controversy over who should be accorded credit as leader of the expedition continued for many years. **$440**

RARE IN 1887

380. [GARNIER, Francis] "FRANCIS, G." La Cochinchine Française en 1864.

Octavo, 48 pp.; various library stamps and annotations, but a good copy in original printed wrappers, spine expertly restored, lower wrapper renewed. Paris, Dentu, 1864.

This important and rare pamphlet, written by Francis Garnier under the pseudonym G. Francis before his great expedition, advocated an expansion of the French position in Vietnam, and portrayed France as a beneficent liberator: "this generous nation… has received from Providence a higher mission, that of emancipation, of bringing into the light and into liberty the races and people still enslaved by ignorance and despotism" (translation from Osborne, *The French Presence in Cochinchina and Cambodia*).

This pamphlet is a known rarity: Edouard Petit, in his biography of Garnier published in 1887, said that it was already very rare thirteen years after publication. Illustrated at end of section. **$1200**

FIRST PUBLICATION ON THE EXPEDITION

381. GARNIER, Francis. Notice sur le Voyage d'Exploration effectué en Indo-chine par une Commission Française pendant 1es années 1866-67-68.

Octavo, 80 pp., with a folding map; a fine uncut copy with the original front wrapper, in contemporary half blue calf, double labels. Paris, Challamel ainé, 1869.

Garnier's first publication on the Mekong River Expedition, and very scarce indeed. A separately-paginated offprint from *La Revue Maritime et Coloniale* for 1869, this was probably issued in only a small quantity for private distribution. This very attractive copy is reputed to have come from Garnier's own library.

Garnier gives quite a detailed account of the Mekong river expedition, with a very good map tracing the expedition's progress and stopping points along the Mekong and then into China and Tibet. Illustrated at end of section.

Cordier, 'Sinica', 330. **$2400**

RARE PAMPHLET

382. GARNIER, Francis. Episode du Voyage d'Exploration dans L'Indo-Chine… Communication adressée à la Société de Géographie dans sa Séance du 30 Avril 1869.

Octavo, the whole printed on stiff paper; minor wear to binding but very good in contemporary crimson half calf with double labels. Paris, E. Martinet, 1869.

Very scarce indeed. This copy is particularly well

ÉPISODE

DU

VOYAGE D'EXPLORATION

DANS

L'INDO-CHINE

PAR

FRANCIS GARNIER

Lieutenant de vaisseau
Chef de la mission scientifique du Mé-kong.

382. Garnier

bound, and matches the previous item in this catalogue, which was said to have come from Garnier's own library. It was published in Paris while Garnier was supervising the publication of the official account of the voyage and at a time when he was being criticised for not giving sufficient credit to Doudart de Lagrée, the expedition's original leader

The text describes the mission under the command of Garnier to visit the Tibetan stronghold at Ti-ly, generally regarded as the most hazardous of all the expedition's undertakings. When Garnier was awarded the Patron's Medal by the Royal Geographical Society it was specifically for leading this mission, and therefore his wish that it also be conferred posthumously on Lagrée was denied.

Illustrated at end of section.

Not in Cordier. $985

THE MEKONG EXPEDITION

383. **GARNIER, Francis. Voyage d'Exploration en Indochine effectué pendant les années 1866, 1867 et 1868, par une Commission Française présidée par M. le Capitaine de Frégate Doudart de Lagrée et publiée par les Ordres du Ministre de la Marine sous la Direction de M. le Lieutenant de Vaisseau Francis Garnier. Avec le Concours de M. Delaporte, Lieutenant de Vaisseau, et de Min. Joubert et Thorel, Médecins de la Marine, Membres de la Commission.**

Two volumes, large quarto, one with portrait frontispiece, 11 maps, a plate and a plan, many wood-engraved illustrations throughout the text; and two atlas volumes, folio, the first "Cartes et plans" with 12 maps, nine plans and an aerial view, (some tinted); the second "Album pittoresque" with 48 plates (including 44bis) on 40 sheets (six double-page), 11 of the plates in full colour; a few leaves or plates spotted, some marginal staining to volume one of the text, but a very handsome set in contemporary half calf. Paris, Librairie Hachette, 1873.

The rare official account of the most important exploration in Indochina in the nineteenth century. The Mekong Expedition followed years of lobbying, mainly on the part of Francis Garnier, who held passionate beliefs about France's "civilising mission" as well as the riches which would flow from the Mekong, in terms of trade with China. He had been a vocal, though pseudonymous, pamphleteer on the subject and became a most energetic second-in-command of the expedition. The French representative in Cambodia, Doudart de Legrée, was chosen to lead the expedition.

The party left Saigon on 5 June 1866 with orders to find a navigable route into south-western China, and to search for any potentially profitable commodities en route. The journey took them through Phnom Penh (then the royal capital), through Vietnam and Cambodia, Laos and into China.

Although the mission was a failure in commercial terms – the Mekong was found to be unnavigable by boats of any size – the party's achievements were immense: mapping over four thousand miles of previously unsurveyed territory and becoming the first Europeans to enter Yunnan province by a southern route.

Whilst Garnier was the expedition's cartographer, the fine views in the atlas are based on the drawings of Louis Delaporte, a young regular naval officer, but also an artist of considerable talent and great technical skill. "Delaporte's illustrations, ranging from detailed architectural drawings to sketches of people and places the Frenchmen encountered, make the record of the Mekong mission one of the most pictorially complete for the exploration of Indochina in the nineteenth century" (Osborne, *River Road to China*, p. 86).

The beauty of the landscape, the great diversity of the peoples of the region and the exotic ancient monuments were all recorded in breath-taking detail by Delaporte, and reproduced in sumptuous lithographs and engravings in Garnier's account. Of particular interest are the intricately detailed plates of the mighty ruins of Angkor, the centre of Cambodian glory between the ninth and fifteenth centuries and the overgrown ruins of Vien Chan, the ancient Laotian capital.

Apart from the very first stage of the expedition, the large party with their mountain of provisions travelled up the river in *pirogues*, narrow dug-out canoes, which were pushed forward by the use of long bamboo poles or, in the treacherous rapids, painstakingly pulled along the shore by means of a boat hook. Whilst the Mekong's fearsome rapids, particularly those of Khon, and spectacular waterfalls dashed any hopes of it as a future trade route, more immediate problems beset

383. Garnier

the expedition, such as encounters with tigers, leeches, quicksand and inevitable illness. Dysentery was a common problem, but much worse was fever. Garnier was lucky to survive a bout of typhoid fever, which rendered him unconscious for eighteen days, and caused him to lose a lot of skin and all his hair. Nevertheless he was fit enough within a couple of months to make a journey of one thousand miles mostly on foot through territory never before explored by Europeans, back to Phnom Penh to secure passports for the party to travel into China.

Much to Garnier's regret the party had to abandon the Mekong after Se-Mao in Yunnan, and follow the Red River to Tong-king. De Lagrée succumbed to fever and remained in Hui-tse (where he died soon after) while the party proceeded under the command of Garnier to the Muslim stronghold of Ta-li across the Yang-tse in Tibet. Here Garnier was instructed by the Sultan to abandon the journey, which he was compelled to do, returning to Saigon by way of Shanghai.

Published originally in an edition of only eight hundred copies, this is now a very rare book when complete and in good condition: most of those copies held in Indochina have been destroyed by subsequent events, and the beautiful plates have made it a target for the print trade.

This set has a fine provenance: it is a presentation copy from the family of the expedition's late leader, inscribed by C. de Legrée and dated 21 January 1873.

Cordier, 'Indosinica', 1012; 'Sinica', 329; Numa Broc, 208; Osborne, 'River Road to China'. **$44,500**

Also illustrated on catalogue cover.

384. Garnier

THE CONDENSED ACCOUNT

384. GARNIER, Francis. Voyage d'Exploration en Indo-Chine.

Quarto, profusely illustrated with wood engravings, two large folding colour maps (each with sealed tears); a very good copy in original quarter red morocco, all edges gilt. Paris, Hachette, 1885.

A later, condensed but very handsome edition of Garnier's famous account by the same publisher. This edition is illustrated by engravings based on the original drawings by Delaporte and has an introduction by Leon Garnier.

Cordier, 'Sinica', 330. $2400

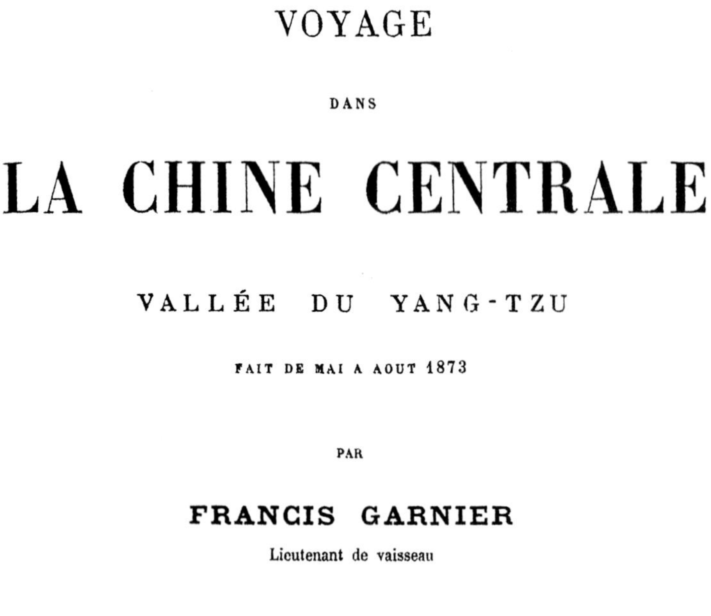
VOYAGE

DANS

LA CHINE CENTRALE

VALLÉE DU YANG-TZU

FAIT DE MAI A AOUT 1873

PAR

FRANCIS GARNIER

Lieutenant de vaisseau

385. Garnier

INTO VIETNAM

385. GARNIER, Francis. Voyage dans la Chine Centrale. Vallée du Yang-Tzu, fait de Mai a Août 1873.

Octavo, large folding map and tables; foxing throughout but an attractive copy in contemporary half calf, original wrappers bound in. Paris, Delagrave, 1874.

Very scarce indeed. Published posthumously, this is a separately-paginated offprint from *Bulletin de la Société de Géographie* of Garnier's account of a journey he undertook to China in an effort to capitalise on his earlier experiences during the Mekong expedition. Although ostensibly a journey to explore the Yang-tse, Garnier was called from Shanghai to Saigon in August 1873 – the point at which his narrative ends – and then became involved in local intrigue between the Governor of Cochinchina, Marie-Jules Dupré, and a French trader, Dupuis. When he reached Hanoi he confronted Vietnamese officials and led a successful attack on the Hanoi citadel. However, in an attempt to defend it against Chinese Black Flag bandits on 21 December 1873, Garnier was killed. The French negotiated a withdrawal from Vietnam in early 1874.

Illustrated at end of section.

Not in Cordier. $1400

386. Garnier

PARIS TO TIBET

386. GARNIER, Francis. De Paris au Tibet. Notes du Voyage.

Octavo, with forty plates and a folding map; modern cloth binding retaining original decorative wrappers. Paris, Hachette, 1882.

Published posthumously, these letters form the journal of Garnier's second expedition to Indochina, during which he was killed. $485

LIFE OF GARNIER

387. [GARNIER] PETIT, Edouard. Francis Garnier: Sa Vie, ses Voyages, son Oeuvre, 1839-1874...

Octavo, folding map; a very good copy in modern buckram, original wrappers bound in. Paris, Maurice Dreyfous, 1887.

Comprehensive account of the life and expeditions in Indochina of the explorer and enthusiastic colonist Garnier. $425

RARE PAMPHLET

388. [RIEUNIER, A.B.L.] "ABEL, M.H." Solution Pratique de la Question de Cochinchine ou Fondation de la Politique Française dans l'Extrême-Orient.

Octavo, 24 pp.; a very good copy in original wrappers, spine expertly restored. Paris, Challamel aîné, 1864.

Abel was the pseudonym of A.B.L. Rieunier, a French Naval officer who had served with Admiral Bonard in Cochinchina. With Francis Garnier, he was one of the most active propagandists for the extension of French control beyond the three southern provinces around Saigon; a call which which led to the Mekong expedition (see Osborne, *French Presence in Cochinchina and Cambodia*, pp. 40-41).

This pamphlet is of particular interest for the publisher's catalogue on the back wrapper which advertises another work by "Abel" as well as one under Rieunier's own name; it also lists G. Francis's *Le Cochinchine Française en 1864*, all of which sold for one franc. This, like all these ephemeral pamphlets, is very rare. $600

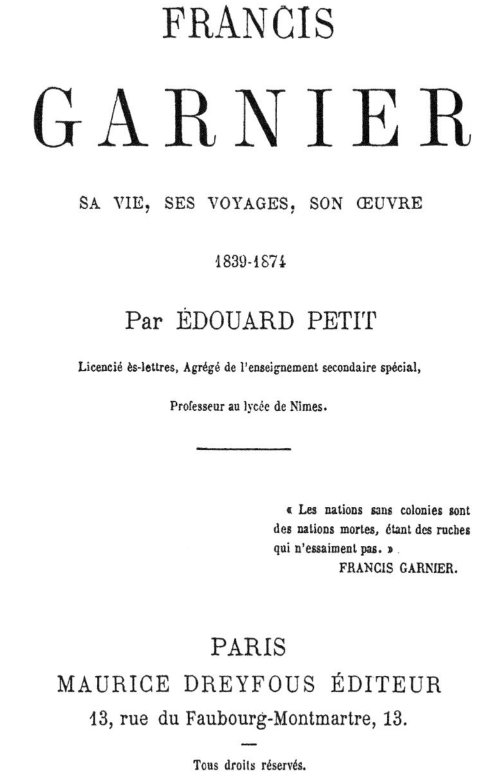

FRANCIS

GARNIER

SA VIE, SES VOYAGES, SON ŒUVRE

1839-1874

Par ÉDOUARD PETIT

Licencié ès-lettres, Agrégé de l'enseignement secondaire spécial,

Professeur au lycée de Nîmes.

« Les nations sans colonies sont des nations mortes, étant des ruches qui n'essaiment pas. »

FRANCIS GARNIER.

PARIS

MAURICE DREYFOUS ÉDITEUR

13, rue du Faubourg-Montmartre, 13.

Tous droits réservés.

388. Rieunier ("Abel")

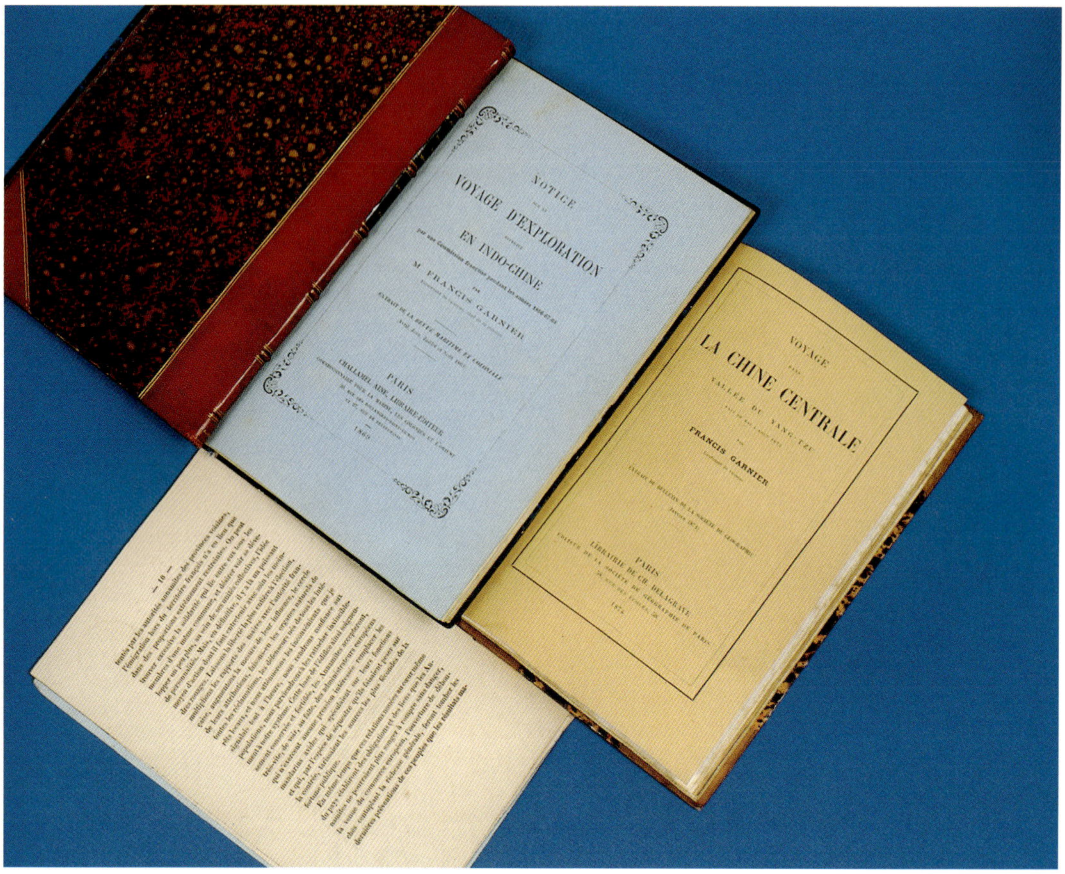

Clockwise from top left: 380, 381, 382, 385 - The rare series of pamphlets by Garnier

V: *The Pacific*

15: THE PACIFIC

CRUISE OF THE BACCHANTE

389. ALBERT VICTOR and GEORGE, Princes. The Cruise of Her Majesty's Ship "Bacchante" 1879-1882.

Two volumes, octavo, illustrations and maps; slight spotting at the start of the first volume, but a good copy in the original blue cloth gilt. London, Macmillan, 1886.

Based on the diaries kept by the Princes on their three-year world tour in the *Bacchante*, including their voyage through the Pacific and a long visit to Australia.

Ferguson, 5795. **$350**

THE OLD SAILOR

390. ALLYN, Gurdon L. The Old Sailor's Story, or a Short Account of the Life, Adventures and Voyages of Capt. Gurdon L. Allyn. Including Three Trips around the World.

Octavo; a good copy in the original black limp cloth. Norwich, Conn., Gordon Wilcox, 1879.

Very scarce: probably privately published.

The author of this first-hand account of three trips around the world on a whaler was a native of Connecticut "in the Seventy-ninth and Eightieth Years of his age". Plying the waters of the Pacific in search of sperm whales, the ship called in at Honolulu, New Zealand and Tasmania, of which the old-timer recollects: "this land produces the finest grains, fruits and vegetables, the people are friendly and hospitable, and there is no finer country in the world for the abode of man".

This is an interesting account not only of whaling, but also of the life of trading commodities around the world in the nineteenth century.

Not in Ferguson, Hill or Judd. **$650**

LADY BRASSEY'S FIRST VOYAGE

391. BRASSEY, Lady Anne. Voyage in the 'Sunbeam'. Our Home on the Ocean for Eleven Months.

Octavo, numerous woodcuts in the text and a number of full-page plates, some printed in blue; a very good copy in original blue decorative cloth, slight edge wear. St. Louis, J.W. Henry, 1891.

In this, the first of a number of extensive cruises in their yacht *Sunbeam*, Lord Thomas Brassey, the Admiralty Secretary and diplomat, and his wife visited Tahiti, Hawaii and other Pacific island destinations, and various ports of call in Asia including Japan, Canton, Macao and Singapore.

Lady Brassey's account of this voyage proved very popular; it was first published in England in 1879 and went through numerous editions, including this American version.

O'Reilly-Reitman, 1315-18 (various editions). **$400**

392. Lady Brassey

LAVISH FRENCH PRODUCTION

392. BRASSEY, Lady Anne. Aux Indes et en Australie dans le yacht le "Sunbeam"...

Quarto, with many illustrations; a few spots but a fine copy in the original ornately pictorial cloth, with decoration in silver and gilt, gilt edges. Tours, Alfred Mame et fils, 1893.

A fine copy of the handsome and splendid first French edition, which puts the English edition of 1890 quite in the shade. It is quite scarce – and is not in fact recorded by Ferguson, although almost half the book is devoted to the Australian portion of the voyage.

Lord and Lady Brassey made several cruises around the world in their yacht *Sunbeam*. Lady Brassey's account of their earlier voyage published in 1879 was very popular and went through numerous editions. Her last voyage was undertaken in 1886-7: she died before the cruise was completed and her grieving husband completed her work for publication. He later authorised this French version, which was translated by Gaston Bonnefont.

Not in Ferguson. **$1400**

SEAFARING

393. BREWER, Charles. Reminiscences [caption-title].

Octavo, 67 pp., an excellent copy in the original maroon cloth. Jamaica Plain, 1884.

Privately published and very scarce; this is a

presentation copy (with a cyclostyled inscription from the author inserted) of these seafaring reminiscences, by an old man, of much earlier days in the Pacific and the Atlantic. Brewer made several visits to Hawaii, described here.

Not in Hill; Judd, 27. **$925**

FROM ALEXANDER BERRY'S LIBRARY

394. BRITTON, Henry. Fiji in 1870...

Octavo, with a folding map; the map lightly spotted, but an excellent copy in the original mauve printed wrappers, sunned. Melbourne, Samuel Mullen, 1870.

First edition: a second edition appeared in the same year. This account of Fiji, with a supplementary gazetteer, is based on thirteen long reports sent back by Britton as Special Correspondent for *The Argus* newspaper.

This copy was once part of the library of the Australian pioneer landowner Alexander Berry, with his signature on the front wrapper. Berry's library was substantial, and its contents reflected the reading and concerns of a nineteenth-century landowner with political interests.

Ferguson, 7436. **$575**

COMMERCIAL ENTERPRISES

395. CLEVELAND, Richard J. A Narrative of Voyages and Commercial Enterprises.

Two volumes, octavo; a few spots but a good copy in the original dark brown cloth, paper labels chipped. Cambridge, John Owen, 1842.

Scarce: the first edition of a work that appeared essentially unchanged under several different titles.

Cleveland, an adventurer and fur-trader, describes his adventures on the northwest coast of America, including a journey to Alaska in 1799, another to California in 1803, and his appearance at the Old Sitka site of the first Russian establishment only the day after Baranov had struck his deal with the Indians allowing the Russians to build an outpost there.

Hill, p. 56; Judd, 36; Lada-Mocarski, 115; Sabin, 13665. **$1100**

SURVEYING

396. COPPINGER, Richard W. Cruise of the "Alert". Four years in Patagonian, Polynesian, and Mascarene Waters (1878-82).

Octavo, with 16 wood-engraved plates; slight spotting to the frontispiece, small stamp removed from title just affecting date, but an excellent copy in the original green pictorial cloth, decorated in gilt. London, W. Swan Sonnenschein, 1883.

First edition of a quite important late voyage, which set out to continue the survey of the Straits of Magellan, to investigate certain doubtful reefs and islands in the South Pacific, and to survey parts of the northern and western coasts of Australia, following on from the work done by the *Adventure* and *Beagle* (1826-36) and later by the *Nassau* (1866-9).

The book describes the expedition's experiences in the Straits of Magellan, Tierra del Fuego, Patagonia, the Tahitian and Fijian islands, Sydney, Torres Strait, the north-east and north-west coasts of Australia, Port Darwin, Singapore, the Seychelles, and South Africa.

The wood-engravings are after drawings by Coppinger as well as photographs taken by Frederick North during the cruise. Coppinger pays particular attention to ornithology in his description of the *Alert*'s voyage, and in the preface he acknowledges his debt to his friend Bowdler Sharpe in persuading him to publish the book.

Ferguson, 8682; not in Hill; O'Reilly-Reitman, 1339. **$625**

GUANO TRADE

397. COWANS, David. Anecdotes of a Life on the Ocean, being a portion of the Experiences of Twenty-seven years' service in Many Parts of the World.

Duodecimo, frontispiece; original red cloth, gilt vignette; spine worn at extremities. Montreal, John Lovell, 1871.

A scarce volume of memoirs by a Scots merchant-mariner who ended up in Canada, including an account of doubling the Horn on a voyage to Peru for guano.

Cowans also gives a description of the coolie trade with a narrative of a voyage from India to the West Indies with four hundred men, women and children: eighty-two died on the voyage. **$325**

TWO OCEANS

398. DELESSERT, Eugène. Voyages dans les Deux Océans Atlantique et Pacifique 1844 à 1847...

Large octavo, with 16 plates, four maps (two folding), and numerous illustrations in the text; a few spots but a very good copy in original quarter morocco. Paris, A. Franck, 1848.

First edition: "an important French illustrated book.

398. Delessert

nids de perroquets, dont ils mangent les œufs. Quant aux petits de ces oiseaux, ils les élèvent pour les vendre dans les villes, et c'est, à ce qu'il paraît, la seule industrie à laquelle ils se livrent. Il arrive souvent qu'un aborigène fera vingt lieues pour aller vendre un perroquet moyennant quelque faible somme d'argent, bientôt dissipée en boissons et en liqueurs fortes.

Ils emploient pour chasser l'émeu ou casoar de la Nouvelle-Hollande, une

Naturels chassant le casoar ou émeu.

TOME I. 18

Most of the main ports of the Orient and Polynesia, as well as Australia, were visited and reported upon by this observant traveler. This appears to be the only edition of this title" (Hill). Delessert has accounts of New Zealand, and Australia as well as the Pacific, and the book is richly illustrated with engravings of native fauna, Aborigines and Sydney views. "During his sojourn in New South Wales the author took great interest in sport. He recounts his experiences and observations" (Ferguson).

Borba de Moraes, 253; Ferguson, 4751; Hill, p. 401; O'Reilly-Reitman, 1013; Sabin, 19429. **$825**

TAHITI

399. ELLIS, William. Polynesian Researches, during a residence of nearly six years in the South Sea Islands; including descriptions of the natural history and scenery of the islands – with remarks on the history, mythology, traditions, government, arts, manners, and customs of the inhabitants.

Two volumes, octavo, with portrait of Pomare, two folding maps, seven engraved plates and 16 wood engravings; frontispiece and title of the second volume a little stained, otherwise in fine condition internally; uncut in the original blue-grey boards, plain linen spines (numbered by hand) worn, small splits to joints. London, Fisher, Son & Jackson, 1830.

First edition: one of the most important works on the history and ethnology of the Society Islands, by one of the most perceptive of the missionary travellers in the Pacific. This is "une des sources les plus abondantes pour l'histoire de l'ancien Tahiti..." (O'Reilly-Reitman).

The work begins with Ellis's account of his voyage in which he visited Rio de Janeiro, Sydney, and New Zealand. He was in Tahiti from 1816 to 1822, during which time very little seems to have escaped his notice. A very popular and influential work, its publication "went far to redeem the character of missionaries in the eyes of those who had thought of them as ignorant and narrow minded men" (Hill).

After his time in Tahiti Ellis went on to Hawaii; he published his works on both islands once he was back in England.

Not in Ferguson; Hill, p. 96; Judd, 61 (later edition); O'Reilly-Reitman, 7556. **$550**

400. Goodridge's sealskins

DRESSED IN SEALSKINS

400. GOODRIDGE, Charles Medyett. Narrative of a Voyage to the South Seas, and the Shipwreck of the Princess of Wales cutter with an account of two years' residence on an uninhabited island.

Small octavo, with four plates including frontispiece portrait of the author; with errata slip and a final advertising slip asking for work; complete with list of subscribers and "Opinions of the Press"; original brown blind-stamped cloth boards, gilt coat of arms on front cover surrounded by the words "Patronised by Her Majesty". Exeter, W.C. Featherstone... and Sold by the Author, 1852.

"Sixth edition" (though not all the first five may actually exist) of this thoroughly eccentric book, whose author seems to have sold it door-to-door judging by the subscribers' list. The frontispiece portrait of Goodridge in his seal skin costume must be one of the oddest confections ever to have appeared in a book. Goodridge's record of his two-year stay on one of the Crozet Islands after being shipwrecked was written as a new *Robinson Crusoe.* The survivors lived on sea-elephant seals, penguins and other sea birds and were in the process of building a boat to escape when they were rescued by an American ship. They immediately fell out with the captain and were left on St. Paul Island for three months before being picked up and taken to Tasmania. Goodridge's eight years there furnished him with material to write his *Statistical View of Van Diemen's Land* as well as the Addenda to this volume which details climate, soil types, river systems, disease, the Aboriginal people, transportation and the penal system (with which he was familiar having been falsely imprisoned for four days in Hobart).

Ferguson, 9986c. Hill, p.126 (third edition only). **$485**

FRENCH CHILDREN'S BOOK

401. GUERIN, Leon. Le Tour du monde, ou les mille et une merveilles des voyages...Océanie...

Duodecimo, with wood-engraved vignette illustrations; original blind-stamped roan gilt, a fine copy. Paris, Langlois et Leclerq, 1841.

Attractive children's book, describing the experiences of Williams (sic) Jarvis on the ship *Aventure* mainly in New Zealand, Tonga, New Caledonia, the Marquesas and Hawaii. The delightful vignette illustrations are mostly sourced from earlier voyage accounts, so that there are interestingly debased images of Maoris from the Parkinson engraving, Webber's death of Cook and the Choris portrait of Kamehameha, for example.

$325

MAYHEM

402. HICKS, Albert W. The Life, Trial, Confession and Execution of Albert W. Hicks, the pirate and murderer, executed on Bedloe's Island... Containing the history of his life (written by himself)... With a full account of his piracies, murders, mutinies, highway robberies, etc... Also, his phrenological character, as described by I. N. Fowler.

Octavo, 68 pp. with eight woodcut plates; a few tears but a good copy in binder's cloth. New York, Robert M. De Witt, [1860].

First edition, first issue, published before the description of the execution but with a pasted-in slip printed in white on blue announcing the sentence. A later issue extended the text from 68 pp. to 84 pp.

This rare "shocking" publication gives all the gory details of the dreadful career of Hicks, finally executed for the murder of the Captain and crew of a New York oyster sloop. Hicks denied his guilt until pronounced guilty at the trial, and then produced a lengthy and detailed confession of his life and substantial misdeeds which makes remarkable reading. He was involved in mutinies and mayhem throughout the high seas, beginning with mutinies aboard the whaleship *Warren* on a voyage to the Northwest Coast, with terrible episodes in Hawaii (he "engaged in every kind of wickedness; I robbed and beat the natives, and was finally taken by the authorities and locked up until the vessel was ready for sea..."). In the Marquesas he got into a fight with the natives ashore, and "thus having a taste for blood, and the demon in me fully aroused", set out to kill the officers and set sail for the Northwest Coast. He and his accomplices were found out and put in irons, but escaped when the ship returned to Honolulu in 1847.

Not recorded by Judd (neither the ship's visits nor the book).

$740

CONFUSION, HORROR & DISMAY

403. HOLDEN, Horace. A Narrative of the Shipwreck, Captivity and Sufferings of Horace Holden & Benj. H. Nute; who were cast away in the American Ship Mentor, on the Pelew Islands, in the year 1832; and for two years afterwards were subjected to unheard of sufferings among the barbarous inhabitants of Lord North's Island.

Duodecimo, with two wood-engraved plates; original cloth. Boston, Russell, Shattuck and Co., 1836.

First Edition: a sad tale of "confusion, horror and dismay". The *Mentor* sailed from New Bedford under Captain Edward Barnard in 1831, and wrecked the following year in the Pelews. About half the crew were killed perished, the remainder being mistreated by the inhabitants of Lord North's Island, who also tattooed them as was their custom. Melville's discussion of tattooing in *Typee* is probably derived from this account. The book also includes a vocabulary of the language of Lord North's Island.

Forster, 56; Hill, p. 147; Huntress, 277C. **$825**

PATAGONIANS

404. HOXSE, John. The Yankee Tar. An authentic narrative of the Voyages and Hardships of John Hoxse, and the cruise of the U.S. frigate Constellation...

Small octavo, a bit spotted and stained; original cloth with paper label, joints repaired. Northampton, John Metcalf, 1840.

Descriptions of a life at sea, the early parts actually eighteenth century: Hoxse first put to sea in 1795. He lost an arm in 1800 when the *Constellation* took on some French shipping, but stayed in the service, later transferring out to run his own ship. Hoxse describes sealing for several seasons in Patagonia: it is interesting that even at this late date a supposed eye-witness description still has the Patagonians at between seven and nine feet tall! Such is the power of myth. **$240**

NOOTKA SOUND

405. JEWITT, John Rodgers. Narrative of the Adventures and Sufferings of John R. Jewitt; only survivor of the crew of the Ship Boston, during a captivity of nearly three years among the Savages of Nootka Sound...

Octavo, with a woodcut frontispiece and several vignettes; printed on poor-quality paper and spotted as usual and lightly stained throughout; original printed boards, cloth spine, rubbed, edges worn. New York, Printed for the Publisher, [1815].

"With the exception of Dana's book, the best known and most popular of the early narratives of adventure and experience on the western coast" (Hill). This is a scarce account of a famous shipwreck. Jewitt was armourer on the *Boston*, which had two officers and a crew of twenty-four; they were seized by Indians at Nootka Sound as a result of an insult to their chief. Jewitt survived, along with apparently another crew member (despite his description in the title as the lone survivor), and was rescued in 1805.

There was also a New York edition in 1815.

Hill, p. 156; Howes, A189. **$745**

SOUTH SEA BUBBLES

406. KINGSLEY, George Henry and George Robert Charles, Earl of PEMBROKE. South Sea Bubbles by the Earl and the Doctor.

Octavo; contemporary blue half calf. London, Richard Bentley, 1872.

First of many editions: a very attractive copy of this South Seas classic, agreeable memoirs mainly set in Tahiti with chapters on Tonga and Samoa. Pembroke, a young nobleman, travelled to the Pacific for his health with his friend Kingsley. They made two cruises in the South Pacific, ending with a gentle shipwreck on the coral. The book is noteworthy for its attack on the missionaries in the South Seas, which had to be toned down for later editions.

O'Reilly-Reitman, 1277. **$425**

UNDER THE HORN

407. LAMSON, Joseph. Round Cape Horn. Voyage of the Passenger-Ship James W. Paige, from Maine to California in the year 1852.

Small octavo; a nice copy in original quarter cloth and boards, printed paper label. Bangor, Maine, O.F. & W.H. Knowles, 1878.

A scarce book, describing a voyage from Maine to California passing under the Horn, with short descriptions of life in California in the 1850s. "This is another narrative of a voyage to San Francisco via Cape Horn; it deals a bit more deeply with the personalities aboard ship and has a "California Scenes" section, which is a series of vignettes on California – Sacramento, the Redwoods, Indians, Yosemite and native animals are among the subjects dealt with" (Hill).

Cowan, p. 341; Hill, p. 170. **$650**

SEAFARING IN THE 1820S

408. MANT, Frederick Woods. The Midshipman: or Twelve Years at Sea.

Duodecimo, inscription on title; contemporary half calf, a fine copy. London, G. Routledge & Co., 1854.

Reminiscences of seafaring in the 1820s and 1830s, including a Pacific crossing and a passage under the Horn. This is a scarce book, unknown to Ferguson despite the chapter devoted to New South Wales in 1826, and the Sydney Aborigines: "with the peculiarity of the animal race may be added the peculiarities of the human, which marked the country from King Bungaree and his tribe of savages, to the bright eyed but ignorant Currency Lass".

Not in Ferguson; not in Hill. **$685**

SHIPWRECKS

409. MARINER'S CHRONICLE. The Mariner's Chronicle: containing Narratives of the Most Remarkable Disasters at Sea.

Duodecimo, 504 pp., plates; publisher's sheep, worn. New Haven, 1834.

A famous collection of shipwreck accounts which includes thirty-five narratives, among them a description of the sinking of the *Essex* after being attacked by a whale (one of the main sources for *Moby Dick*). There is also a chapter on the loss of the *Hibernia* bound for Van Diemen's Land, with an engraving – but it is not recorded by Ferguson.

Hill, p. 192; Sabin, 44588. **$850**

THE COMMUNE & THE CANAQUES

410. MICHEL, Louise. Legendes et Chants de Gestes Canaques.

Duodecimo, with four double-page wood-engravings by Louise Michel; slight spotting, but an excellent copy in the original printed blue wrappers, spine restored; library stamps. Paris, Keva et Cie., 1885.

First edition: with striking wood-engravings by Michel herself. Louise Michel, or "La Bonne Louise", had arrived in New Caledonia in 1873, one of a number of exiles deported from France for their involvement in the Commune. When the Canaques of New Caledonia rebelled against the French occupation in 1878, the majority of the communard exiles sided with the French army. Michel, however, sympathised with the Canaque cause, recognising that in fighting injustice and oppression they were attacking the same enemy as were the communards in France. She gave practical help, showing the rebels how to cut telegraph wires and intercept communications. Her friendship with the Canaque Daoumi taught her the songs and legends which she publishes here. When she eventually returned to France, she remained an important figure in the anti-colonialist movements.

O'Reilly-Reitman, 1805. **$585**

CRUISE OF THE COLUMBIA

411. MURRELL, William Meacham. Cruise of the Frigate Columbia around the world...

Duodecimo, with some spotting; old signature on title-page, another cut from first leaf of text; original cloth with gilt vignette, respined in buckram, endpapers renewed. Boston, Benjamin B. Mussey, 1840.

A scarce book: "strangely enough, this work is a largely unknown source for this important voyage" (Hill). Murrell was a crew member of the *Columbia*, part of the United States East India Squadron, which undertook a long voyage at the end of the 1830s. The only other printed description of the voyage was F.W. Taylor's *The Flag Ship* (1840).

The *Columbia* arrived at Hawaii at an important time in the history of the islands, when the French expedition under Laplace was insisting upon French rights. Murrell gives a good description of Hawaii; he calls his work "a plain and succinct account of our cruise, with short descriptions of the places we have visited. I am told there may be some who will carp and cavil at my little work. Well, let them...". O'Reilly and Reitman were impressed by the book: "Souvenirs écrits d'une plume alerte et délurée".

Hill, p. 507; Judd; 133; O'Reilly-Reitman, 990. **$275**

WRECK OF THE GLIDE

412. OLIVER, James. Wreck of the Glide; with an account of life and manners at the Fijii Islands.

Duodecimo; original grey boards, spine repaired but defective at top; otherwise a good copy. Boston, Wm. D. Ticknor, 1846.

First edition of this rare Pacific narrative of a shipwreck in the Fiji islands, with a good account of Tahiti and a longer description of a visit to Hawaii in the 1830s. This is a book that because of its paper stock is more usually seen browned and stained, like many American voyage books of the period.

Hill, p. 218; Judd, 136; O'Reilly-Reitman, 901. **$700**

TRADING ON THE NORTHWEST COAST

413. PATTERSON, Samuel. Narrative of the Adventures and Sufferings... experienced in the Pacific Ocean, and many other parts of the world, with an account of Feegee, and Sandwich Islands.

Duodecimo, browned and stained as always; final two leaves defective and restored with the loss of about fifteen words; modern plain quarter calf. Palmer, Mass., 1817.

"Very important" (Hill). This is the first edition (though Howes refers to this as second issue, citing a Rhode Island version of the same year as the first; but no Rhode Island issue is recorded by any other authority consulted and it may well be a ghost).

This is a rare Pacific narrative, of importance for its material on Hawaii (1806) and Fiji (1808), at early dates for both groups of islands, and for the short description of an 1808 visit to Port Jackson. Patterson also made three trips to the Northwest coast – "interesting details of the early experiences of New England traders on the Northwest coast, including Alaska" (Lada- Mocarski).

Ferguson, 692 ("very rare"); Hill, p. 227; Judd, 141; Lada-Mocarski, 74; Sabin, 59145. **$465**

ROUND THE WORLD

414. SIMPSON, Sir George. Narrative of a Journey round the World, during the Years 1841 and 1842…

Two volumes, octavo, frontispiece portrait and folding map (slight wear to one fold); an excellent copy, bound without the advertisement leaves in contemporary half calf, slightly rubbed. London, Henry Colburn, 1847.

"A model record of travels by an exceedingly able man and keen observer…" (Cowan). Simpson made two visits to Sitka, and his material on Alaska is significant. He made stops at San Francisco, Monterey, and Santa Barbara, and gives a detailed account of his time in California, as well as a good description of Hawaii.

Cowan, p. 589; Hill, p. 274; Judd, 162; Lada-Mocarski, 129; Sabin, 81343. **$750**

JEMMY BUTTON

415. SNOW, William Parker. A Two Years' cruise off Tierra del Fuego, the Falkland Islands, Patagonia, and in the River Plate: a Narrative of Life in the Southern Seas…

Two volumes, small octavo, with six fine coloured lithograph plates and three folding maps; original blue cloth, slightly rubbed. London, Longman, 1857.

A scarce and attractive two-volume narrative in the original cloth. Snow gives a long account of travels in South America and under the Horn, mainly describing experiences with the Patagonian Missionary Society, ministering to far-flung populations. The coloured plates are fine and include a wonderfully atmospheric lithograph of the small vessel beating to windward under the Horn.

Snow's descriptions of Tierra del Fuego are good, constantly referring to Darwin and Fitzroy, and including descriptions of meetings with Jemmy Button – still speaking English – and his people.

Snow's other book – the *Voyage of the Prince Albert* (1851), describing one of the Franklin search expeditions in the Arctic – is much better known.

Not in Abbey; not in Hill; Spence, 1141. **$3850**

415. Snow

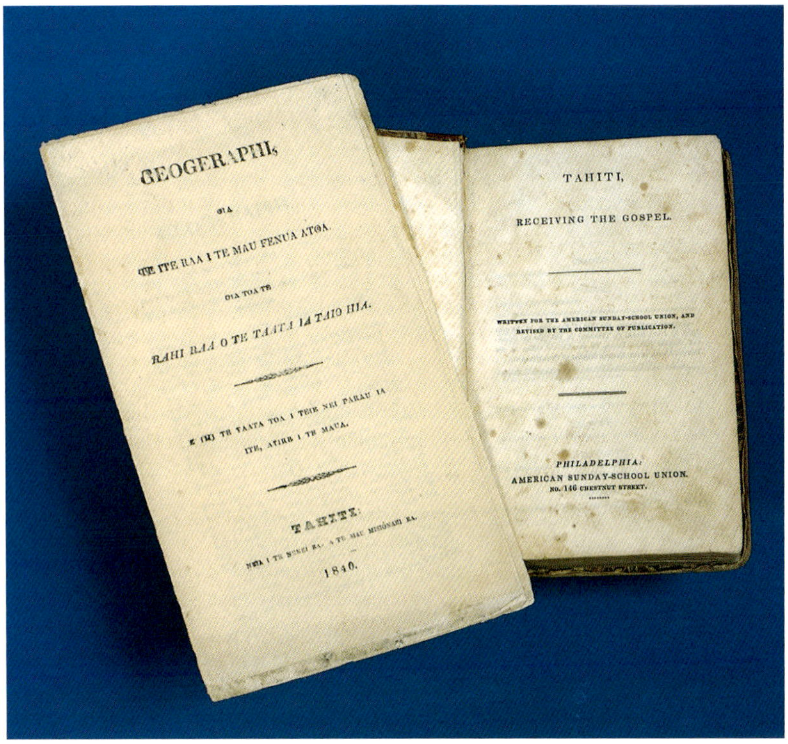

416 & 417. Tahiti

EARLY TAHITI

416. [TAHITI] AMERICAN SUNDAY-SCHOOL UNION. Tahiti, receiving the Gospel. Written for the American Sunday-School Union, and revised by the Committee of Publication.

Duodecimo; a few spots and tears but good in the original quarter roan. Philadelphia, 1832.

A very scarce early account of missions in the South Pacific, with a brief history of Tahiti from the *Duff* voyage to about 1820. The book is not, however, recorded by the encyclopaedic O'Reilly-Reitman. **$785**

TAHITIAN PRINTING

417. [TAHITI: MISSIONARY PRESS] Geogeraphi, oia te ite raa i te mau fenua atoa… [Geography, or the knowledge of all nations as well as the importance of their people…]

Duodecimo, 12 pp.; folded as issued, uncut and unopened. Tahiti, Missionary Press, 1840.

A fine copy of the first and only edition of this geography primer; crudely printed, it is a rare ephemeral survival from the early period of missionary printing in Tahiti. "The early Missionary Presses in Tahiti and the Society Islands never became of importance outside the fields they were meant to cover. The printers were missionaries with only a very scanty knowledge of printing. The types and materials used were deficient, but nevertheless these early presses rank high because of their importance to the native population of the islands…" (Harding and Kroepelien, *Tahitian Imprints of the London Missionary Society*, pp. 37-8).

Kroepelien, 1006; O'Reilly, 5097. **$1500**

THE TEA PARTY

418. [TAHITI] TOPLIFF, R. The King of Otaheite's Tea Party.

11 pp., engraved music; original illustrated wrappers. London, n.d. but c. 1840.

Rare comical song about the cannibalistic King of Tahiti: "each of my neighbours. Tell 'em to come to me, at half past four to tea, see you get everything fit for my table – Human flesh, herbs and rice, everything else that's nice…".

O'Reilly-Reitman, 10250. **$725**

RACING OFF NEW ZEALAND

419. TAYLOR, Joshua N. Sea Yarns: being the reminiscences of Capt. Joshua N. Taylor.

Octavo, 47 pp., with illustrations; original printed wrappers. N.p., n.d., ?Orleans, Mass., c. 1915.

Rare only edition, probably for private distribution. These are the reminiscences of a Cape Cod sea-captain who first went to sea in 1850. Taylor gives an account of an 1860s visit to New Zealand, including a yacht race between his vessel *Charmer*, a trading schooner, and the former mail packet *Sylph*. Later he was master of the *Otago*, which wrecked on the South African coast in 1867, and he sailed in a number of other ships.

Not in Bagnall. **$485**

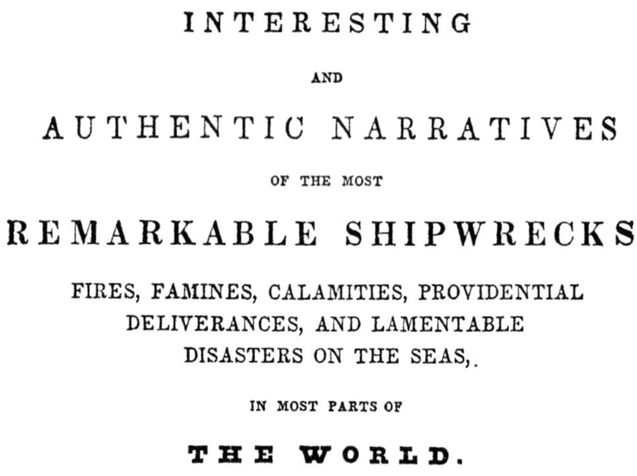
INTERESTING

AND

AUTHENTIC NARRATIVES

OF THE MOST

REMARKABLE SHIPWRECKS,

FIRES, FAMINES, CALAMITIES, PROVIDENTIAL DELIVERANCES, AND LAMENTABLE DISASTERS ON THE SEAS,.

IN MOST PARTS OF

THE WORLD.

420. Thomas

SHIPWRECKS

420. THOMAS, R. Interesting and Authentic Narratives of the Most Remarkable Shipwrecks, Fires, Famines, Calamities, Providential Deliverances, and lamentable Disasters on the Seas, in Most Parts of the World.

Octavo, with wood-engraved plates and text illustrations; some foxing, title-page a little frayed; otherwise fine in the original blindstamped cloth, pictorial gilt spine. Hartford, Silas Andrus & Son, c. 1850.

A collection of maritime disaster stories, many of them Pacific voyages. As Huntress points out, much of the material is filched from the *Mariner's Chronicle* and Dalyell's collection published in Edinburgh in 1812. Some 43 marine disasters are described, the latest dating from 1835. There are accounts of the convict ship the *Amphitrite*, wrecked whilst bound for New South Wales in 1833, the *Sidney*, wrecked when sailing from Port Jackson to Bengal in 1806, and the loss of the whaleship *Essex* in 1820.

Not in Ferguson; Forster, 742; Huntress, 280C; Sabin, 95444. **$685**

FROM THE ROCKIES TO HAWAII

421. TOWNSEND, John K. Narrative of a Journey Across the Rocky Mountains, to the Columbia River, and a Visit to the Sandwich Islands, Chile, &c. With a Scientific Appendix…

Octavo, original brown grained cloth, joints neatly repaired; a good copy. Philadelphia, Henry Perkins, 1839.

First edition: Townsend, an ornithologist, travelled on Wyeth's second Oregon expedition in 1834, which set out from Missouri to establish trading posts beyond the Rocky Mountains. His narrative of the journey includes much on the life and customs of both Indians and fur-traders. It is followed by a long description of their two visits to Hawaii, and their return by way of Tahiti and Juan Fernandez Island. Streeter calls his narrative "one of the best early ones."

Hill, p. 293; Judd, 175; not in O'Reilly-Reitman; Sabin 96381. **$550**

MISSIONARIES

422. TYERMAN and BENNET. MONTGOMERY, James, Editor. Journal of Voyages and Travels, by Daniel Tyerman and George Bennet, deputed from the London Missionary Society, to visit their various stations in the South Seas, China, India, & c., between the years 1821 and 1829. Compiled from original documents…

Two volumes, octavo, with an engraved frontispiece and plates; some spotting; modern quarter calf. London, Frederick Westley and A.H. Davis, 1831.

A famous account of missionary activities, mainly set in the Pacific. "The first part was written in conjunction with George Bennet, but the latter part was entirely [Tyerman's] own. It affords a graphic picture of the state of the London Society's missions at the period" (DNB). "In 1821, Tyerman and George Bennet of Sheffield were appointed by the London Missionary Society to visit their southern stations. They sailed from London in the whaler *Tuscan*, and proceeding round Cape Horn, visited Tahiti, the Leeward and Hawaiian Islands, and other mission stations in the South Seas. In 1824 they visited New South Wales, and on the way, narrowly escaped from the Maoris of New Zealand. From Sydney, in September 1824, they sailed through the Torres Strait to Java, and from there to Singapore, Canton and Calcutta…" (Hill).

Ferguson, 1487; Hill, p. 295; O'Reilly-Reitman, 802. **$420**

423. Waterhouse

BAXTER PRINT

423. [WATERHOUSE] BAXTER, George. The Rev. J. Waterhouse Superintending the Landing of the Missionaries at Taranaki, New Zealand.

Original Baxter print, 320 x 430 mm.; one spot and lacks margins but otherwise in good condition, mounted. London, G. Baxter, 1844.

The Reverend Waterhouse was general superintendent of the Wesleyan Missionary Society's missions in Australia and Polynesia. This print shows him conversing with a Maori chief, whilst the wife of the new missionary Rev. Charles Creed, the first European woman to arrive at Taranaki, is carried ashore. In the background is the missionary ship *Triton*, and in the distance Mount Edgcombe.

The print is significant for its version of events which was intended to communicate to English audiences the success of the missionary enterprises in the Pacific: the Maori chief has thrown down his weapons and appears to be humbly welcoming the reverend to their shores. The native women – dressed as vestal virgins – are carrying Mrs Creed with awe and reverence. The Maoris on the shore appear to be overjoyed at the prospect of embracing the Europeans and their civilising ways.

Courtney Lewis, 'George Baxter – The Picture Printer', London, 1924. **$1500**

WHALES, SEALIONS, CAMELS

424. [WHALES] History of Wonderful Animals; Containing an Account of the Most Remarkable Fishes, Beasts, and Birds...

24mo., with a wood-engraved frontispiece picturing a whale, and numerous full-page illustrations; early owner's inscription; extremities rubbed and joints very worn, but an attractive, unsophisticated copy in publisher's quarter roan. London, J. Arliss, 1818.

An early and charming natural history text for children. With descriptions and engravings of a good number of animals – from sea lions to camels – it contains a 19-page chapter dealing with whales in surprising detail. This is the eighth edition of this popular title. **$240**

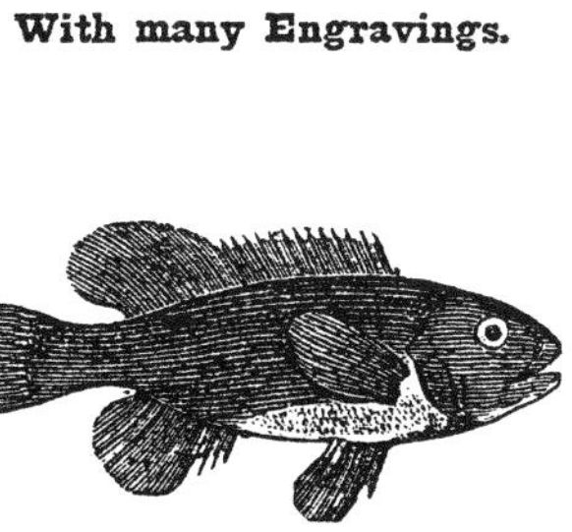

425. Whaling

SPERMACETI & AMBERGRIS

425. [WHALING] BOYS AND GIRLS LIBRARY. Natural History of Fishes [and] The Peace Maker.

16mo., with wood-engravings; slight damage to spine but a very good copy in the original stiff pictorial wrappers. Worcester, Lazell, c. 1830.

Two works in one. The first, *The Natural History of Fishes*, is a primer for children describing twelve species of fish, each of which is illustrated by a wood engraving. The "Cachelot" or "Spermaceti Whale" – "by far the most valuable, and the most sought after by man, as it contains two very valuable drugs, spermaceti and ambergris" – is described as well the various methods of hunting the whale. **$275**

426. *Whaling*

DUTCH JUVENILE

426. [WHALING] In het ruime sop. Reis- en Jachtavonturen ter zee benevens het leven en de Bijzonderheden van eenige Zeedieren...

Small octavo, with nine fine hand-coloured engravings; original black cloth, tiny snags to spine but a fine copy. Leiden, D. Noothoven van Goor, c. 1870.

A rare and most attractive Dutch whaling juvenile, which includes descriptions of voyages in the Pacific and to Batavia and the Dutch East Indies. **$985**

COLOUR PLATES

427. [WHALING] MELVILLE, Herman. Moby Dick or The White Whale.

Octavo, colour plates tipped in, pictorial endpapers; very good in original cloth with gilt vignette. London, Hodder & Stoughton, c. 1940.

A beautifully-illustrated edition of what is arguably *the* American novel. Published in countless editions over the last 150 years, Melville's challenging tale of a profound and symbolic struggle between man and his fate which takes place during a tempestuous whaling voyage has never lost its immense power. This edition is illustrated by twelve colour plates after Mead Schaeffer. **$275**

DANGERS OF WHALING

428. [WHALING] Pictures of Travel & Adventure [cover title].

Octavo; twelve hand-coloured plates, folded accordion-style; original pictorial boards; spine reinforced with paper tape; a few tears, folds reinforced on versos with tape. N.p., c. 1860.

A collection of exotic scenes from around the world, from Lapland to Egypt to China. One of the scenes, entitled "Dangers of the Whale Fishery", shows a whale-hunt gone wrong: a boat has been thrown in the air by a whale and the sailors are falling into the sea. There is also a dramatic "Escape from a Shark". **$225**

CULTIVATION OF THE TASTE

429. [WHALING] The Ocean and its Inhabitants, with their uses to man.

Small quarto, colour lithograph frontispiece and half-title and eight handcoloured lithograph plates; original gilt cloth, skilfully rebacked. London, Darton and Clark, c. 1845.

Excellent copy of a scarce and particularly beautiful children's book on whales and the sea. The sperm and common whales are described and illustrated along with the products which were made from them: corsetry, candles, perfumes, umbrellas, and oil for lamps and cooking. Of particular interest is the manner in which this book is designed for a young readership, by a publisher concerned with aesthetics: "the cultivation of the taste has also been kept in view, by regard to correct drawing and careful colouring" (Preface).

Ingalls, 127-8, 427. **$1250**

429. *The Ocean*

MISSIONARY ENTERPRISES

430. WILLIAMS, John. A Narrative of Missionary Enterprises in the South Seas.

Octavo, with a colour Baxter print ("To Po a Chief of Rarotonga") as frontispiece, a folding map and several wood-engraved plates and vignettes; title-page a little spotted; a fine copy in contemporary maroon morocco, ornately gilt. London, John Snow, 1841.

A particularly good copy, in a fine contemporary binding, of this enormously popular book, whose success was probably due to its mix of piety with the thrill of the exotic. Now seen as the classic account of missionary activities in the South Seas, it went through numerous editions; this is an example of the "fourteenth thousand", in fact a relatively early issue. It is the main source for the life and works of Williams, and the insights it gives into missionary and native life are important. **$485**

DEATH AT ERROMANGA

431. [WILLIAMS] BAXTER, George. The Reception of the Revd. J. Williams at Tanna in the South Seas the Day before he was Massacred [and] The Massacre of the Lamented Missionary the Rev. J. Williams and Mr. Harris at Erromanga South Seas.

Pair of sepia Baxter Prints, each 240 x 340 mm.; in fine condition, framed. London, G. Baxter, 1841.

News of Williams's violent death at Erromanga in the New Hebrides in November 1839 was received with much public sorrow. George Baxter produced this pair of prints to commemorate the event. These prints are "extremely rare" according to Mitzman and particularly so as a pair, and in such fine condition.

C.T. Courtney Lewis, 'George Baxter – The Picture Printer', London, 1924; Max E. Mitzman, 'George Baxter and the Baxter Prints', London, 1978. **$2850**

THRILLING INCIDENTS

432. ZOLLERS, George D. Thrilling Incidents on Sea and Land: the prodigal's return.

Octavo, with a frontispiece and vignette illustrations; original purple cloth, small stain on upper cover but an excellent copy. Mount Morris, Illinois, 1892.

Scarce: the first of many editions of a quite surprisingly popular and enduring book (at least seven editions before 1910), which "gives an interesting account of early day whaling adventures in the Pacific. Included are sections copied from a diary of a whaling voyage to Margarita Bay, Baja California, in January of 1865" (Hill).

Hill, p. 626. **$300**

431. The Reception

431. The Massacre

MODE OF BUILDING HOUSES AMONG THE INHABITANTS OF NORTHERN CLIMES.

432. Zollers

16: HAWAII

TWICE TO HAWAII

433. **ALLEN, Miss A.J. Ten Years in Oregon. Travels and adventures of Doctor E. White and Lady, west of the Rocky Mountains; with incidents of two sea voyages via Sandwich Islands around Cape Horn...**

Octavo, original brown cloth, spine decorated in gilt; spine chafed at top but a good copy. Ithaca, N.Y., Andrus, Gauntlett, 1850.

Second and best edition. Miss Allen's account includes two interesting voyages to Hawaii, arriving in the *Hamilton* in 1836 and leaving on the *Diana* in 1837, returning on the *Maryland* in 1840. "Dr. Elijah White, a Presbyterian missionary, went overland to Oregon in 1842 and returned in 1845 by sea. This is the best edition because of the addition of passages from Frémont's *Narrative*, which extended the text from 399 pages to 430..." (Hill).

Hill, p. 339; Judd, 2. **$650**

KAMEHAMEHA'S FUNERAL

434. **AMERICAN SUNDAY SCHOOL UNION. History of the Sandwich Islands; with an account of the American mission established there in 1820...**

Duodecimo, with a frontispiece map and three wood-engraved plates; general ageing and staining, owner's scribbles on front endpaper; original quarter roan, boards worn. Philadelphia, Sunday School Union, 1831.

A scarce early history of the islands and the missions, with a plate surprising in this context of Kamehameha's funeral procession. **$440**

PUAAIKI

435. **[BARTIMEUS] BINGHAM, Hiram. Bartimeus, of the Sandwich Islands.**

16mo, with a frontispiece; a few stains and scribbles but a good copy in the original brown blind-stamped cloth. New York, American Tract Society, 1844.

Very scarce uplifting biography of one of the most ideal early converts to Christianity in the islands. Blind, lame, nicknamed Puaaiki (the little hog), poor and a hopeless drunk, he was redeemed by conversion. In passing Bingham gives some local history of the 1820s and 1830s. **$275**

MISSIONARIES IN HAWAII

436. **BARTLETT, Prof. S.C. Historical Sketch of the Hawaiian Mission, and the Missions to Micronesia and the Marquesas Islands.**

Octavo, with a map printed on the inside of each wrapper; original printed wrappers, spine splitting and a few spots on front wrapper, otherwise fine. Boston, American Board of Commissioners for Foreign Missions, 1869.

Short but detailed history of missionary activities in Hawaii, as well as other islands in the Pacific, including some mention of exploring voyages etc.

Not in Hill. **$325**

THE HAOLE

437. **[BATES, George Washington] Sandwich Island Notes. By a Häolé.**

Duodecimo, with twenty-two fine steel-engraved plates and vignettes of Hawaii; original brown cloth, old bookplates; an excellent, bright copy. New York, Harper & Bros., 1854.

An excellent copy of this famous memoir of Hawaii by a "Häolé" or outsider, which gives a thorough and considered account of the islands based on travels there in the early 1850s. Bates urged annexation by the United States.

Hill, p. 18; Judd, 14. **$300**

AN AUSTRIAN VIEW

438. **BECHTINGER, Dr. J. Ein Jahr auf den Sandwich-Inseln...**

Octavo, with eight plates and a folding map; original boards, linen back, lacking the spine label; a little loose and with old library stamps, but a good copy. Wien, 1869.

Bechtinger sailed from San Francisco in 1867, and spent about a year in Hawaii. His book includes detailed observations of people and manners of the islands, with some notice of matters economic and medical. **$750**

TWENTY-ONE YEARS

439. **BINGHAM, Hiram. A Residence of Twenty-One Years in the Sandwich Islands...**

Octavo, with steel-engravings; original dark brown cloth, chafing to joints with a few pieces missing. Hartford, Hezekiah Huntington, 1847.

"Bingham arrived in Honolulu in 1820 and spent the next 21 years in missionary work. He soon learned the Hawaiian language and created a written language as well. Along with Rev. William Ellis, Bingham printed the first Hawaiian book, a hymnal, in 1823. With his associates, Bingham translated the Bible into Hawaiian. The Hawaiian King and Queen supported Bingham's missionary work, and he was able to travel extensively throughout the islands gaining impressions which appeared in this book written after he finally left for the United States in 1840. There were additional Hartford editions in 1848 and 1849 and one published at Canandaigua, New York, in 1855. Sabin said that this was probably the best work on the Sandwich Islands" (Hill).

Hill, p. 354; Judd, 20. **$220**

MRS. BIRD

440. **BIRD, Isabella L. The Hawaiian Archipelago. Six months among the Palm Groves, Coral Reefs, and Volcanoes of the Sandwich Islands...**

Octavo, clean tear to frontispiece; original green cloth, gilt vignette on upper cover; a fine copy. London, John Murray, 1880.

Third edition of this very successful memoir of travels in the Hawaiian islands. **$225**

ORIGINAL PRINTED BOARDS

441. BOELEN, J. Reize naar de Oost-en Westkust van Zuid-Amerika en, van daar, naar de Sandwichs-en Philippijnsche Eilanden, China...

Three volumes, octavo, with five folding maps, six engravings and four lithograph views (two folding); some pencil notes in the second volume; a few spots or stains, but free from the usual foxing; some of the plates expertly cleaned, a few marginal repairs; original printed boards entirely uncut, a bit soiled and spotted, spines wearing and one restored; with the original printed leaves of spine labels and directions to the binder, boxed. Amsterdam, Ten Brink & De Vries, 1835.

A rare Pacific and Hawaiian piece, most unusual in original bindings. The narrative of this little known Dutch commercial circumnavigation is significant as an extensive portion of the text relates to the author's trading in ports on both sides of the South American coast including Rio de Janeiro, Montevideo, Valparaiso and Guyaquil.

The book is particularly important for its narrative of Boelen's 1828 visit to Hawaii in the *Wilhelmina en Maria*; stopping first at Kealakekua Bay, the ship continued on to Honolulu, and the account contains important descriptions of both ports, with perceptive comments on the King, the various chiefs and foreigners with whom he came into contact. About half of the third volume relates to Hawaii, and one of the folding lithographs shows Kealakekua Bay, while one chart is of the Hawaiian Islands, and a Hawaiian vocabulary is included.

At Canton, Boelen reported on Dutch activities in the area, and the voyage continued to the Philippines and Batavia, returning to Europe via the Cape of Good Hope.

The plates include a view of Rio, five costume plates of South American interest, an engraved map of the Hawaiian Islands, a view of the *Wilhelmina en Maria* at Kealakekua Bay, two maps of the China coast, an engraving of Canton merchants and a view of the Hongs at Whampoa.

Hill, p. 357; Judd, 23; Sabin, 6116; not in Hunnewell. **$5500**

FIFTY FAMILIES

442. CHEEVER, Henry T. Life in the Sandwich Islands: or, The Heart of the Pacific as it was and is.

Duodecimo, frontispiece and plates; original blue cloth, slight wear to head of spine, joints repaired. New York, A.S. Barnes & Co., 1851.

"This interesting work relates mainly to Cheever's wide travels among the Hawaiian Islands and to his observations of them. Much interesting data is included on the history and native culture of the islands. Missionary endeavour in Hawaii is another favourite subject. Chapters are given on the first Hawaiian college and on Hawaiian literature. Cheever predicted eventual American rule of the islands because fifty missionary families had settled there and owned extensive lands" (Hill).

Hill, p. 51. **$300**

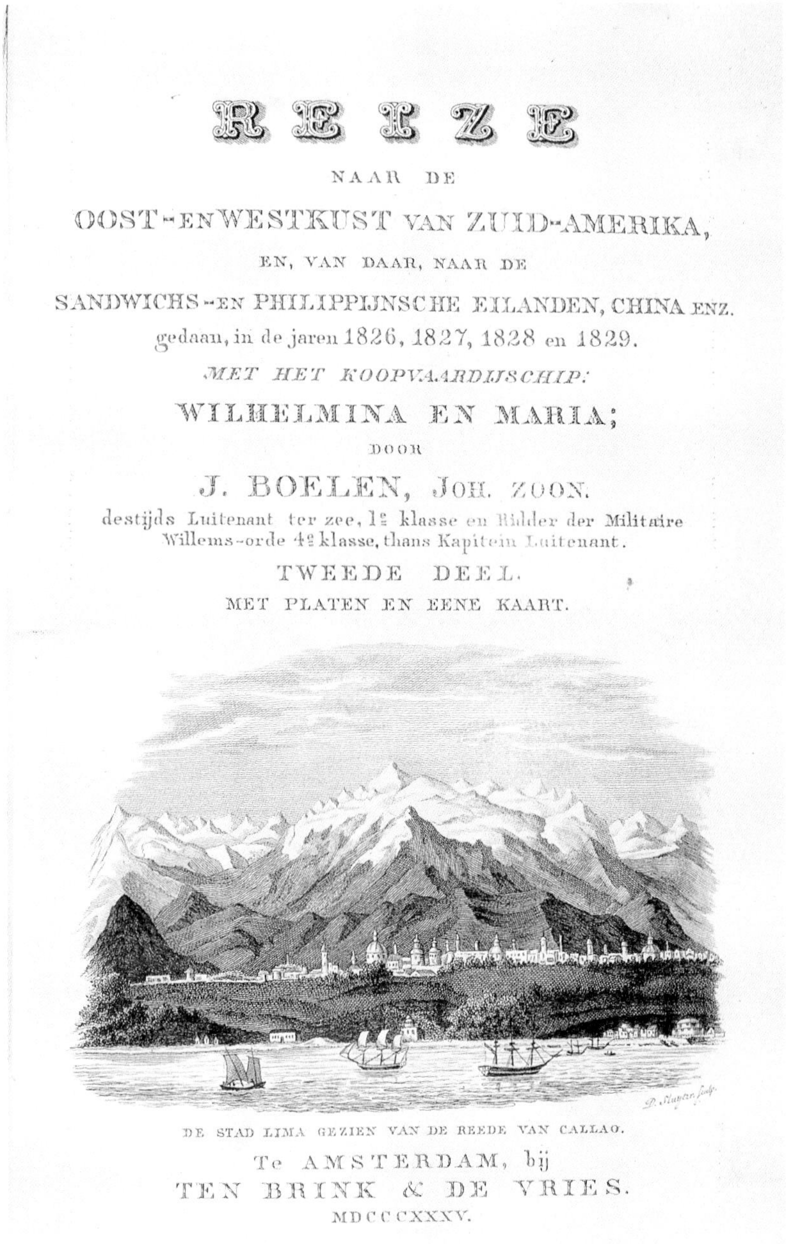

REIZE

NAAR DE

OOST-EN WESTKUST VAN ZUID-AMERIKA,

EN, VAN DAAR, NAAR DE

SANDWICHS-EN PHILIPPIJNSCHE EILANDEN, CHINA ENZ.

gedaan, in de jaren 1826, 1827, 1828 en 1829.

MET HET KOOPVAARDIJSCHIP:

WILHELMINA EN MARIA;

DOOR

J. BOELEN, JOH. ZOON.

destijds Luitenant ter zee, 1e klasse en Ridder der Militaire Willems-orde 4e klasse, thans Kapitein Luitenant.

TWEEDE DEEL.

MET PLATEN EN EENE KAART.

DE STAD LIMA GEZIEN VAN DE REEDE VAN CALLAO.

Te AMSTERDAM, bij

TEN BRINK & DE VRIES.

MDCCCXXXV.

441. Boelen

SPANISH VISITS BEFORE COOK?

443. DAHLGREN, Erik. Were the Hawaiian Islands visited by the Spaniards before their discovery by Captain Cook in 1778? A contribution to the geographical history of the North Pacific Ocean especially of the relations between America and Asia in the Spanish period.

Quarto, with six folding maps, illustrations; binder's cloth, preserving the original green printed wrappers; a fine copy. Stockholm, Almquist & Wiksells, 1916.

One of only 650 copies printed: an analysis which concludes that "no historical fact proves, nor is there any sort of probability, that the Hawaiian Islands were ever visited, or even seen, by the Spaniards before their discovery by Captain Cook in 1778...".

Beddie, 1644; Hill, p. 388; Kroepelien, 244; Palau, 68084.

$450

A MISSIONARY ON COOK

444. DIBBLE, Rev. Sheldon. History and General Views of the Sandwich Islands' Mission.

Octavo; a little spotting but a good copy in the original ribbed black cloth. New York, Taylor & Dodd, 1839.

An important early history of Hawaii by one of the most influential of the missionaries. Although preoccupied with conversion, Dibble gives quite a good account of the early history, although he is unusual for his attitude to the usually sainted Cook. As Cook's bibliographer Holmes writes, it "contains an account of Cook's discovery of the islands and of his subsequent death at the hands of the natives. It is highly critical of his conduct, particularly in respect of the divine honours which were accorded to him and his alleged immorality. Cook's reputation in the United States suffered for some years from the biased attitude of Dibble and other missionaries. Both this account and the same author's 'History of the Sandwich Islands', 1843 were based on replies by the oldest inhabitants to a questionnaire prepared by him…". Cook scholarship, historical method, and the study of the original inhabitants of Hawaii have all advanced since Dibble wrote and Holmes cavilled.

Holmes, 88. $725

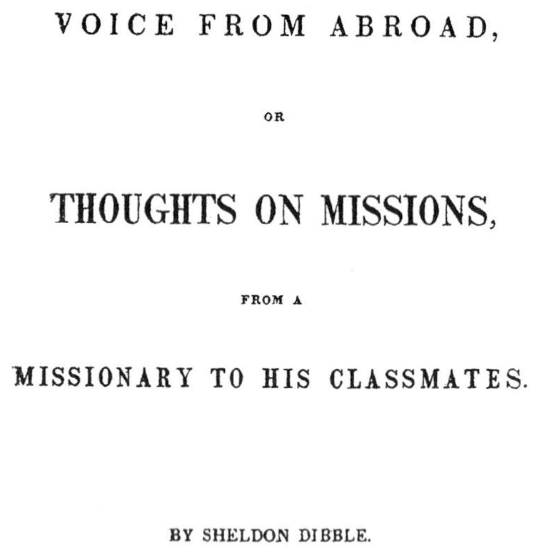
VOICE FROM ABROAD,

OR

THOUGHTS ON MISSIONS,

FROM A

MISSIONARY TO HIS CLASSMATES.

BY SHELDON DIBBLE.

445. Dibble

PRINTED AT LAHAINALUNA

445. DIBBLE, Rev. Sheldon. A Voice from Abroad, or Thoughts on Missions, from a missionary to his classmates.

Octavo; original quarter roan and marbled boards; slight wear to extremities, but a fine copy. Lahainaluna, Press of the Missionary Seminary, 1844.

Rare: a lovely copy of this book by the influential missionary, printed on the missionary press at Lahainaluna. This is a thoughtful book about missionaries and their role, and as such must be among the earliest printings of original material, rather than straight educational or religious matter, in Hawaii. $1100

446. Duhaut-Cilly

CALIFORNIA & HAWAII

446. DUHAUT-CILLY, Auguste. Viaggio intorno al globo principalmente alla California ed alle isole Sandwich negli anni 1826, 1827, 1828 e 1829…

Two volumes bound together, octavo, four wood-engraved plates; inoffensive library number on the verso of the title but a very good copy in contemporary half morocco, gilt press marks on the spine. Turin, Fontana, 1841.

Very scarce: a most important book for California and Hawaii. Duhaut-Cilly was a French trader who reached Yerba Buena in 1827, and in the next two years visited most of the missions, presidios, and pueblos of Alta California. No other traveller had visited so many of the Californian establishments, and his is the best contemporary account of the region. He also describes at some length his 1828 visit to the Hawaiian Islands, and includes a vocabulary of the Hawaiian language.

The plates show San Francisco, San Luis Rey, Fort Ross and Hawaii. This is the second edition of the book, much expanded with additional chapters from its first appearance in French in 1834, which is now extremely rare (John Howell Books conjectured in their catalogue 50 that only about a dozen copies of the French edition survive); there has never been an English edition.

Hill, p. 23; Judd, 57; Sabin, 21164; Zamorano 80, 31. $3300

ORIGINAL SHEEP BINDING

447. ELLIS, Rev. William. A Journal of a Tour Around Hawaii, the largest of the Sandwich Islands. By a deputation from the Mission on those Islands.

Duodecimo, with a frontispiece, four plates and a folding map; some light browning (as usual); an excellent copy in original sheep, joints repaired. Boston, Crocker & Brewster. New York, John P. Haven, 1825.

Very scarce: "the first book written about the Hawaiian Islands and the first serious notice of the Islands since Captain Cook's discovery of them forty-eight years earlier. It may be called the first scientific book on Hawaii because although its main purpose was to report on mission progress, much of the work was devoted to observations on volcanoes, antiquities, and the anthropology of the Islands. Although as a missionary Ellis was mainly interested in the local manners and traditions, he was sufficiently broad-minded to include observations in his book on the natural and cultural history of the Islands. The book became quite popular almost immediately..." (Hill catalogue, of later editions).

The lithographed plates are after Ellis's own drawings. He shows himself to be an acute observer in his text as well: there are very good descriptions of landscape and quite detailed accounts of conversations with Hawaiians. Ellis gives an account of the manufacture of *tapa* cloth, and there are numerous mentions of Cook, his arrival in the islands and his death. Subsequent enlarged editions were entitled *Narrative of a Tour* rather than *Journal*; the Hill Catalogue records (pp. 95, 413) editions of 1826 and 1827, but not this first appearance, which is rare on the market.

Not in Hill or Sabin; Holmes, 83. **$2100**

FIRST DESCRIPTION OF SURFING

448. ELLIS, Rev. William. Narrative of a Tour Through Hawaii, or Owhyhee;. with remarks on the history, traditions, manners, customs and language of the inhabitants of The Sandwich Islands.

Octavo, with a folding map and seven engraved views; old owner's stamp, a large, clean copy, uncut in the original boards, worn, joints repaired. London, for the author, 1826.

First edition in this form, extensively revised and enlarged from the rare Boston publication of the previous year (see previous item). The engraved plates include views of the City of Refuge, a Hula performance and the Volcano.

In this edition the material on Cook has been extended, with pp. 99-106 containing "a valuable account, derived from eye-witnesses, of the circumstances of Cook's death and the honours rendered to his remains" (Holmes). This version also includes one of the earliest detailed descriptions of surfing in chapter 13, where Ellis describes the sport itself, its popularity amongst the islanders, the use of a *papa he naiu* or "wave sliding board" five or six feet long which is greased with coconut oil. "Those who are expert frequently change their position on the board, sometimes sitting and sometimes standing erect in the midst of the foam... we have seen Karaimoku and Kakioeva, some of the highest chiefs in the island, both between 50 and 60 years of age, and large corpulent men, bouncing themselves on their narrow board, or splashing about in the foam, with as much satisfaction as youths of sixteen...".

Hill, p. 413; Holmes, 84. **$950**

THE HULA

449. EMERSON, Nathaniel B. Unwritten Literature of Hawaii. The Sacred Songs of the Hula collected and translated, with notes and an account of the hula.

Octavo, with numerous photographic illustrations; a fine copy in the original dark green cloth. Washington, Government Printing Office, 1909.

The standard early work on the *hula* was the first Hawaiian work funded by the Bureau of American Ethnology, whose ambit had been restricted to American Indians until 1906. This was published as Smithsonian Institution, Bureau of American Ethnology, Bulletin 38. **$215**

BINGHAM'S MARRIAGE

450. GALLAUDET, Thomas H. An Address, delivered at a meeting for prayer, with reference to the Sandwich Mission, in the Brick Church in Hartford, October 11, 1819.

Octavo, 16 pp., stain to title, inner margin taped, disbound. Hartford, Lincoln & Stone, 1819.

A rare pamphlet. This was the address given by Gallaudet after the marriage of Reverend Hiram Bingham and Sybil Moseley "prior to their departure as members of the pioneer mission to the Sandwich Islands. Rev. Gallaudet was also interested in female education and children's books, and was a scholar of the Greek and Hawaiian languages" (Hill).

Hill, p. 427. **$785**

HAWAIIAN CLUB

451. HUNNEWELL, James. [Bibliography of Hawaiiana, contained in:] Hawaiian Club Papers. Edited by a committee of the Club, October 1868.

Octavo; an excellent copy in the original black cloth, title in gilt on upper cover; extremities worn, some chipping to spine. Boston, Abmer A. Kingman, 1868.

Very scarce: an ambitious production by a club of Hawaiian "expatriates" in Boston, noted for its important bibliography of the Hawaiian Islands, the earliest serious attempt at such a project. There are additional articles including printing history, the Hawaiian translation of the Scriptures, Volcanoes, fragments of legends, current events, a comprehensive list of islands in the North Pacific, Hawaiian flora, and obituaries of pioneer missionary Asa Thurston and rancher John P. Parker. Most articles are signed, the majority being by James Hunnewell, Sanford B. Dole, E. W. Clark, and William T. Brigham, all of whom left their mark on Hawaiian history.

The Hawaiian Club remained in existence for many years, but this was its only publication. This is a presentation copy, inscribed by J.F.B. Marshall to Warren L. Brigham. **$435**

HE

HOIKEHONUA,

HE MEA IA E HOAKAKA'II

KE ANO O KA HONUA NEI,

A ME NA MEA

MALUNA IHO.

UA UNUHIIA NOLOKO MAI O KA HOIKEHONUA A W. C. WOODBRIDGE.

Ua haumia ka honua i ka poe e noho ana maluna iho.—*Isaia.*

Oahu:
MEA PAI PALAPALA A NA MISIONARI.
1845.

452. He Hoikehonua

THE JUMPING KANEGARU

452. [HAWAIIAN IMPRINT] WHITNEY, S., translator. He Hoikehonua, he mea ia a hoakaka'ii ke ano o ka honua nei, a me na mea maluna iho... [A Geography. A work describing the form of the earth and things on it...].

Duodecimo; two leaves browned; original mottled boards, leather spine renewed in cloth; an excellent copy. Oahu, Mea Pai Palapala a na Misionari, 1845.

A particularly good copy of this scarce educational text produced by the Oahu missionary press, translated by Whitney from W.C. Woodbridge's original, with charming woodcuts of peoples and places of the world. This is not recorded by Ferguson despite a New Holland section including a delightful *kanegaru.*

Judd, 278. **$1200**

ALMANAC

453. HONOLULU ALMANAC. The Honolulu Almanac and Directory, 1886. Containing complete statistical and general information relating to the Hawaiian Islands. Third year of publication.

Octavo, photographic illustrations; original printed pink wrappers; a few chips, but essentially a fine copy. Honolulu, Pacific Commercial Advertiser Steam Print, 1886.

Very scarce in this condition. This is a detailed source for the minutiae of Honolulu life in the late 1880s, not least for the sake of its series of informative advertisements. **$275**

SECOND & BEST EDITION

454. HOPKINS, Manley. Hawaii. The Past, Present, and Future of its Island Kingdom. A historical account of the Sandwich Islands (Polynesia).

Octavo, with plates; original green cloth gilt, slight damage to edges. London, Longmans, 1866.

Second and best edition, with considerable additions in the form of five new chapters reflecting recent events in Hawaii's history especially the death of Kamehameha IV and the constitutional changes brought about under Kamehameha V. "Hopkins' book covers the natural and cultural history of the Hawaiian Islands, their discovery by Captain Cook and his death there, the lives of many of the Hawaiian rulers, and the relations of the Islands with foreign countries. This augmented version is the best edition; it was quoted by most later writers and is considered to be prime source material" (Hill).

This is a presentation copy, inscribed by the author to the Bishop of Lichfield.

Hill, p. 147. **$360**

PRESENTATION COPY

455. HUNT, T. Dwight. The Past and Present of the Sandwich Islands, being a series of lectures to the First Congregational Church, San Francisco.

Duodecimo, a few pages browned, small defect to lower margin of first few leaves; a good copy in the original brown cloth. San Francisco, Whitton, 1853.

Very scarce: this is a presentation copy, inscribed by the author to his daughter in 1880, with a couple of manuscript corrections by him in the text. The text is an examination of Christian good done in the islands which gets at times into the nitty-gritty of politics. Hunt is a partisan writer, sometimes straying far from his (missionary) brief. **$400**

FINE COPY

456. JARVES, James Jackson. History of the Hawaiian or Sandwich Islands...

Octavo, with a folding map and six plates; original black blind-stamped cloth; a fine, bright copy. Boston, Tappan & Dennet, 1843.

This was the first really substantial history of Hawaii, and is still one of the most important and influential books on the history of the islands. It is also an important Cook item: Jarves describes at length Cook's visits to and death in Hawaii, while most of the illustrations are derived from images by the third voyage artist Webber.

Hill, p. 154 (the later London edition only); see Holmes, 119; Hunnewell, p. 45; Sabin, 35796. **$600**

FIRST HAWAIIAN EDITION

457. JARVES, James Jackson. History of the Hawaiian Islands...

Octavo, with a full-page wood-engraved plate and a number of vignettes; very slight worming to first and last four leaves, a few stains; contemporary half calf, a bit worn. Honolulu, C.E. Hitchcock, 1847.

The third Edition overall, and the first Hawaiian Edition of the standard early history. **$450**

458. Jarves

PHOTOGRAPH OF KAMEHAMEHA V

458. [HAWAII] JARVES, James Jackson. History of the Hawaiian Islands...

Octavo, with a portrait photograph as frontispiece, wood-engravings in the text; a fine copy in modern quarter calf. Honolulu, Henry M. Whitney, 1872.

Fourth edition, with the rare albumen photograph of Kamehameha V (added to this edition only and missing in many copies) in fine, unfaded condition.

See Holmes, 119; Hunnewell, p. 45; Sabin, 35796. **$850**

LARGER THAN LIFE

459. LITTLE, George. Life on the Ocean, or Twenty Years at Sea; being the personal adventures of the author.

Large duodecimo, wood-engraved frontispiece; front flyleaf cut away but an excellent copy in the original dark brown cloth. Baltimore, Armstrong & Berry, 1843.

First edition, and scarce. Little is described on the title-page as "For many years Captain in the Merchant Service out of the Port of Baltimore, but now entirely blind". His memoirs may not always have the ring of absolute truth (did he really almost tunnel his way to freedom from Dartmoor Prison?) but the voyaging material is clearly authentic, and based on his experiences at sea between 1807 and 1826.

Of particular interest is his account of a visit to Hawaii, including a description of Cook's burial place and an interview with natives who told him that "every year all the natives assembled here to perform a religious rite in memory of his lamentable death...". Little also met John Young, by now a man of power in the islands, who would not discuss his connection with the mutiny on the *Bounty*, claiming only that he had arrived in Hawaii on a whaling ship.

Not in Hill; not in Ferguson; Borba de Moraes, p. 496; Judd, 111; Sabin, 41513. **$750**

SCARCE SECOND EDITION

460. LITTLE, George. Life on the Ocean, or Twenty Years at Sea...

Large duodecimo, woodcut frontispiece; slight browning but a good copy in the original faded purple cloth, slightly rubbed. Aberdeen, George Clark and Son, Ipswich, J.M. Burton, 1847.

Second, first English, edition: scarce (Ferguson records only his own and the Mitchell Library copies; there is no copy in the Hill catalogue).

Borba de Moraes, p. 496; Ferguson, 4575; Judd, 111; Sabin, 41513. **$475**

NOT IN BEDDIE, HILL OR SABIN

461. LUCAS, Samuel. The Sandwich Islands: a Prize Poem, recited in the Theatre, Oxford; June 15, 1841...

Small octavo, 16 pp.; a disbound copy. Oxford, J. Vincent, 1841.

First edition, surprisingly rare: this was the prize-winning poem in the 1841 Oxford University competition for a long poem. The poem discusses the early history of the Hawaiian islands and their inhabitants, superstition and sacrifice and the role of the missionaries, and alludes to Cook. It is not recorded by Beddie, Hill or Sabin. **$325**

HAWAIIANS IN NEW HAVEN

462. NARRATIVE OF FIVE YOUTH. A Narrative of Five Youth from the Sandwich Islands, now receiving an Education in this Country. Published by order of the Agents appointed to

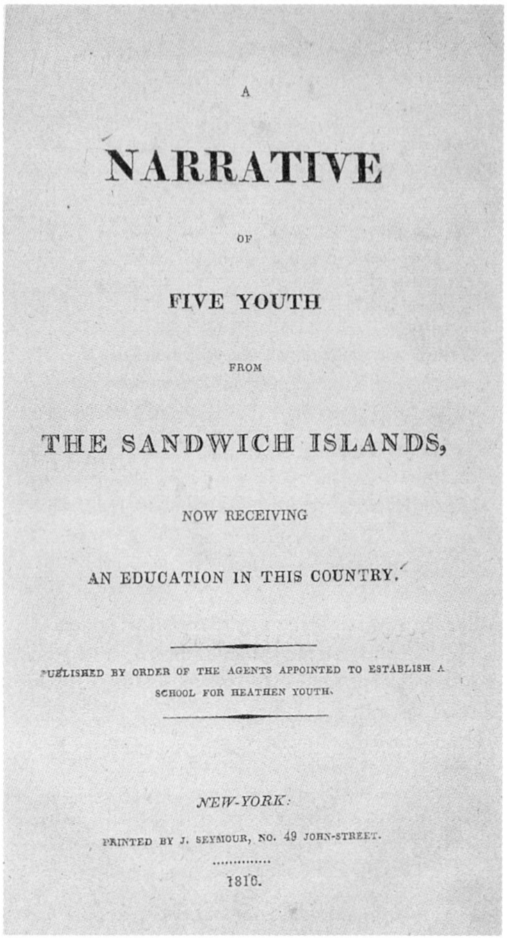

A

NARRATIVE

OF

FIVE YOUTH

FROM

THE SANDWICH ISLANDS,

NOW RECEIVING

AN EDUCATION IN THIS COUNTRY.

PUBLISHED BY ORDER OF THE AGENTS APPOINTED TO ESTABLISH A SCHOOL FOR HEATHEN YOUTH.

NEW-YORK:

PRINTED BY J. SEYMOUR, NO. 49 JOHN-STREET.

....

1816.

462. Narrative of Five Youth

establish a School for Heathen Youth.

Octavo, 44 pp.; modern wrappers. New York, J. Seymour, 1816.

Very scarce pamphlet, giving the full stories of several Hawaiians brought to mainland America for religious and other education, beginning with Obookiah and Hopoo who were allowed by Kamehameha to sail with Captain Brintnal of New-Haven in 1809. The experiment of their education, at first in the Brintnal household, was seen as hugely successful, and the narrative ends with accounts of the first attempts to found a full formal educational establishment for "heathen children".

Sabin, 51790. **$1350**

PHOTOMEZZOTYPES

463. NOTTAGE, Charles G. In Search of a Climate.

Octavo, with a map and thirty photographic plates, some printed in coloured tints, as well as numerous illustrations in the text; the half-title lightly spotted and upper fore-corner clipped; a very good copy in the original cloth, ornately decorated with a volcano vignette in gilt and colours. London, Sampson Low, 1894.

A lengthy account of travels, centring on the Hawaiian Islands, but also dealing with visits to Melbourne and Sydney, California and Mexico. The Hawaiian section is by far the most interesting, and the illustrations are quite wonderful: they are "photomezzotypes", printed by the London Stereoscopic & Photographic Company.

Ferguson, 13496. **$650**

OBOOKIAH

464. [OBOOKIAH] [DWIGHT, Edwin Welles] Memoirs of Henry Obookiah, a native of Owhyhee, and a member of the Foreign Mission School, who died at Cornwall, Conn. Feb. 17, 1818, aged 26 years.

Duodecimo, with a frontispiece portrait, rather foxed throughout (as usual); uncut in the original plain boards, worn, spine renewed. New Haven, Conn., Religious Intelligencer, 1818.

Scarce first edition of this much reprinted memoir of the young Hawaiian, one of five Hawaiians taken to New Haven for their education in 1809; the experiment is described in *A Narrative of Five Youth from the Sandwich Islands* (see catalogue number 462). Hill records only later editions.

See Hill, pp. 91-2 & 410-11. **$325**

FIRST VERSION IN HAWAIIAN

465. [HAWAIIAN IMPRINT] OBOOKIAH. Ka Moolelo o Heneri Opukahaia… [The Story of Henry Obookiah].

Duodecimo, 103 pp., engraved portrait; a fine copy in the original tan blind-stamped cloth. New York, American Tract Society, 1867.

The first Hawaiian language version of the life of Henry Obookiah. Scarce in this fine condition.

Judd, 439. **$250**

HAWAIIAN LEGENDS

466. PATTON, Beatrice Ayer. Légendes Hawaiiennes.

Quarto, with fine border and vignette illustrations by J.M. Fraser; original decorated wrappers. Paris, Les Belles Lettres, 1932.

A fine unopened copy, one of an edition of only 200 numbered copies of this beautifully decorated book of Hawaiian legends. **$450**

DUBLIN EDITION

467. STEWART, C.S. The Private Journal… abridged from the American edition.

Duodecimo, with a frontispiece; a fine copy in original quarter roan. Dublin, Religious Tract Society, 1831.

Scarce edition of this classic account of Hawaii, written by one of the first permanent preachers in the islands. In 1822 Stewart "was sent to Hawaii by the American Board of Commissioners for Foreign Missions accompanied by his wife… During their stay in Hawaii, they were joined by the Rev. and Mrs. William Ellis. Due to a serious illness of Mrs. Stewart, both she and Rev. Stewart departed from Hawaii. Stewart later was appointed chaplain in the U.S. Navy, and during his visits to all parts of the world he collected material for his works" (Hill).

Hill, p. 283; Judd, 168 (earlier edition).. **$275**

REVISED & ENLARGED

468. STEWART, C.S. A Residence in the Sandwich Islands...

Duodecimo, a little spotting to title-page and elsewhere, but an excellent copy in the original green blind-stamped cloth. Boston, Weeks, Jordan, 1839.

Fifth edition, revised and enlarged, and intended by Stewart as the first publication in a series of four titles, which does not seem to have eventuated.

Hill, p. 283 (other editions). **$325**

NEARLY COMPLETE

469. THRUM, Thomas G., compiler. Hawaiian Almanac and Annual for 1875 [-1932]

11 volumes, octavo, 57 (of 58) numbers, lacking only the issue for 1886 (here substituted at an early date by the Creighton Almanac); with all original wrappers 1875-1885, and 1902-1932 (the others removed at an early date by the binder), some issues trimmed with no loss of text; with the folding maps of Honolulu (1891 and 1903) and the 1885 and 1899 maps of the Islands, and numerous illustrations and tables; an excellent set, uniformly bound in old half brown morocco. Honolulu, Thomas G. Thrum, 1875-1932.

A very scarce and nearly complete run of the most important almanac of the Pacific, probably the most frequently quoted source of information on Hawaii. Thrum, the editor, compiler, and publisher (who was born in Newcastle, New South Wales, where his father and grandfather were shipwrights), arrived in Honolulu as a boy in 1833, and became one of the three great nineteenth-century historians of Hawaii. He claimed that his great-grandfather, John Thrum, also a shipwright, had sailed with Cook on his third voyage.

The amount of information is staggering, and indicative of Thrum's wide range of interests. Included are business histories, customs house statistics, government documents, the Chinese question, Japanese and Portuguese immigration, and an annual retrospect of business and social events. Early issues contain an annual list of marine casualties, and articles on maritime history. Agricultural and natural history articles include much on the sugar, coffee, rice, and pineapple industries, cattle and sheep ranching, the sandalwood trade, and forestry. Pieces on Hawaiian legends, linguistics, and folklore record much that would otherwise have been lost, and there are frequent travel articles for the benefit of tourist and resident readers, including good material on the volcano of Kilauea.

Issues of the 1880s have always been extremely difficult to obtain. The missing number here was replaced (in about 1915) by the equivalent issue of Creighton's *Honolulu Almanac and Directory* (published only between 1884 and 1886, and itself scarcer than the Thrum issue). **$6000**

IN OUR SEPULCHRES

470. THURSTON, Lucy Goodale. The Missionary's Daughter: A Memoir of Lucy Goodale Thurston.

16mo., frontispiece; a few spots but a good copy in the original brown blind-stamped cloth. New York, Dayton and Newman, 1842.

A mawkish volume, commemorating Lucy Thurston who was born in Hawaii in 1823, one of the earliest native-born settlers; she was the child of Rev. Asa Thurston, one of the first two foreign missionaries to arrive in Hawaii. At the age of seventeen she travelled back to the mainland: "she landed upon our shores... but within three weeks she found a place in our sepulchres".

This devotional book deals with the history of the islands and their conversion to Christianity, not pulling any missionary punches in its historical view: "Chapter II. Character of the inhabitants – Despotic government – Habitations – Dress – Filthiness – Vicious habits – Idolatry – Kabu system. Chapter III Remarkable events of Divine Providence which prepared the way for the mission to the Sandwich Islands...".

See Hill, p. 592 for more on the Thurstons but not this title.

$200

ORIGINAL WRAPPERS

471. WHITNEY, Henry M. The Hawaiian Guide Book, for Travellers: containing a brief description of the Hawaiian Islands, their harbors, agricultural resources, plantations, scenery, volcanoes, climate, population, and commerce.

Duodecimo, 144 pp., interspersed with advertisements, some on coloured paper; a fine copy in the original orange printed wrappers, just wearing at spine. Honolulu, H.M. Whitney, 1875.

First edition of the first guide book to Hawaii: the second issue, with the advertisement for Thrum and Oat, who started business in 1875.

This was the first traveller's guide to the Hawaiian Islands containing the most comprehensive descriptions for this period. Whitney, a prominent Honolulu publisher and newspaperman, anticipating a tourist rush to Hawaii, noted that his object was to present "such information as travellers require... and at the same time supply all the facts which intelligent settlers may wish to know". Ports and towns, natural wonders and the cost of meals and transportation are included, and the glowing account of the newly completed Hawaiian Hotel informs the reader that all 42 sleeping rooms were lighted by gas "manufactured on the premises" and connected to the office with an "Annunciator". "No better kept public house can be found in any port of the Pacific".

Despite the large printing of 4000 copies, the entire edition was soon distributed, and the work has had a very low survival rate. **$750**

15: SHIPS & SHIP-BUILDING

GLOBES

472. **FERGUSON, James. Lectures on Select Subjects in Mechanics, Hydrostatics, Pneumatics, and Optics. With the Use of the Globes, the Art of Dialing, and the Calculation of the Mean Times of New and Full Moons and Eclipses...**

Quarto, with 23 folding engraved plates; a few spots but an excellent copy in contemporary calf, rebacked. London, A. Millar, 1764.

The use of scientific instruments, with some relevance for navigation, and two long chapters on terrestrial and celestial globes from a technical point of view.

Not in Crone. $850

LONGITUDE AT SEA

473. **GRANT, Major General, Viscount de Vaux. The Means of Finding the Longitude at Sea, gradually developed, discovered, and demonstrated in four astronomical, geographical, nautical, historical, mathematical, and mechanical Dissertations.**

Quarto, frontispiece portrait of the author (engraved in France) and eight other engraved plates (four coloured, one folding); an excellent copy in contemporary full green polished calf, spine gilt; bookplate of the Earl of Derby. London, Printed for the author by D.N. Shury, 1808.

Scarce semi-private publication by a gifted scientific enthusiast of private means.

Grant also wrote the standard early history of Mauritius as well as a number of scientific or cosmographical works. The four coloured circular maps here are particularly attractive.

Not in Crone. $1650

LITHOGRAPHS OF SMALL CRAFT

474. **GRUNDY, W.M. The Sketch Book of Shipping and Craft.**

Oblong octavo, with 24 lithographed plates; light foxing but a very good copy in the original gilt and blind decorated cloth, a little wear at head of spine. London, Charles Tilt, [c. 1838-1840].

Rare: the only edition of this detailed sketchbook of English and European small craft, including some naval vessels.

The work includes handsome sketches of a wide variety of ships, from very small craft such as row boats, a Cologne market boat, schooners, fishing boats, canal boats, up to the impressive brigs and Indiamen, such as the *Duke of Argyle*. This very scarce book is not recorded in any of the standard references. No copy is listed by NUC, and it is not in the British Library Catalogue. $2100

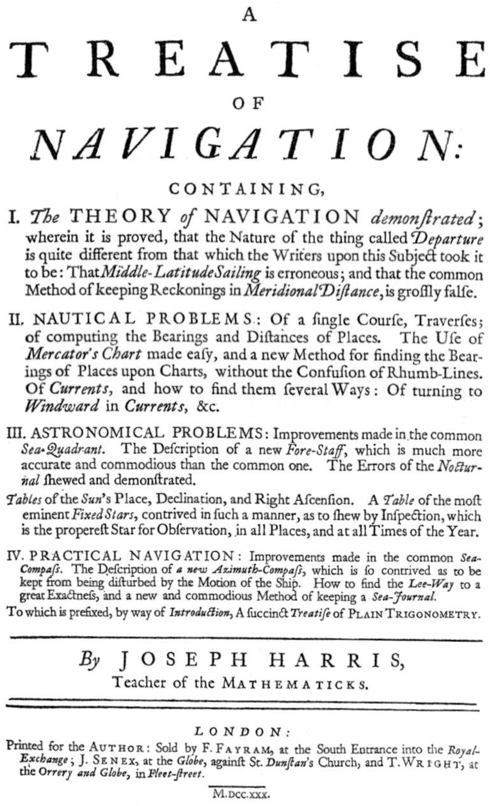
A

TREATISE

OF

NAVIGATION:

CONTAINING,

I. *The* THEORY *of* NAVIGATION *demonstrated*; wherein it is proved, that the Nature of the thing called *Departure* is quite different from that which the Writers upon this Subject took it to be: That *Middle-Latitude Sailing* is erroneous; and that the common Method of keeping Reckonings in *Meridional Distance*, is grossly false.

II. NAUTICAL PROBLEMS: Of a single Course, Traverses; of computing the Bearings and Distances of Places. The Use of *Mercator's Chart* made easy, and a new Method for finding the Bearings of Places upon Charts, without the Confusion of Rhumb-Lines. Of *Currents*, and how to find them several Ways: Of turning to *Windward* in *Currents*, &c.

III. ASTRONOMICAL PROBLEMS: Improvements made in the common *Sea-Quadrant*. The Description of a new *Fore-Staff*, which is much more accurate and commodious than the common one. The Errors of the *Nocturnal* shewed and demonstrated.

Tables of the *Sun*'s Place, Declination, and Right Ascension. A *Table* of the most eminent *Fixed Stars*, contrived in such a manner, as to shew by Inspection, which is the properest Star for Observation, in all Places, and at all Times of the Year.

IV. PRACTICAL NAVIGATION: Improvements made in the common *Sea-Compass*. The Description of *a new Azimuth-Compass*, which is so contrived as to be kept from being disturbed by the Motion of the Ship. How to find the *Lee-Way* to a great Exactness, and a new and commodious Method of keeping a *Sea-Journal*.

To which is prefixed, by way of *Introduction*, A succinct *Treatise* of PLAIN TRIGONOMETRY.

By JOSEPH HARRIS,
Teacher of the MATHEMATICKS.

LONDON:
Printed for the AUTHOR: Sold by F. FAYRAM, at the South Entrance into the *Royal-Exchange*; J. SENEX, at the *Globe*, against St. *Dunstan*'s Church, and T. WRIGHT, at the *Orrery and Globe*, in *Fleet-street*.

M.DCC.XXX.

475. Harris

HARRIS ON NAVIGATION

475. **HARRIS, Joseph. A Treatise of Navigation.**

Large quarto, with twelve folding plates; contemporary mottled calf; bookplate of Thomas Darby; an excellent fresh and crisp copy. London, for the Author, 1730.

Very rare first edition of this important treatise on navigation, which describes among other things a new forestaff, a new and improved azimuth compass, and a new measuring scale to be used with Mercator's chart.

Harris deals extensively with trigonometry, problems of method, problems of nautical practice (how to keep a single course, the simplified use of Mercator's chart), astronomical problems and the use of instruments, and further improvements on some navigational instruments, including his new forestaff. He lambasts earlier nautical guides for limiting the mathematics sailors should know to rote-learning, ascribing their difficulty to poor teaching methods.

He also maintains that nothing of real worth has been written since Edmund Wright's *Certain Errors in Navigation* of 1599, and pokes fun at "the Longitude

Finders", who are more interested in claiming the lucrative "Longitude Prize" than producing anything generally useful.

The rapid rise of the Cornish-born Harris (1702-64) attests to the great increase in social mobility afforded those who developed the technologies of eighteenth-century Britain. Trained as a blacksmith, he already had a considerable reputation as an instrument-maker when he arrived in London in about 1724. He is known to have sailed to Vera Cruz in 1727 and four years later to have accompanied the astronomer Colin Campbell to Kingston to found an observatory there. His measuring scale for simplifying the use of Mercator's chart – described here – involved him in a lengthy dispute with Haselden.

Harris's *Description and the Use of the Globes and Orrery* of 1731 was one of the most popular of such guides, going through at least ten editions. He ended his life as Master of the London Mint.

Houzeau & Lancaster, 10730; Taylor, 'Mathematical Practitioners', 1714-1840, 185. **$8850**

SHEET ANCHOR

476. LEVER, Darcy and George W. BLUNT. American Edition. The Young Sea Officer's Sheet Anchor; or a Key to the Leading of Rigging, and to Practical Seamanship. By Darcy Lever Esq. with Additions by George W. Blunt.

Quarto, engraved title-page and 114 engraved plates, two folding; a few stains; original speckled sheep, worn. New York, E. & G.W. Blunt, 1858.

The classic and generously illustrated (a plate to every page of text) guide to practical seamanship, and particularly to rigging, here in a specially arranged American edition. The editor, George Blunt, explains that thirty-four years earlier, "his vocation then being upon the deep", he had used an English edition of Lever's *Sheet Anchor*. Modern changes in rigging have been included, and the new material includes Tables on Rope, Canvas, Guns, Spars, Clippers, and Forbes's New Rig. **$3000**

LAUNCHING THE ORWELL

477. ORWELL, East-Indiaman. The Launching of the Orwell, East-Indiaman, at the Halifax Ship-Yard, Ipswich, on Thursday the 28th August, 1817: a New Song.

Octavo, 8 pp.; cover vignette, self-wrappers, slipcase. Ipswich, J. Raw, 1817.

The launch of a new ship (the *Orwell* contract was initiated by Jabez Bayley with Capt. M. Isacke) deserved a new sea song. This ditty is to be sung to the tune of "Come sweep up the Planchers, and fill up your Gotch". **$275**

476. Lever

COLOUR PLATES

478. PACINI, Eugène. La Marine, arsenaux, navires, équipages, navigation, atterrages, combats…

Large octavo, heavily illustrated with plates and text-illustrations, some of the plates coloured; some spotting or browning but a handsome copy overall in the splendid original blue cloth, ornately gilt. Paris, L. Curmer, 1844.

A lavish production – typical of the publisher Curmer – with very good pictorial material, some (uniforms, flags, etc.) in colour: but this is also a serious and extensive work on French naval history and tradition. **$1400**

479. *Marine binding*

SHIP IN A BINDING

479. **[MARINE BINDING] WILLIAMS, M. The Adventures of a Post Captain: by a Naval Officer. With characteristic engravings by Mr. Williams.**

Large octavo, with a coloured frontispiece and 24 coloured engravings; text a little aged and frayed (and possibly missing four leaves, but the text apparently complete); the coloured plates in fine bright condition; in a contemporary binding of marbled calf, flat spine panelled in gilt; with a stippled design on the covers of each side of a ship under full sail within double gilt borders; the binding's joints and endpapers renewed. London, J. Johnston, n.d., c. 1830.

A most unusual and attractive early nineteenth-century binding with a fine marine design executed directly onto each cover. The series of plates is an amusing picaresque tale of the adventures of a young officer in the typical highly-coloured style of the period. **$1500**

JIGSAW PUZZLE

480. **[SHIPBUILDING PUZZLE] PHILIP, Son, and Nephew. The Building of the Ship. A new dissected puzzle.**

Children's jigsaw puzzle, 47 pieces, colour lithograph paper laid down on wood, one piece repaired; with colour lithograph guide, 380 x 280 mm., some wear to original folds; and original booklet, 24 pp. in original green wrappers; preserved in a modern case. London & Liverpool, Philip, Son, & Nephew, c. 1880.

Very scarce: the complete puzzle with the original guide and booklet. This is a delightful puzzle showing the entire process of ship construction from initial design, to laying out the plans, building the framework of the hull, raising the masts, sail-making, and forging the anchor, to the completed ship being loaded at the docks and majestically setting off under full sail. Amusingly one of the pennant flags reads P. S. & N., a subtle advertisement for Philip, Son and Nephew, the publishers of the puzzle.

Present here is the very rare explanatory booklet – these are "almost invariably lost" (Hannas, *The English Jigsaw Puzzle*). It describes in some detail the processes of ship building, and also the roles of the various trades associated in the process, in order that "our young readers will then be able to understand the reason for many of the peculiarities of form etc.". This puzzle was evidently originally published by John Betts, and formed part of his extensive catalogue of educational puzzles and maps. It seems that when he went out of business in about 1875 his goodwill was taken over by Philip, Son & Nephew, who continued to sell the stock under the Betts banner. This puzzle forms part of the publisher's catalogue of fifty-five dissected puzzles in the back of the booklet, from which we learn that it originally sold for 5 shillings.

Not in Hannas. **$4250**

NAVAL ARCHITECTURE

481. **STALKART, Marmaduke. Naval Architecture or the Rudiments and Rules of Ship Building exemplified in a series of Draughts and Plans...**

Folio text, and large folio atlas with 17 engraved plates (11 double-page and some folding); some marginal tears and restorations and other signs of use to the plates, light spotting at the start of the text, but an excellent set, the text in contemporary crimson morocco ornately gilt, neatly rebacked, atlas in half crimson morocco to match. London, Printed for the Author, 1781.

480. *Shipbuilding puzzle*

First edition: a splendid series of shipbuilding plates, with a long and detailed practical text. One of the more comprehensive eighteenth-century shipbuilding manuals, and a richly illustrated textbook on marine architecture, this is "a neglected and important work" (McDonald); Roeding compares its importance with that of Chapman's famous book on the same subject.

So accurate are the engravings that even today they could be used to plan and build a sailing-ship. Published at the time of the great English voyages to the Pacific, and in the decade of the First Fleet's voyage, this book provides a direct connection between the activities of the eighteenth-century practical shipyard, and the exploring ships that they produced. This copy has the bookplate of Henry Beaufoy, Fellow of the Royal Society and evidently the father of Mark Beaufoy, the naval architect.

McDonald, 269; Roeding, I, 166; Scott Collection, 380; Stratico, IV, 124. $9500

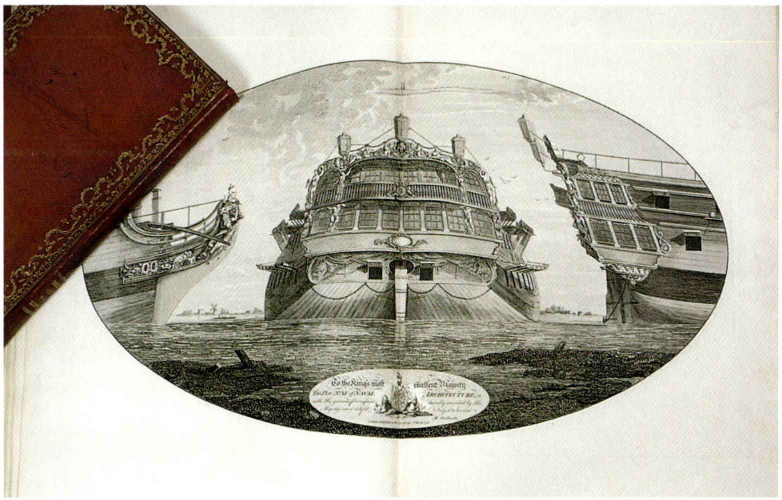

481. *Stalkart*

HORDERN
HOUSE

PRINCIPAL REFERENCES CITED IN THIS CATALOGUE

ABBEY, J.R. *Travel in Aquatint and Lithography.* London, 1956-57.

ADB. *Australian Dictionary of Biography.* Melbourne, 1966-91.

ALDEN, John, ed. *European Americana.* New York, 1980-88.

ALLEN, Joel Asaph. *Preliminary List of Works and Papers relating to the Mammalian Orders Cete & Sirenia 1494-1840.* Hythe, 1977.

ATKINSON, Geoffroy. *The Extraordinary Voyage in French Literature from 1700 to 1720.* Paris, 1922.

BARBIER, G. *Dictionnaire Des Ouvrages Anonymes.* Paris, 1872-89.

BEAGLEHOLE, J.C., ed. *The Journals of Captain James Cook.* Hakluyt Society, 1955-67.

BEDDIE, M.K. *Bibliography of Captain James Cook.* Sydney, 1970.

BORBA DE MORAES, Rubens. *Bibliographia Brasiliana.* Revised edition. Los Angeles, 1983.

BUCK, Peter H. *Explorers of the Pacific.* Honolulu, 1953.

CARLSMITH. *Pacific Voyages and Exploration from the Carlsmith Collection.* Sydney, Hordern House, 1987.

CHURCH, E.D. *A Catalogue of Books relating to the Discovery and Early History of North and South America.* New York, 1907.

CORDIER, Henri. *Bibliotheca Indosinica*, Paris, 1912-32; *Bibliotheca Japonica*, Paris, 1912; *Bibliotheca Sinica*, Paris, 1904-24.

COWAN, R.N. & R.G. *A Bibliography of the History of California 1510-1930.* San Francisco, 1933.

COX, E.G. *Reference Guide to the Literature of Travel.* Seattle, 1935-49.

DAVIDSON, R. *A Book Collector's Notes. On Items relating to the Discovery of Australia.* Melbourne, 1970.

DAY, A. Grove. *Books about Hawaii.* Honolulu, 1977.

DNB. *Dictionary of National Biography.* London, 1908- .

DSB. *Dictionary of Scientific Biography.* New York, 1970-1980.

DUNMORE, John. *French Explorers in the Pacific. Vol. I. The Eighteenth Century; Vol. II. The Nineteenth Century.* Oxford, 1965-69.

DUVIOLS, Jean-Paul. L'Amérique espagnolevue et revée. Les livres de voyages. Paris, 1986.

FERGUSON, J.A. *Bibliography of Australia.* Sydney, 1941-86.

FORD, E. *Bibliography of Australian Medicine.* 1790-1900. Sydney, 1976.

FORSTER, Honore. *The South Sea Whaler.* Sharon, Massachusetts, 1985.

FORSTER, Honore. *More South Sea Whaling.* Canberra, 1991.

FREEMAN, R.B. *The Works of Charles Darwin: an Annotated Bibliographical Handlist.* Second edition. London, 1977.

FRIEDERICH, Werner P. *Australia in Western Imaginative Prose Writings 1600-1960.* Chapel Hill, 1967.

GIBSON, R.W. *St. Thomas More: a Preliminary Bibliography... with a Bibliography of Utopiana.* New Haven, 1961.

GOVE, P.B. *The Imaginary Voyage in Prose Fiction.* London, 1941.

GREAT FLOWER BOOKS. SITWELL & BLUNT. *Great Flower Books.* London, 1957.

GUMUCHIAN, & Cie. *Les livres d'enfance du XVe au XIXe siècle.* Paris, n.d.

HASKELL, Daniel C. *The United States Exploring Expedition, 1838-1842, and its publications.* New York Public Library, 1942.

HILL. *The Hill Collection of Pacific Voyages.* San Diego, 1974-1983.

HOCKEN, T.M. *A Bibliography of New Zealand.* Wellington, 1909.

HOLMES, Sir M. *Captain James Cook, R.N., F.R.S. A Bibliographical Excursion.* London, 1952.

HOWES, Wright. *U.S.Iana (1650-1950).* Newberry Library, 1962.

HUNNEWELL, James F. *Bibliography of the Hawaiian Islands.* Boston, 1869.

HUNTRESS, Keith. *A Checklist of Narratives of Shipwrecks and Disasters at Sea to 1860.* Ames, 1979.

JAMES FORD BELL. *Annotated Catalog... James Ford Bell Library.* Minnesota, 1994.

JOPPIEN, R. & B. SMITH. *The Art of Captain Cook's Voyages.* Melbourne, 1985-1987.

JUDD, Bernice, et al. *Hawaiian Language Imprints, 1822-1899.* Honolulu, 1978.

JUDD, Bernice. *Voyages to Hawaii before 1860.* Honolulu, 1974.

KAEPPLER, Adrienne L. *Artificial Curiosities.* Honolulu, 1978.

KELLY, Celsus. *Calendar of Documents: Spanish voyages in the South Pacific.* Madrid, 1965.

KROEPELIEN. *Bibliotheca Polynesiana. A Catalogue of some of the books in the Polynesiana Collection formed by the late Bjarne Kroepelien, by R. du Rietz.* Oslo, 1969.

LACH, Donald F. *Asia in the Making of Europe.* Chicago, 1965-93.

LADA-MOCARSKI, V. *Bibliography of books on Alaska published before 1868.* New Haven, 1969.

LANDWEHR, John. *VOC: bibliography of publications relating to the Dutch East India Company.* Utrecht, 1991.

LUST, John. *Western books on China published up to 1850 in the library of SOAS.* University of London, 1987.

MARIA, Ron. *The Burke and Wills Exploring Expedition. Catalogue.* Sydney, 1986.

McLAREN, Ian F. *Lapérouse in the Pacific.* University of Melbourne, 1993.

MEDINA, J.T. *Bibliografia española de las Islas Filipinas*, Santiago de Chile, 1897; *Biblioteca Hispano-Americana*, 1898-1907; *Biblioteca Hispano-Chilena*, 1897-99.

MENDELSSOHN, Sydney. *South African Bibliography.* London, 1910-79.

MUIR, Marcie. *Australian Children's Books. A Bibliography.* Melbourne, 1992.

NAN KIVELL, R. & S. SPENCE. *Portraits of the Famous and Infamous.* London, 1970.

NISSEN, Claus. *Die Botanische Buchillustration*, Stuttgart, 1966; *Die Zoologische Buchillustration*, 1969-72.

NMM. *National Maritime Museum Catalogue of the Library.* London, 1972.

NUC. *National Union Catalog, Pre 1956 Imprints.* Washington, 1983.

NZNB. BAGNALL, A.G. *New Zealand National Bibliography. Volume I: to 1889.* Wellington, 1980.

O'REILLY-REITMAN. O'Reilly, P. & E. REITMAN. *Bibliographie de Tahiti et de la Polynesie Française.* Musée de l'Homme, 1967.

OSBORNE, Milton. *The French Presence in Cochinchina and Cambodia*, Cornell, 1969; *River Road To China: The Mekong River Expedition*, London, 1975; *Southeast Asia*, 1995.

PALAU Y DULCET, Antonio. *Manual del Librero Hispanoamericano.* Second edition. Barcelona, 1948-87.

PARKER, John. *Books to build an Empire. A bibliographical history of English overseas interests to 1620.* Amsterdam, 1965.

PENROSE, Boies. *Travel and Discovery in the Renaissance.* Harvard, 1955.

PEOPLE, PRINT AND PAPER. RICHARDS, Michael. *People, print and paper. Catalogue of a travelling exhibition.* National Library of Australia, 1988.

POLAK, Jean. *Bibliographie Maritime Française.* Grenoble, 1976-83.

PRINTING AND THE MIND OF MAN. CARTER, J. & P. H. MUIR. *A Descriptive Catalogue illustrating the Impact of Print on the Evolution of Western Civilisation.* London, 1967.

QUINN, D.B. *The Hakluyt Handbook.* Hakluyt Society, 1974.

RETANA, W.E. *Aparato bibliografico de la historia general de Filipinas.* Madrid, 1906.

ROBERT, Willem. *Contributions to a Bibliography of Australia and the South Sea Islands.* Amsterdam, 1969.

ROSCOE, S. *John Newbery and his Successors 1740-1814.* Wormley, 1973.

SABIN, Joseph. *Dictionary of Books relating to America.* New York, 1868-1936.

SCHILDER, Gunter. *Australia Unveiled: the share of the Dutch navigators in the discovery of Australia.* Amsterdam, 1976.

SHARP, Andrew. *The Discovery of the Pacific Islands.* Oxford, The Clarendon Press, 1960.

SHIRLEY, R.W. *The mapping of the world. Early printed world maps 1472-1700.* Holland Press, 1993.

SPENCE, S.A. *Antarctic Miscellany.* London, 1980.

STAFLEU, F.A. & R.S. COWAN. *Taxonomic Literature.* Utrecht, 1976-86.

STC. *Short-Title Catalogue of English Books 1475-1640,* by A.W. Pollard & G.R. Redgrave. Revised edition. Bibliographical Society, 1986-1976-1991.

STEELE, Colin. *English Interpreters of the Iberian New World (1603-1726).* Oxford, 1975.

STREETER. *The celebrated collection of Americana formed by the late Thomas Winthrop Streeter.* New York, Parke-Bernet Galleries, 1966-69.

STREIT, Robert. *Bibliotheca Missionum.* Aachen, 1916-55.

TIELE, P.A. *Mémoire Bibliographique sur les Journaux des Navigateurs Neerlandais, la plupart en la possession de Frederik Muller.* Amsterdam, 1867.

TIELE, P.A. *Nederlandsche Bibliographie van Land- en Volkenkunde.* Amsterdam, 1884.

TOOLEY, R.V. *English Books with Coloured Plates 1790 to 1860.* Folkestone, 1979.

TOOLEY, R.V. *The Mapping of Australia.* Folkestone, 1979.

WAGNER, Henry R. *The Cartography of the Northwest coast of America.* Berkeley, 1937.

WAGNER, Henry R. *The Spanish Southwest 1542-1794.* Berkeley, 1924.

WANTRUP, J. *Australian Rare Books 1788-1900.* Sydney, Hordern House, 1987.

WICKERSHAM, James. *Bibliography of Alaskan literature 1724-1924.* Cordova, 1927.

WING, Donald. *Short Title Catalogue of English Books 1641-1700.* Second edition. New York, 1994-1982-1988.

YAKUSHI, Yoshimi. *Catalogue of the Himalayan Literature.* Tokyo, 1984.

ZAMORANO 80. *The Zamorano 80. A selection of distinguished California books made by members of the Zamorano Club.* Los Angeles, 1945.

ZIMMER, John Todd. *Catalogue of the Edward E. Ayer Ornithological Library.* Chicago, 1926.

AUTHOR AND TITLE INDEX

A

Aa: Reizen naar Nieuw-Guinea, 1879 360

Account of a Geographical, An (Billings), 1802 227

Account of a Geographical, An (Sauer), 1802 227

Account of a Voyage, An (Barrington), 1810 148

Account of an Embassy, An (Symes), 1827 373

Account of an Expedition (Fox), 1837 87

Account of an Expedition (Fox), 1849 88

Account of Capt. Ross's Voyage, An (Ross), 1835 266

Account of Several, An (Narborough), 1711 51

Account of some specimens, An (Fitton), 1826 181

Account of Discoveries, An (Dalrymple), 1996 64

Account of the Colony, An (Collins), 1798 150

Account of the Colony, An (Collins), 1802 151

Account of the Colony, An (Collins), 1804 152

Account of the Pelew Islands, An (Keate), 1788 258

Account of the Pelew Islands, An (Keate), 1789 259-60

Account of the Russian Discoveries (Coxe), 1787 126

Account of the Voyage, An (Tuckey), 1805 171

Account of the Voyages (Hawkesworth), 1773 106

Acosta: Tractado de las drogas, 1578 338

Address, An (Gallaudet), 1819 450

Address, An (Roscoe), 1802 120

Adventures of a Post Captain (Williams), 1830 479

Aguiar: Viagem ao interior, 1841 86

Albert: The Cruise of HMS Bacchante, 1886 389

Aleck, and the Mutineers (Fiske), 1854 221

Allen: Ten Years in Oregon, 1850 433

Allom: China, in a Series of Views, 1843 312

Allyn: The Old Sailor's Story, 1879 390

Alvares: Historiale description, 1558 17

American: History of the Sandwich Islands, 1831...434

American: Tahiti, receiving the Gospel, 1832 416

Anderson: Sketches of Canton, 1852 325

Anecdotes of a Life (Cowans), 1871 397

New Discoveries Concerning the World, 1778 78

Portuguese account of the Moluccas, 1750 371

Anson: A Voyage Round the World, 1748 1

Anson: Voyage autour du Monde, 1750 4

Anson: A Voyage Round the World, 1785 2

Anson: The Voyage, 1818 3

Appointment to an expedition (Nodal), 1618 38

Arago: Narrative of a Voyage, 1823 297

Arago: Souvenirs d'un aveugle, 1840 298

Arago: Souvenirs d'un aveugle, 1868 299

Argensola: Conquista de las Islas Molucas, 1609 6

Argensola: The discovery and conquest, 1708 7

Asia and its Islands (Laurie), 1799 329

Authentic Account, An (Staunton), 1797 335

Authentic Narrative of a Voyage, An (Ellis), 1782 127

Authentic Narrative of a Voyage, An (Ellis), 1783 128

Aux Indes et en Australie (Brassey), 1893 392

B

Balaguns (Webber), 1809 136

Barrington: A Voyage to Botany Bay, 1801 147

Barrington: An Account of a Voyage, 1810 148

Barrington: History of New South Wales, 1802......149

Barrow: A Voyage to Cochinchina, 1806 339

Barrow: The Eventful History, 1831 220

Barrow: The Life of George Lord Anson, 1839..........5

Barrow: Travels in China, 1804 313

Barrow: Travels in China, 1806 314

Bartimeus (Bingham), 1844 435

Bartlett: Historical Sketch, 1869...........................436

Bates: Sandwich Island Notes, 1854......................437

Baxter: Reception of the Revd. J. Williams, 1841....431

Baxter: Rev. J. Waterhouse, 1844 423

Bayldon: HMS Endeavour, 1968 110

Beauvais: Louis Delaporte, 1929...........................378

Bechtinger: Ein Jahr,1869....................................438

Bedelet: Monde en Estampes, Le, 1858...................56

Behrens: Histoire de l'expedition, 1739 45

Beknopte Beschrijving (Laar), 1867.........................349

Belcher: The Botany of HMS Sulphur, 1846..........245

Belcher: The Zoology of HMS Sulphur, 1846........245

Bennett: Narrative of a Whaling Voyage, 1840 246

Benyowsky: Voyages et Mémoires, 1791 226

Betagh: A Voyage round the World, 1728..................8

Bibliography of Hawaiiana (Hunnewell), 1868 451

Bibliothèque Géographique (Campe), 1807 142

Bibliothèque portative (Breton), 1807 57

Billings: An Account of a Geographical, 1802........227

Bingham: A Residence, 1847 439

Bingham: Bartimeus, 1844 435

Bird: The Hawaiian Archipelago, 1880..................440

Blaxland: A Journal of a Tour, 1870 175

Bligh Notebook, The (Bligh), 1986 214

Bligh: A Voyage to the South Sea, 1792 213

Bligh: Memoir to accompany a chart, 1803............218

Bligh: Minutes of the Proceedings, 1794 215

Bligh: Proceedings of a Court-Martial, 1811..........216

Bligh: Reports on Dublin Harbour, 1799..............217

Bligh: The Bligh Notebook, 1986.........................214

Boelen: Reize naar de Oost-en Westkust, 1835.......441

Boke of Idrography, The (Rotz), 1981 46

Bonaparte: Medal for Baudin expedition, 1801.......274

Borget: Fragments d'un voyage, 1850 276

Botany of HMS Sulphur (Belcher), 1846 245

Bougainville: Voyage autour du Monde, 1771.........277

Bougainville: Voyage round the World, 1772..........278

Bougainville: Journal de la navigation, 1837 279

Bourne: Journal, 1862 191

Brambila: Vista de la Ciudad de Santiago, 1795......237

Brassey: Aux Indes et en Australie, 1893 392

Brassey: Voyage in the Sunbeam, 1891 391

Breton: Bibliothèque portative, 1807 57

Brewer: Reminiscences, 1884 393

Brief Statement of Facts, A (Hume), 1874................187

Brierly: HMS Galatea, 1870 257

Brierly: HMS Mæander, 1852..............................247

Brierly: HMS Rattlesnake, 1849 201

Britton: Fiji in 1870, 1870 394

Brosses: Histoire des Navigations, 1756...................58

Brosses: Vollständige Geschichte, 1767....................59

Bugnon: Relation exacte, 1707 315

Building of the Ship, The (Philip), 1880...................480

Burford: Description of a View of Nanking, 1845...316

Burke and Wills Expedition, 1861...........................176

Burke and Wills: Report, 1862.............................178

Burke and Wills: The Expedition, 1861.................176

Burnes: Travels into Bokhara, 1834.......................317

Burney: A Chronological History, 1807...................60

Burney: Memoir, Explanatory of a Chart, 1811 318

Byron: Narrative of the Hon. John Byron, 1767........10

Byron: Narrative of the Hon. John Byron, 1768..........9

Byron: Narrative of the Hon. John Byron, 1788........11

Byron: Voyage of HMS Blonde, 1826 248

C

Callander: Terra Australis Cognita, 1768 61

Calvert: The Discovery of Australia, 1902.................13

Cameron: Our Tropical Possessions, 1865.............340

Campe: Bibliothèque Géographique, 1807.............142

Carte de l'Ile de Java (Java), 1750 348

Catalogue of the Specimens, A (Shaw), 1787..............133

Chappe d'Auteroche: Journey into Siberia, 1770.....280

Characteristics of Volcanoes (Dana), 1890 244

Cheever: Life in the Sandwich Islands, 1851..........442

China Sea Directory, The (King), 1868....................328

China, in a Series of Views (Allom), 1843................312

Choris: Vues et Paysages, 1826 231
Chronological History, A (Burney), 1807 60
Clark: Observations on the Diseases, 1775 341
Clercq: Ethnographische beschrijving, 1893 361
Clerke: Viage del Comandante Byron, 1769 12
Cleveland: A Narrative of Voyages, 1842 395
Cleveley: Views, 1788 123
Cleveley: View of Huaheine, 1787 124
Cleveley: View of Moorea, 1787 125
Cochinchine Française (Lemire), 1884 352
Cochinchine Française (Garnier), 1864 380
Coleridge: Lives of northern worthies, 1852 143
Collection of Voyages, A (Dampier), 1729 22
Collingridge: Discovery of Australia, 1895 14
Collingridge: First Discovery of Australia, 1906 15
Collins: An Account of the English Colony, 1798 150
Collins: An Account of the English Colony, 1802 151
Collins: An Account of the English Colony, 1804 152
Complete Account, A (Tench), 1793 170
Conquista de las Molucas (Argensola), 1609 6
Consequences of Invasion (Gillray), 1850 153
Cook: Set of the three voyage accounts, 1784 141
Cook: A Voyage towards the South Pole, 1777 114
Cook: A Voyage to the Pacific Ocean, 1785 121
Cook: Troisième voyage de Cook, 1785 122
Cooke: Scenery of China, Turkey, 1813 63
Cooke: Scenery of the East India Islands, 1813 62
Coppinger: Cruise of the Alert, 1883 396
Corals and Coral Islands (Dana), 1875 243
Coronelli: Mare del Sud, 1696 16
Costume, Manners, The (Harris), 1830 74
Cowans: Anecdotes of a Life on the Ocean, 1871 397
Coxe: Account of the Russian Discoveries, 1787 126
Crozet: Nouveau Voyage à la Mer du Sud, 1783 281
Cruise of HMS Bacchante, The (Albert), 1886 389
Cruise of the Alert (Coppinger), 1883 396
Cruise of the Frigate Columbia (Murrell), 1840 411
Cruise of the Marchesa, The (Guillemard), 1886 363

D

D'Entrecasteaux: Voyage, 1808 211
Dahlgren: Were the Hawaiian Islands, 1916 443
Dalrymple: An Account of the Discoveries, 1996 64
Dalrymple: An Historical Collection, 1771 65
Dalrymple: Voyages dans la mer du Sud, 1774 66
Dampier: A Collection of Voyages, 1729 22
Dampier: A New Voyage, 1697 18
Dampier: A New Voyage, 1698 19
Dampier: A Voyage to New Holland, 1703 21
Dampier: Voyages and Descriptions, 1699 20
Dana: Characteristics of Volcanoes, 1890 244
Dana: Corals and Coral Islands, 1875 243
Darwin: Journal of Researches, 1839 249
Darwin: Journal of Researches, 1845 250
Darwin: Journal of Researches, 1846 251
Davis: Tracks of McKinlay, 1863 196
De la Chine (Grosier), 1820 323
De La Perouse (Playbill), 1825 94
De Mensch (Stuart), 1807 81
De optimo reipublicae statu (More), 1777 98
De Paris au Tibet (Garnier), 1882 386
De situ orbis libri III (Mela), 1722 35
De: Doudart de la Grée, 1875 379
De: Novae Guineae Forma & Situs, 1593 23
De: The Philippine Islands, 1868 42
Death of Captain Cook (Webber), 1784 137
Décisions. Précaution (Toulon), 1777 119
Découvertes dans la Mer du Sud, 1795 93
Découvertes des François (Fleurieu), 1790 69
Delano: A Narrative of Voyages, 1817 223
Delaporte: Voyage au Cambodge, 1880 377
Delessert: Voyages dans les Deux Océans, 1848 398
Deleuze: Notice historique sur M. Péron, 1810 275
Dellon: A Voyage to the East-Indies, 1698 342
Dellon: The History of the Inquisition, 1750 343
Descripcion de las Indias (Herrera), 1730 26
Description Géographique (Marchal), 1824 350
Description of Nanking (Burford), 1845 316
Détails nouveaux (Samwell), 1786 132
Diarium Nauticum (Stockmans), 1598 29
Diary of a Journey Overland (R., J), 1822 337
Dibble: A Voice from Abroad, 1844 445
Dibble: History and General Views, 1839 444
Dillon: Voyage aux Iles de la Mer du Sud, 1830 212
Directions for sailing (Horsburgh), 1811 347
Directions to Accompany Charts (Lindsay), 1796 353
Discoveries & Surveys (Moresby), 1876 364
Discoveries of the French (Fleurieu), 1791 70
Discovery and conquest, The (Argensola), 1708 7
Discovery of Australia, The (Calvert), 1902 13
Discovery of Australia, The (Collingridge), 1895 14
Dixon: A Voyage Round the World, 1789 252
Doudart de la Grée (De), 1875 379
Duff, The (Memorial), 1796 254
Duff: A Missionary Voyage, 1799 253
Duhaut-Cilly: Viaggio intorno al globo, 1841 446
Dumont d'Urville: Astrolabe Medal, 1826 282
Dumont d'Urville: Rapport, 1838 287
Dumont d'Urville: Voyage au Pole Sud, 1842 288
Dumont d'Urville: Voyage de l'Astrolabe, 1833 283-6
Duperrey: Medal for the Coquille voyage, 1822 291
Duperrey: Voyage autour du monde, 1826 289
Dwight: Memoirs of Henry Obookiah, 1818 464

E

Earl: The Native Races, 1853 362
Early Voyages to Terra Australis (Major), 1861 33
East-India Register, The (East), 1805 344
Ein Jahr (Bechtinger), 1869 438
Ekelöf: Ett år i stilla hafvet reseminnen, 1872 229
Ellis: A Journal of a Tour, 1825 447
Ellis: An Authentic Narrative of a Voyage, 1782 127
Ellis: An Authentic Narrative of a Voyage, 1783 128
Ellis: Journal of the Proceedings, 1817 319
Ellis: Narrative of a Tour, 1826 448
Ellis: Polynesian Researches, 1830 399
Emerson: Unwritten Literature of Hawaii, 1909 449
En Upptäckts-Resa (Sparrman), 1801 271
Endeavour, HMS (Bayldon), 1968 110
Engelmann: Porte-feuille géographique, 1820 67
English Hermit, The (Longueville), 1822 96-7
Entdeckungs-Reise (Kotzebue), 1821 230
Episode du Voyage (Garnier), 1869 382
Epistola (Manuel), 1513 34
Esquemeling: The History of the Bucaniers, 1785 68
Ethnographische beschrijving (Clercq), 1893 361
Ett år i stilla hafvet reseminnen (Ekelöf), 1872 229
Eventful History, The (Barrow), 1831 220
Expedition to Borneo, The (Keppel), 1846 351

F

Facsimile-Atlas (Nordenskiold), 1889 40
Ferguson: Lectures on Select Subjects, 1764 472
Field: Geographical Memoirs, 1825 180
Fiji in 1870 (Britton), 1870 394
First Discovery of Australia (Collingridge), 1906 15
First Night of the Races (Paltock), 1828 100
Fiske: Aleck, and the Mutineers, 1854 221
Fitton: An Account of some specimens, 1826 181
Fitzroy: Passage Table, 1859 255
Fleurieu: Découvertes des François, 1790 69
Fleurieu: Discoveries of the French, 1791 70

Flinders: Matthew Flinders' Narrative, 1946.......182-3
Fore-edge view (Sydney), 1830.............................165
Forrest: A Voyage to New Guinea, 1780................256
Forster: A Voyage round the World, 1777..............115
Forster: Tagebuch, 1781......................................140
Fortune: A Journey to the Tea Countries, 1852.......320
Fortune: A Residence among the Chinese, 1857.....321
Fortune: Yedo and Peking, 1863...........................322
Founding of the Settlement (Gosse), 1799................154
Fox: Account of an Expedition, 1837......................87
Fox: Account of an Expedition, 1849......................88
Fox: The Southlanders, 1860.................................89
Fragments d'un voyage (Borget), 1850....................276
Fragments of Voyages (Hall), 1834............................72
Francis Garnier (Petit), 1887...............................387
Fregatten Eugenies Resa (Skogman), 1855...............241
Freycinet: Voyage autour du monde, 1826..............295
Freycinet: Voyage autour du monde, 1844..............294
Frezier: A Voyage to the South-Sea, 1717................24

G

Galatea, HMS (Brierly), 1870...............................257
Gallaudet: An Address, 1819................................450
Garnier: Cochinchine Française, La, 1864.............380
Garnier: De Paris au Tibet, 1882..........................386
Garnier: Episode du Voyage d'Exploration, 1869....382
Garnier: Notice sur le Voyage, 1869......................381
Garnier: Voyage d'Exploration, 1873.....................383
Garnier: Voyage d'Exploration, 1885.....................384
Garnier: Voyage dans la Chine Centrale, 1874........385
General Collection, A (Pinkerton), 1814....................79
General Collection, A (Richardson), 1789..................80
Geogeraphi, oia te ite raa (Tahiti), 1840...................417
Geographical Memoirs (Field), 1825.......................180
Geographical Panorama, A (Harvey), 1840.................75
Gillray: Consequences of Invasion, 1850................153
Giolo, Prince (Savage), 1692................................345
Goodrich: Peter Parley's Tales of the Sea, 1848.......206
Goodridge: Narrative of a Voyage, 1852................400
Gosse: Founding of the Settlement, 1799..............154
Gosse: Transplanting the Bread-Fruit, 1796...........222
Grant: Means of Finding the Longitude, 1808.......473
Grant: The Narrative of a Voyage, 1803.................185
Grey: Journals of Two Expeditions, 1841...............186
Grosier: De la Chine, 1820...................................323
Grundy: The Sketch Book of Shipping, 1840.........474
Guerin: Tour du monde, Le, 1841........................401
Guignes: Voyages à Peking, 1808..........................324
Guillemard: The Cruise of the Marchesa, 1886......363

H

Hakluyt: The Principal Navigations, 1600................71
Hall: Fragments of Voyages, 1834...........................72
Hall: Mundus alter et idem, 1606...........................90
Hall: Mundus alter et idem, 1607...........................91
Hall: Mundus alter et idem, 1648...........................92
Hamilton: New Account of the East Indies, 1930...346
Harris: A Treatise of Navigation, 1730..................475
Harris: Navigantium atque Itinerantium, 1764.........73
Harris: The Costume, Manners, 1830.....................74
Harvey: A Geographical Panorama, 1840................75
Hawaii. The Past, Present (Hopkins), 1866.............454
Hawaiian Almanac and Annual (Thrum), 1932.......469
Hawaiian Archipelago, The (Bird), 1880..................440
Hawaiian Guide Book, The (Whitney), 1875...........471
Hawkesworth: An Account of the Voyages, 1773....106
He Hoikehonua (Whitney), 1845...........................452
Heath: New Panorama, 1829..............................155
Hechos de Don García Hurtado (Suarez), 1613...........50
Heeres: The Part Borne by the Dutch, 1899............25
Henry: An Historical Account, 1774......................76
Hermit, The (Longueville), 1768.............................95
Herrera: Descripcion de las Indias, 1730..................26
Hicks: The Life, Trial, Confession, 1860...............402
Histoire de l'Expédition (Pallu), 1888......................368
Histoire de l'expedition, (Behrens), 1739...................45
Histoire des Navigations (Brosses), 1756...................58
Histoire du Prince Lee Boo (Lee boo), 1820.............263
Historia de la vida del Cook (Kippis), 1795..............145
Historia general de la Yndia (San Roman), 1603.........47
Historiale description (Alvares), 1558........................17
Historica Relacion del Chile (Ovalle), 1646................41
Historical Account, An (Henry), 1774.......................76
Historical Collection, An (Dalrymple), 1771...............65
Historical Journal, An (Hunter), 1793.................156-7
Historical Sketch (Bartlett), 1869...........................436
History and General Views (Dibble), 1839...............444
History of New South Wales (Barrington), 1802........149
History of Prince Lee Boo, The (Newbery), 1793.......261
History of Sandwich Islands (American), 1831.........434
History of the Bucaniers (Esquemeling), 1785............68
History of the Hawaiian Islands (Jarves), 1843.........456
History of the Hawaiian Islands (Jarves), 1847.........457
History of the Hawaiian Islands (Jarves), 1872.........458
History of the Inquisition, The (Dellon), 1750...........343
History of Wonderful Animals (Whales), 1818..........424
Hodges: Travels in India, 1793.............................326
Hodgson: Reminiscences of Australia, 1846...........193
Holden: A Narrative of the Shipwreck, 1836..........403
Hondius: Polus antarcticus, 1650............................27
Honolulu Almanac, The (Honolulu), 1886...............453
Honolulu: The Honolulu Almanac, 1886..............453
Hopkins: Hawaii. The Past, Present, 1866............454
Horsburgh: Directions for sailing, 1811.................347
Hoxse: The Yankee Tar, 1840...............................404
Huc: Souvenirs d'un voyage, 1850........................327
Hulk at Sheerness, A (Rowlandson), 1800................164
Hume: A Brief Statement of Facts, 1874..............187
Hunnewell: Bibliography of Hawaiiana, 1868.........451
Hunt: The Past and Present, 1853........................455
Hunter: An Historical Journal, 1793.................156-7
Hunter: Resa til Nya södra Wallis, 1797................158

I

Iles Taïti (Vincendon-dumoulin), 1844..................290
In het ruime sop (Whaling), 1870..........................426
In Search of a Climate (Nottage), 1894...................463
Interesting and Authentic (Thomas), 1850...............420

J

Jackson: Robert O'Hara Burke, 1862....................177
Jarves: History of the Hawaiian Islands, 1843........456
Jarves: History of the Hawaiian Islands, 1847........457
Jarves: History of the Hawaiian Islands, 1872........458
Java: Carte de l'Ile de Java, 1750...........................348
Jenkins: Voyage of the Squadron, 1850..................242
Jeoly, Prince (Savage), 1692.................................345
Jewitt: Narrative of the Adventures, 1815.............405
Johan Hunters Resa (Hunter), 1797........................158
John Whites Resa till Nya Holland (White), 1793.....113
Journal d'un Baleinier (Thiercelin), 1866................311
Journal d'un Voyage (Magra), 1772.........................105
Journal de la navigation (Bougainville), 1837...........279
Journal of a Tour, A (Blaxland), 1870......................175
Journal of a Tour, A (Ellis), 1825............................447
Journal of a Voyage (White), 1790......................172-3
Journal of a Voyage, A (Magra), 1771......................103
Journal of a Voyage, A (Magra), 1772......................104
Journal of a Voyage, A (Parkinson), 1773.............107-8
Journal of a Voyage, A (Parkinson), 1784.................109
Journal of an Expedition (Mitchell), 1848...............198
Journal of an Expedition (Leichhardt), 1847...........194

Journal of Cook's Voyage, A (Ledyard), 1783 129
Journal of Exploration (Mckinlay), 1862 195
Journal (Bourne), 1862 191
Journal (Landsborough), 1862 190
Journal of Researches (Darwin), 1839 249
Journal of Researches (Darwin), 1845 250
Journal of Researches (Darwin), 1846 251
Journal of the Proceedings (Ellis), 1817 319
Journal of Voyages and Travels (Tyerman), 1831........422
Journals of Two Expeditions (Grey), 1841 186
Journals of Two Expeditions (Oxley), 1820...............199
Journey into Siberia, A (Chappe), 1770 280
Journey to the Tea Countries, A (Fortune), 1852 320
Jukes: Narrative of the Surveying Voyage, 1847 188

K

Ka Moolelo o Heneri Opukahaia, 1867 465
Kaart van de Zuyd-Ooster (Valentijn), 1726 55
Kathay: A Cruise (Macauley), 1852 331
Keate: An Account of the Pelew Islands, 1788 258
Keate: An Account of the Pelew Islands, 1789..259-60
Keppel: The Expedition to Borneo, 1846 351
King of Otaheite's Tea Party, The (Topliff), 1840.......418
King: Narrative of a Survey, 1826 189
King: The China Sea Directory, 1868 328
Kingsley: South Sea Bubbles, 1872 406
Kippis: Historia de la vida del Cook, 1795 145
Kippis: The Life of Captain James Cook, 1788.......144
Kotzebue: A New Voyage, 1830 232
Kotzebue: Entdeckungs-Reise, 1821 230
Kotzebue: Resa omkring Jorden, 1830 233
Krusenstern: Reise um die Welt, 1812 234

L

La Pérouse: A Voyage Round the World, 1798.......203
La Pérouse: Rélation abrégée, 1799 204
La Pérouse: Voyage autour du Monde, 1797 202
Laar: Beknopte Beschrijving, 1867 349
Labillardière: Novæ Hollandiæ Plantarum, 1804 210
Labillardière: Rélation du Voyage, 1800 207
Labillardière: Voyage, 1800 208
Labillardière: Voyage, 1800 209
Lamson: Round Cape Horn, 1878 407
Landsborough: Journal of Expedition, 1862 190
Landsborough's Exploration (Laurie), 1866...............192
Langsdorff: Voyages and Travels, 1817 235
Laplace: Reis rondom de wereld, 1836 301
Launching of the Orwell, The (Orwell), 1817 477
Laurie: Asia and its Islands, 1799 329
Laurie: Landsborough's Exploration, 1866 192
Lecomte: Pèche de la Baleine, 1833 306
Lectures on Select Subjects (Ferguson), 1764.............472
Ledyard: A Journal of Cook's Voyage, 1783............129
Lee Boo: Histoire du Prince Lee Boo, 1820...........263
Lee Boo: The History of Prince Lee Boo, 1825......262
Legendes et Chants (Michel), 1885 410
Légendes Hawaiiennes (Patton), 1932 466
Leichhardt: Journal of an Expedition, 1847............194
Lemire: Cochinchine Française, 1884.....................352
Les Isles Philippines (Sanson), 1654 370
Lesson: Voyage autour du monde, 1839 293
Lesson: Voyage Médical, 1829 292
Lettres du Tonkin (Lyautey), 1920 355
Lever: The Young Sea Officer, 1858 476
Life and Times of Patuone, The (Patuone), 1876 111
Life in the Sandwich Islands (Cheever), 1851 442
Life of Captain James Cook, The (Kippis), 1788 144
Life of George Lord Anson, The (Barrow), 1839.............5
Life of William Bligh, The (Mackaness), 1951 219
Life on the Ocean (Little), 1843 459
Life on the Ocean (Little), 1847 460
Life, Trial, Confession, The (Hicks), 1860 402
Lindley: The Log of Fortuna, 1880.......................330
Lindsay: Directions to Accompany Charts, 1796 353
Link with Magellan, A (Penrose), 1929 32
Little: Life on the Ocean, 1843 459
Little: Life on the Ocean, 1847 460
Lives of Travellers, The (St. John), 1832 131
Lives of northern worthies (Coleridge), 1852............143
Lodewyckszoon: Prima pars descriptionis, 1598........28
Log of Fortuna, The (Lindley), 1880 330
Longueville: The Hermit, 1768 95
Longueville: The English Hermit, 1822..............96-7
Louis Delaporte (Beauvais), 1929 378
Low: Sarawak, 1848 354
Lucas: The Sandwich Islands, 1841 461
Luillier: Nouveau voyage, 1726..............................30
Lyautey: Lettres du Tonkin, 1920 355

M

Macauley: Kathay: A Cruise, 1852.........................331
MacGillivray: Narrative of the Voyage, 1852 200
MacGregor: Through the Buffer State, 1896..........356
Mackaness: The Life of William Bligh, 1951 219
Mæander, HMS (Brierly), 1852 247
Magnum Mare Del Zur (Wit), 1680 52
Magra: A Journal of a Voyage, 1771......................103
Magra: A Journal of a Voyage, 1772 104-5
Major: Early Voyages to Terra Australis, 1861 33
Mant: The Midshipman, 1854 408
Manuel: Epistola Emanuelis, 1513 34
Map of Eastern China (Stanford), 1904 334
Map of the Malay Peninsula (Stanford), 1910 372
Marchal: Description Géographique, 1824.............350
Mare del Sud (Coronelli), 1696 16
Marine, La (Pacini), 1844 478
Mariner's Chronicle, The (Mariner's), 1834 409
Mariner's: The Mariner's Chronicle, 1834.............409
Maritime Geography (Tuckey), 1815 82
Matthew Flinders' Narrative (Flinders), 1946 182-3
McClintock: A Narrative of the Discovery, 1859 264
McKinlay: Journal of Exploration, 1862 195
McLeod: Voyage of HMS Alceste, 1818 357
Means of Finding Longitude, The (Grant), 1808 473
Meares: Tvänne Resor från Ostindien, 1797...........265
Medal for Astrolabe voyage (d'Urville), 1826 282
Medal for Baudin expedition (Bonaparte), 1801........274
Medal for Coquille voyage (Duperrey), 1822..............291
Medal for Uranie voyage (Freycinet), 1817...............296
Megiser: Septentrio Novantiquus, 1613...................43
Mela: De situ orbis libri III, 1722...........................35
Memoir to accompany a chart (Bligh), 1803..............218
Memoir, Explanatory of a Chart (Burney), 1811 318
Mémoires du Capitaine Péron (Peron), 1824.............160
Memoirs of Henry Obookiah (Dwight), 1818 464
Memoirs of Henry Taylor (Taylor), 1811 146
Memoirs of John Ledyard (Sparks), 1828 130
Memorial: The Duff, 1796 254
Michel: Legendes et Chants, 1885........................410
Michelena: Viajes cientificos, 1843 238
Midshipman, The (Mant), 1854...............................408
Minutes of the Proceedings (Bligh), 1794 215
Missionary Voyage, A (Duff), 1799..........................253
Missionary's Daughter, The (Thurston), 1842 470
Mitchell: Journal of an Expedition, 1848 198
Mitchell: Three Expeditions, 1839........................197
Moby Dick or The White Whale (Whaling), 1940......427
Moeurs, Usages et Costumes (Wahlen), 1844...............84
Moll: A System of Geography, 1701 77
Monde en Estampes, Le (Bedelet), 1858 56
Monument à la Pérouse (Touanne), 1826.................205
Moor: Sailing Directions, 1801 358

More: Utopia, 1684 99
More: De optimo reipublicae statu, 1777 98
Morea, or Eimo (Cleveley), 1788 123
Moresby: Discoveries & Surveys, 1876 364
Morin: Voyages de Gulliver, 1850 102
Mort de Cook (Webber), 1790 138
Mortier: Partie orientale du monde, 1730 36
Muller: Ueber Alterthuemer, 1863 359
Mundus alter et idem (Hall), 1606 90
Mundus alter et idem (Hall), 1607 91
Mundus alter et idem (Hall), 1648 92
Murray: Pitcairn, 1857 224
Murrell: Cruise of the Frigate Columbia, 1840 411
Mutiny of the Bounty (Young), 1894 225

N

Narborough: Account of Several Voyages, 1711 51
Narrative of a Journey (Simpson), 1847 414
Narrative of a Journey (Townsend), 1839 421
Narrative of a Survey (King), 1826 189
Narrative of a Tour (Ellis), 1826 448
Narrative of a Voyage (Arago), 1823 297
Narrative of a Voyage (Goodridge), 1852 400
Narrative of a Voyage, The (Grant), 1803 185
Narrative of a Whaling Voyage (Bennett), 1840 246
Narrative of Five Youth, A (Narrative), 1816 462
Narrative of Missionary, A (Williams), 1841 430
Narrative of the Adventures (Jewitt), 1815 405
Narrative of the Adventures (Patterson), 1817 413
Narrative of the Discovery, A (Mcclintock), 1859 264
Narrative of the Expedition, A (Tench), 1789 166-7
Narrative of John Byron, The (Byron), 1767 10
Narrative of John Byron, The (Byron), 1768 9
Narrative of John Byron, The (Byron), 1788 11
Narrative of the Mission, A (Yule), 1858 376
Narrative of the Shipwreck, A (Holden), 1836 403
Narrative of the Surveying Voyage (Jukes), 1847 188
Narrative of the Voyage (Macgillivray), 1852 200
Narrative of Voyages, A (Cleveland), 1842 395
Narrative of Voyages, A (Delano), 1817 223
Narrative: A Narrative of Five Youth, 1816 462
Native Races, The (Earl), 1853 362
Natural History of Fishes (Whaling), 1830 425
Naval Architecture (Stalkart), 1781 481
Navigantium atque Itinerantium (Harris), 1764 73
New Account, A (Hamilton), 1930 346
New Discoveries (Anonymous), 1778 78
New Light on the Discovery (Stevens), 1930 44
New Map of Asia, A (Ogilby), 1673 366
New Panorama (Heath), 1829 155
New Voyage, A (Dampier), 1697 18
New Voyage, A (Dampier), 1698 19
New Voyage, A (Kotzebue), 1830 232
Newbery: The History of Prince Lee Boo, 1793 261
Nodal: Appointment to an expedition, 1618 38
Nodal: Relación del Viage, 1769 37
Nordenskiold: Periplus an Essay, 1897 39
Nordenskiold: Facsimile-Atlas, 1889 40
Notice historique sur M. Péron (Deleuze), 1810 275
Notice sur le Voyage (Garnier), 1869 381
Nottage: In Search of a Climate, 1894 463
Nouveau Voyage à la Mer du Sud (Crozet), 1783 281
Nouveau voyage (Luillier), 1726 30
Novae Guineae Forma & Situs (De), 1593 23
Novæ Hollandiæ Plantarum (Labillardière), 1804 210
Novo: Viaje político-científico, 1885 236

O

Obookiah: Ka Moolelo, 1867 465
Observations on the Diseases (Clark), 1775 341
Ocean and its Inhabitants, The (Whaling.), 1845 429
Ogilby: A New Map of Asia, 1673 366
Old Sailor's Story, The (Allyn), 1879 390
Oliver: Wreck of the Glide, 1846 412
Oriental Navigator, The (Oriental), 1794 367
Oriental: The Oriental Navigator, 1794 367
Original Astronomical, The (Wales), 1777 117
Orwell: The Launching of the Orwell, 1817 477
Oud en Nieuw Oost-Indiën (Valentijn), 1726 83
Our Tropical Possessions (Cameron), 1865 340
Ovalle: Historica Relacion del Chile, 1646 41
Oxley: Journals of Two Expeditions, 1820 199

P

Pacini: Marine, La, 1844 478
Pages: Voyages autour du monde, 1782 300
Pallu: Histoire de l'Expédition, 1888 368
Paltock: First Night of the Races, 1828 100
Parker: A Voyage round the World, 1795 159
Parkinson: A Journal of a Voyage, 1773 107-8
Parkinson: A Journal of a Voyage, 1784 109
Part Borne by the Dutch, The (Heeres), 1899 25
Partie orientale du monde (Mortier), 1730 36
Passage Table (Fitzroy), 1859 255
Past and Present (Hunt), 1853 455
Patterson: Narrative of the Adventures, 1817 413
Patton: Légendes Hawaiiennes, 1932 466
Patuone: The Life and Times of Patuone, 1876 111
Penrose: A Link with Magellan, 1929 32
Periplus an Essay (Nordenskiold), 1897 39
Péron: Mémoires du Capitaine Péron, 1824 160
Péron: Voyage de Découvertes, 1816 272
Péron: Voyage of Discovery, 1809 273
Peter Parley's Tales of the Sea (Goodrich), 1848 206
Petit: Francis Garnier, 1887 387
Philip: The Building of the Ship, 1880 480
Philippine Islands, The (De), 1868 42
Phillip: The Voyage, 1789 161
Phillip: The Voyage, 1790 162
Phillip: Voyage, 1791 163
Philosophical and Political (Raynal), 1777 369
Pickering: The Races of Man, 1848 85
Pictorial History of China (Sears), 1853 333
Pictures of Travel & Adventure (Whaling), 1860 428
Pigafetta: Relation du premier voyage, 1923 31
Pinkerton: A General Collection, 1814 79
Pitcairn (Murray), 1857 224
Playbill: De La Perouse, 1825 94
Polus antarcticus (Hondius), 1650 27
Polynesian Researches (Ellis), 1830 399
Port Lincoln (Westall), 1835 184
Porte-feuille géographique (Engelmann), 1820 67
Portuguese account of the Moluccas, 1750 371
Pratique de la pèche (Lecomte), 1833 306
Prima pars descriptionis (Lodewyckszoon), 1598 28
Primera parte (Torquemada), 1723 54
Prince Giolo (Savage), 1692 345
Principal Navigations, The (Hakluyt), 1600 71
Private Journal, The (Stewart), 1831 467
Proceedings of a Court-Martial (Bligh), 1811 216
Progress Reports and Final Report (Royal), 1863 179

R

R., J: Diary of a Journey Overland, 1822 337
Races of Man, The (Pickering), 1848 85
Rapport (Dumont d'Urville), 1838 287
Rattlesnake, HMS (Brierly), 1849 201
Raynal: Philosophical and Political, 1777 369
Reception of Revd. J. Williams (Baxter), 1841 431
Recréations Instructives (St. Aulaire), 1840 308
Recréations Instructives (St. Aulaire), 1845 309
Reis rondom de wereld (Laplace), 1836 301

Reise der öesterreichischen (Scherzer), 1862 239
Reise um die Welt (Krusenstern), 1812 234
Reistochten (Rosenberg), 1875 365
Reize naar de Oost-en Westkust (Boelen), 1835 441
Reizen naar Nieuw-Guinea (Aa), 1879 360
Relación del último viage (Vargas), 1793 228
Relación del Viage (Nodal), 1769 37
Rélation abrégée (La Pérouse), 1799 204
Rélation du premier voyage (Pigafetta), 1923 31
Rélation du Voyage (Labillardière), 1800 207
Relation exacte (Bugnon), 1707 315
Remarks on Mr. Forster's Account (Wales), 1778 118
Reminiscences (Brewer), 1884 393
Reminiscences of Australia (Hodgson), 1846 193
Reminiscences of Travel (Sales), 1892 332
Report (Burke and Wills), 1862 178
Reports on Dublin Harbour (Bligh), 1799 217
Resa omkring Jorden (Kotzebue), 1830 233
Residence among the Chinese, A (Fortune), 1857 321
Residence, A (Stewart), 1839 468
Residence, A (Bingham), 1847 439
Reynolds: Voyage of the Potomac, 1835 240
Reys-Togten (Schouten), 1708 49
Richardson: A General Collection, 1789 80
Rickman: Troisième voyage de Cook, 1783 139
Rieunier: Solution Pratique, 1864 388
Rivière de Derwent (Sainson), 1833 305
Robert O'Hara Burke (Jackson), 1862 177
Robertson: Voyage de Robertson, 1767 101
Roscoe: An Address, 1802 120
Rosenberg: Reistochten, 1875 365
Ross: A Voyage of Discovery, 1847 267
Ross: An Account of Capt. Ross's Voyage, 1835 266
Rotz: The Boke of Idrography, 1981 46
Round Cape Horn (Lamson), 1878 407
Rowlandson: A Hulk at Sheerness, 1800 164

S

Sailing Directions (Moor), 1801 358
Sainson: Rivière de Derwent, 1833 305
Sainson: Sydney. Etablissement principale, 1833 303
Sainson: Vooloo-moloo, 1833 302
Sainson: Vue de L'Entrée de Port Jackson, 1833 304
Sales: Reminiscences of Travel, 1892 332
Samwell: Détails nouveaux, 1786 132
San Roman: Historia general de la Yndia, 1603 47
Sandwich Island Notes (Bates), 1854 437
Sandwich Islands, The (Lucas), 1841 461
Sanson: Les Isles Philippines, 1654 370
Sarawak (Low), 1848 354
Sarmiento: Viage al Estrecho, 1768 48
Sauer: An Account of a Geographical, 1802 227
Savage: Prince Giolo, 1692 345
Scenery of China, Turkey (Cooke), 1813 63
Scenery of the East India Islands (Cooke), 1813 62
Scherzer: Reise der öesterreichischen, 1862 239
Schouten: Reys-Togten, 1708 49
Scott-Waring: A Second Letter, 1774 112
Sea Yarns (Taylor), 1915 419
Sears: Pictorial History of China, 1853 333
Second Letter, A (Scott-Waring), 1774 112
Septentrio Novantiquus (Megiser), 1613 43
Set of the three voyage accounts, A (Cook:), 1784 141
Shaw: A Catalogue, 1787 133
Simpson: Narrative of a Journey, 1847 414
Sketch Book of Shipping, The (Grundy), 1840 474
Sketches of Canton (Anderson), 1852 325
Skogman: Fregatten Eugenies Resa, 1855 241
Snow: A Two Years' cruise, 1857 415
Solution Pratique (Rieunier), 1864 388
Sonnerat: Voyage à la Nouvelle Guinée, 1776 307
South Sea Bubbles (Kingsley), 1872 406
Southlanders, The (Fox), 1860 89
Souvenirs d'un aveugle (Arago), 1840 298
Souvenirs d'un aveugle (Arago), 1868 299
Souvenirs d'un voyage (Huc), 1850 327
Sparks: Memoirs of John Ledyard, 1828 130
Sparrman: En Upptäckts-Resa, 1801 271
Sparrman: Tal, om den tilväxt och nytta, 1778 116
St. Aulaire: Recréations Instructives, 1840 308
St. Aulaire: Recréations Instructives, 1845 309
St. Aulaire: Voyage Autour du Monde, 1845 310
St. John: The Lives of Travellers, 1832 131
Stalkart: Naval Architecture, 1781 481
Stanford: Map of Eastern China, 1904 334
Stanford: Map of the Malay Peninsula, 1910 372
Staunton: An Authentic Account, 1797 335
Staunton: Voyage en Chine, 1804 336
Stevens: New Light on the Discovery, 1930 44
Stewart: A Residence, 1839 468
Stewart: The Private Journal, 1831 467
Stockmans: Diarium Nauticum, 1598 29
Stuart: De Mensch, 1807 81
Suarez: Hechos de Don García Hurtado, 1613 50
Sydney. Etablissement principale (Sainson), 1833 303
Sydney: Fore-edge view, 1830 165
Symes: An Account of an Embassy, 1827 373
System of Geography, A (Moll), 1701 77

T

Tagebuch (Forster), 1781 140
Tahiti, receiving the Gospel (American), 1832 416
Tahiti: Geogeraphi, oia te ite raa, 1840 417
Tal, om den tilväxt och nytta (Sparrman), 1778 116
Tasman's Kaart (Wieder), 1942 53
Taylor: Memoirs of Henry Taylor, 1811 146
Taylor: Sea Yarns, 1915 419
Ten Years in Oregon (Allen), 1850 433
Tench: A Complete Account, 1793 170
Tench: A Narrative of the Expedition, 1789 166-7
Tench: Voyage à la Baie Botanique, 1789 168
Terra Australis Cognita (Callander), 1768 61
Terry: A voyage to East-India, 1777 374
Thiercelin: Journal d'un Baleinier, 1866 311
Thomas: Interesting and Authentic, 1850 420
Thomson: The Voyage of the Challenger, 1878 268
Three Expeditions (Mitchell), 1839 197
Thrilling Incidents (Zollers), 1892 432
Through the Buffer State (Macgregor), 1896 356
Thrum: Hawaiian Almanac and Annual, 1932 469
Thurston: The Missionary's Daughter, 1842 470
Tidemarks (Tomlinson), 1924 375
Tomlinson: Tidemarks, 1924 375
Topliff: The King of Otaheite's Tea Party, 1840 418
Torquemada: Primera parte, 1723 54
Touanne: Monument à la Pérouse, 1826 205
Toulon: Précaution contre le Scorbut, 1777 119
Tour du monde, Le (Guerin), 1841 401
Townsend: Narrative of a Journey, 1839 421
Tracks of McKinlay (Davis), 1863 196
Tractado de las drogas (Acosta), 1578 338
Transplanting the Bread-Fruit (Gosse), 1796 222
Travels in China (Barrow), 1804 313
Travels in China (Barrow), 1806 314
Travels in India (Hodges), 1793 326
Travels into Bokhara (Burnes), 1834 317
Treatise of Navigation, A (Harris), 1730 475
Troisième voyage de Cook (Cook), 1785 122
Troisième voyage de Cook (Rickman), 1783 139
Tuckey: An Account of the Voyage, 1805 171
Tuckey: Maritime Geography, 1815 82

Turnbull: A Voyage Round the World, 1813 269
Tvänne Resor från Ostindien (Meares), 1797 265
Two Years' cruise, A (Snow), 1857 415
Tyerman: Journal of Voyages and Travels, 1831.......422

U

Ueber Alterthuemer (Muller), 1863..........................359
Unwritten Literature of Hawaii (Emerson), 1909.....449
Utopia (More), 1684 99

V

Valentijn: Kaart van de Zuyd-Ooster, 1726..............55
Valentijn: Oud en Nieuw Oost-Indiën, 1726............83
Vancouver: Voyage de Découvertes, 1800...............270
Vargas: Relacion del último viage, 1793.................228
Viage al Estrecho (Sarmiento), 1768.........................48
Viage del Comandante Byron (Clerke), 1769 12
Viagem ao interior (Aguiar), 1841............................86
Viaggio intorno al globo (Duhaut), 1841..................446
Viaje político-científico (Novo), 1885 236
Viajes científicos (Michelena), 1843........................238
View of Pulo Condore (Webber), 1809 135
View of Huaheine (Cleveley), 1787 124
View of Moorea, A (Cleveley), 1787 125
Vincendon-Dumoulin: Iles Taïti, 1844..................290
Vista de la Ciudad de Santiago (Brambila), 1795.......237
Voice from Abroad, A (Dibble), 1844.......................445
Vollständige Geschichte (Brosses), 1767......................59
Vooloo-moloo (Sainson), 1833................................302
Voyage (D'Entrecasteaux), 1808 211
Voyage (Phillip), 1791..163
Voyage à la Baie Botanique (Tench), 1789................168
Voyage à la Nouvelle Galles (White), 1795...............174
Voyage à la Nouvelle Guinée (Sonnerat), 1776 307
Voyage au Cambodge (Delaporte), 1880...................377
Voyage au Pole Sud (Dumont d'Urville), 1842..........288
Voyage autour du Monde (Anson), 1750 4
Voyage autour du Monde (Bougainville), 1771 277
Voyage autour du monde (Duperrey), 1826...............289
Voyage autour du monde (Freycinet), 1826................295
Voyage autour du monde (Freycinet), 1844 294
Voyage autour du Monde (La Pérouse), 1797............202
Voyage autour du monde (Lesson), 1839 293
Voyage Autour du Monde (St. Aulaire), 1845............310
Voyage aux Iles de la Mer du Sud (Dillon), 1830........212
Voyage d'Exploration (Garnier), 1873 383
Voyage d'Exploration (Garnier), 1885 384
Voyage dans la Chine Centrale (Garnier), 1874 385
Voyage de Découvertes (Peron), 1816.......................272
Voyage de Découvertes (Vancouver), 1800.................270
Voyage de l'Astrolabe (Dumont d'Urville), 1833 283-6
Voyage de Robertson (Robertson), 1767 101
Voyage en Chine (Staunton), 1804..........................336
Voyage in search (Labillardière), 1800..................208-9
Voyage in the Sunbeam (Brassey), 1891....................391
Voyage Médical (Lesson), 1829.............................292
Voyage of Discovery (Peron), 1809 273
Voyage of Discovery, A (Ross), 1847........................267
Voyage of HMS Alceste (Mcleod), 1818 357
Voyage of HMS Blonde (Byron), 1826......................248
Voyage of the Challenger, The (Thomson), 1878 268
Voyage of the Potomac (Reynolds), 1835...................240
Voyage of the Exploring (Jenkins), 1850...................242
Voyage Round the World, A (Anson), 1748 1
Voyage Round the World, A (Anson), 1785 2
Voyage round the World, A (Betagh), 1728...................8
Voyage round the World, A (Bougainville), 1772 278
Voyage Round the World, A (Dixon), 1789................252
Voyage round the World, A (Forster), 1777...............115
Voyage Round the World, A (La Pérouse), 1798.........203
Voyage round the World, A (Parker), 1795 159
Voyage Round the World, A (Turnbull), 1813............269
Voyage to Botany Bay, A (Barrington), 1801.............147
Voyage to Cochinchina, A (Barrow), 1806.................339
Voyage to East-India, A (Terry), 1777......................374
Voyage to New Guinea, A (Forrest), 1780................256
Voyage to New Holland, A (Dampier), 1703...............21
Voyage to the East-Indies, A (Dellon), 1698..............342
Voyage to the Pacific Ocean, A (Cook), 1785 121
Voyage to the South Sea, A (Bligh), 1792..................213
Voyage to the South-Sea, A (Frezier), 1717 24
Voyage towards the South Pole, A (Cook), 1777 114
Voyage, The (Anson), 1818 3
Voyage, The (Phillip), 1789 161
Voyage, The (Phillip), 1790 162
Voyages à Peking (Guignes), 1808 324
Voyages and Descriptions (Dampier), 1699 20
Voyages and Travels (Langsdorff), 1817 235
Voyages autour du monde (Pages), 1782 300
Voyages dans la mer du Sud (Dalrymple), 1774 66
Voyages dans le pays (Tench), 1790 169
Voyages dans les Deux Océans (Delessert), 1848.........398
Voyages de Gulliver (Morin), 1850 102
Voyages et Mémoires (Benyowsky), 1791..................226
Vue de L'Entrée de Port Jackson (Sainson), 1833........304
Vues et Paysages (Choris), 1826 231

W

Wahlen: Moeurs, Usages et Costumes, 1844 84
Wales: Remarks on Mr. Forster's Account, 1778 118
Wales: The Original Astronomical, 1777 117
Waterhouse, The Rev. J. (Baxter), 1844....................423
Webber: Balaguns or Summer Habitations, 1809....136
Webber: Death of Captain Cook, 1784 137
Webber: Mort de Cook, 1790 138
Webber: View of Pulo Condore, 1809...................135
Were the Hawaiian Islands (Dahlgren), 1916 443
Westall: Port Lincoln, 1835 184
Whales: History of Wonderful Animals, 1818........424
Whaling.: The Ocean and its Inhabitants, 1845......429
Whaling: In het ruime sop, 1870..........................426
Whaling: Moby Dick or The White Whale, 1940 ..427
Whaling: Natural History of Fishes, 1830 425
Whaling: Pictures of Travel & Adventure, 1860 428
White: John Whites Resa till Nya Holland, 1793 ...113
White: Journal of a Voyage, 1790......................172-3
White: Voyage à la Nouvelle Galles, 1795.............174
Whitney: He Hoikehonua, 1845 452
Whitney: The Hawaiian Guide Book, 1875...........471
Wieder: Tasman's Kaart, 1942...............................53
Williams: A Narrative of Missionary, 1841 430
Williams: Adventures of a Post Captain, 1830........479
Wit: Magnum Mare Del Zur, 1680 52
Wreck of the Glide (Oliver), 1846 412

Y

Yankee Tar, The (Hoxse), 1840 404
Yedo and Peking (Fortune), 1863 322
Young Sea Officer, The (Lever), 1858 476
Young: Mutiny of the Bounty, 1894......................225
Yule: A Narrative of the Mission, 1858 376

Z

Zollers: Thrilling Incidents, 1892 432
Zoology of HMS Sulphur (Belcher), 1846 245

- Print Post Approved PP 242800/00015 -